Accounting with QuickBooks Pro® 2000

for Home Builders and Remodelers

Diane C.O. Gilson, CPA, CIA

Home Builder Press®
National Association of Home Builders
1201 15th Street, NW
Washington, DC 20005-2800
(800) 223-2665
www.builderbooks.com

Accounting with QuickBooks Pro® for Home Builders and Remodelers
ISBN 0-86718-506-6

Cover Design by Tim Kaage, Laurel Graphx
Digital Imagery copyright ©2000PhotoDisc, Inc.

Library of Congress Cataloging-in-Publication Data

Gilson, Diane.
Accounting with QuickBooks Pro for Home Builders and Remodelers/Diane Gilson.
p. cm.
ISBN 0-86718-506-6 (pbk. : alk. paper)
1. Construction industry—Accounting. 2. Construction industru—Finance—Computer programs. 3. Building—Estimates-Computer programs. 4. QuickBooks. I. National Association of Home Builders (U.S.) II. Title.

HF5686.B7 G5 2000
657'.869'02855369—dc21 00-046555

For further information, please contact:
Home Builder Press®
National Association of Home Builders
1201 15th Street, NW
Washington, DC 20005-2800
(800) 223-2665
Check us out online at: www.builderbooks.com

12/00 Circle Graphics/Data Reproductions 2500

Contents

About the Author

Diane C.O. Gilson, CPA, CIA is President of Info Plus Accounting. Diane has lived in the Ann Arbor/Ypsilanti area for the last twenty-eight years, is a graduate of Eastern Michigan University, a Certified Public Accountant, and a Certified Internal Auditor. She has more than twenty-five years' hands-on experience in accounting, including public and internal auditing, tax, and budgeting. For the last twelve years she has focused on the design and delivery of management and cost accounting systems.

Diane created the CPA firm of Info Plus Accounting in 1994 with the intent of providing current and future-oriented management accounting information to small to medium-sized business owners. Since the firm's inception, Diane has worked exclusively in QuickBooks Pro®. She is a Certified QuickBooks® Professional Advisor and an ALS-certified QuickBooks® consultant.

Although her firm has a number of clients in the service, technology, retail and other assorted industries, approximately 60% of Info Plus Accounting's clients are builders, developers, and remodelers. Diane has chosen to provide enhanced services for the construction industry because QuickBooks Pro® is especially suited to the industry's needs for understandable, useful job costing and profitability reports.

Info Plus Accounting provides accounting systems design, training, trouble-shooting, and business support services. Their website address is www.infoplusacct.com.

Acknowledgments

I'd like to extend special thanks to the editors at Home Builder Press who made this possible, and especially to Charlotte McKamy, who went to bat to help bring the specifics of QuickBooks® Pro to the people who most needed it. I'd also like to thank John Goucher for his hours of involvement and feedback, and Barbara Minich for her organizational and editing skills and encouragement. I'd especially like to thank Paul LaRoe, CGR, of LaRoe Residential Remodeling who helped to coach me through my first remodeling company job-costing project, and, due to his concern for fellow remodelers, has been a consistent source of support and encouragement. My special thanks go to Linda, who took care of so many details behind the scenes at work so that I could continue with this project and to Stephanie who came to my rescue during the last critical phases of the project.

This book is produced under the general direction of Jerry Howard, NAHB Executive Vice President and CEO, in association with NAHB staff members Rob Pflieger, Senior Staff Vice President, Public Affairs; Greg French, Assistant Staff Vice President, Public Affairs; Charlotte McKamy, Publisher, Home Builder Press; John Goucher, Acquisitions Editor; David Rhodes, Art Director; and Toral Patel, Assistant Editor.

Introduction

I am continually amazed at the variety of skills and abilities owners of remodeling and custom building companies must possess. From the hours and energy required to effectively run a company and produce a service, to the financial and legal risks, and the "I got a lower bid from a guy with a dog and a truck" price competition, you are truly entrepreneurs. Surely such energy, talent, hard work, and risk should ensure an excellent income! Yet, after working with my clients and attending a number of seminars and conferences, I believe far too many remodelers and custom builders are highly skilled, but underpaid.

I sincerely believe that custom builders' and remodelers' profit-margin woes can be traced to the "numbers" part of their business. Like many small-business owners, remodelers and custom builders prefer to focus their time and energy on their strengths: designing, remodeling, building, and successfully managing projects and people.

Often what is lacking is a properly designed accounting and financial information system. Homegrown databases, spreadsheets, and interim financial systems that do not cover internal financial reporting, job costing, payroll, estimates, or invoice record keeping may be costing your business more than you can imagine! Just as buying a hammer and saw and watching a few "Hometime" episodes won't turn *me* into a remodeler, a self-created "on-the-fly" approached to accounting won't give *you* professional, useful management information and results.

Creating an effective system

This book was written to help you, the remodeler, developer, or custom builder, create an effective accounting and financial information system for your business. Although your CPA and tax advisor are valuable resources for tax-oriented questions, if you wait for quarterly tax reports or year-end financial information, the data may be too old to help you make meaningful operating decisions. An accounting system is more than a tax record. It also is a record of the costs and profitability of your individual jobs and, ultimately, a tool to help you improve your business as a whole.

As a remodeler or custom builder you have specialized financial and management information needs that can range from the moderately simple to the very complex. A properly designed accounting and financial information system can have a significant and favorable impact on your operations and profitability by providing you with critical, current operating information.

My goal with this book is to help you *create* and *access* timely management and financial information that will provide you with information you can use effectively to *understand* and *manage* your business in order to reach your daily, short-term, and long-term goals.

Although part of the financial measurement game involves recording *historical* information, you'll find that the approaches illustrated in this book will help you become more knowledgeable about the *present.*

The basic management and measurement controls I outline in this book will help you shape your future success. You will learn how to implement basic management controls and measurement systems that are designed to help you enhance your profitability, enabling you to turn your numbers into your servants, rather than your masters. Take advantage of my experience as a construction-oriented management accountant and use the tips and business control techniques that I've learned from working with remodelers and custom builders just like you.

Taking action step by step

The book is organized in sections. Section 1 helps you determine your requirements. These chapters give you an overview of the management decisions and controls you will need to make in order to set up an effective accounting system. Like your customers, you will want to explore your options and make some key decisions about what you need to effectively build a financial management system before you "sign on the bottom line."

Section 2 outlines the design process. QuickBooks Pro 2000 is composed of a variety of subsystems. Lists are a critical element of these subsystems, helping QuickBooks Pro integrate your information and minimizing data-entry chores. The lists within the subsystems also allow you to customize reports to get the information you need to make management and pricing decisions. Each of the chapters in this section walks you through a list, including its purpose, organization, entry, and data. Once you are finished with your lists, you have a customized blueprint for the structure and subsystems of your financial management system.

In section 3 I'll help you build your system. Using the actual QuickBooks Pro windows, I'll show you how to enter your company's data to accomplish the tasks that are key to day-to-day financial management, job costing, and year-end requirements. Trade contractor time and the special transactions associated with 1099 vendors also are discussed.

Employees are the primary topic of the chapters in section 4. QuickBooks Pro has an extensive payroll system that integrates with other job-cost information. In addition, the system will record payroll taxes and deductions, vacation and sick leave, and other employee-related records.

During the final walk-through in section 5, we'll look at customizing reports to meet your specific needs. We'll also explore troubleshooting processes, month-end procedures, file management issues, and security processes that will help you keep your files safe and your clean-up work to a minimum.

Setting up a customized accounting system is not a lighthearted undertaking. You will need to invest time and energy to develop your own special blend of approaches, techniques, and philosophies. You may even make some mistakes along the way. Developing a solid, workable, customized accounting and information system won't be easy, but I believe you will be rewarded many times over for your efforts.

Introduction to QuickBooks Pro 2000

For me, the key to an effective in-house accounting system is the QuickBooks Pro 2000 accounting software. Over the years, I have designed and implemented a variety of accounting systems for remodelers, developers, and custom home builders. I have found QuickBooks Pro to be a complete financial package. It is robust enough to process and report on all of your financial transactions, eliminating the need to employ multiple software packages.

Integrating data

QuickBooks Pro is composed of a variety of integrated subsystems, including accounts payable, accounts receivable, payroll, and general ledger that are able to access and carry data across the accounting "cycles." For example, when you enter a purchase order in QuickBooks Pro, you select the cost items and indicate the job. The data then flows from the purchase order process to the bill entry process to the bill payment process without you having to re-enter the same cost and job data. Here's another example. If an employee's time is entered by job and job stage in the payroll time tracking process, the data doesn't need to be entered again when it is time to pay the employee. The time data will be available for time and material invoicing and payroll costs will flow through to reports for the job. QuickBooks Pro minimizes the amount of data entry work and helps to maintain the integrity of your data so that you have fewer month-end balancing and reconciling steps.

Among the many features incorporated into QuickBooks Pro are:

- Invoicing and accounts receivable detail and tracking, statements, and sample collection letters
- Accounts payable, reminders, and bill payment features
- Payroll features and time tracking
- Job costing, including job-process costing and percentage-of-completion capacity
- Converting estimates to invoices and purchase orders to bills
- Extensive reporting capacity and flexibility
- Ability to view or correct underlying detail

Customizing to your needs

I believe that an accounting-reporting system should be dynamic and flexible. Today you may want to see the information organized one way. Tomorrow you may require a new perspective and prefer to see a report in a slightly different order. QuickBooks Pro gives you this flexibility. For example, in QuickBooks Pro you can *move* an account from one place to another within its account type by using the mouse to drag it from its old location and drop it in its new location. You also can alter an account name by clicking on the Edit button and changing it, or you can merge two accounts by giving one of them the same number as the other.

Every company is unique. In addition to the needs of company owners, individual users of the accounting system and readers of various financial reports typically want different information, formats, and processes from the system. Once you understand how the program's subsystems work, you can customize *within* QuickBooks Pro to meet your company's varied needs. For example, QuickBooks Pro allows you to:

- Create reports showing all manner of detail or summary information regarding your job costs and estimates
- Prepare financial reports that you can use for banking or tax purposes
- Filter reports to extract only certain data
- Create templates for multiple customized versions of invoices, estimates, customer statements, purchase orders, cash sale receipt documents, and other forms
- Show overhead costs in as detailed or summarized a manner as you wish, such as splitting out your pager costs from your telephone costs, and segregating and subtotaling your occupancy costs
- Include individualized memos on forms sent to customers and vendors
- Add special data fields to customers', vendors', or employees' individual records, such as e-mail addresses, birthdays, or spouses' and children's names
- Include your logo and a special notice on your invoices and forms
- Edit collection letters or tie into Word® to generate customized letters and mailing lists

Making navigation simple

QuickBooks Pro is designed for both new and experienced users. It presents information and activities in a visual, user-friendly format. The Navigator windows especially are useful to new users because individual Navigator windows for each accounting cycle present active icons that show the structural components of each of the major steps in that accounting cycle. The icons are arranged in a flowchart format and prebuilt reports relating to the cycle are accessed through the Report Finder feature at the top of each Navigator window, as shown in Exhibit I.1.

Another set of icons at the bottom of the Navigator window shows related lists and activities commonly associated with the illustrated transactions. This is a logical presentation of accounting functions for anyone with a visual or information-flow mindset.

For users who like pull-down menus, the accounting cycles also are presented across the top of the window. Clicking on each cycle presents a pull-down menu with the same options as the active icons on the Navigator windows. In the following chapters, both the Navigator window and the pull-down menus are given as ways to access the various functions and processes in QuickBooks Pro. A pull-down menu is shown in Exhibit I.2.

QuickBooks Pro 2000 also has a tool bar along the left-hand side of the window. The toolbar includes active icons for the different Navigator windows. I recommend that you hide the toolbar because it is easier to use the Navigator windows or the pull-down menus. To hide the toolbar, click on Tools and select Hide Toolbars.

QuickBooks Pro Centers are a third way to access information. The centers provide a quick glance at each business function by collecting and grouping data for specific areas. The information is organized in a manner that makes it easy for you to see and analyze the data, and take action on it. For example, the Company Center window, shown in Exhibit I.3, displays account balances, an income and expense graph, a reminder or alert list, and summary statements in the Customer Center Highlights section.

User-friendly aids

The forms you see on the screen in QuickBooks Pro closely resemble real-life paper purchase orders, checks, invoices, and so on. Window choices and features are consistent, and transaction-

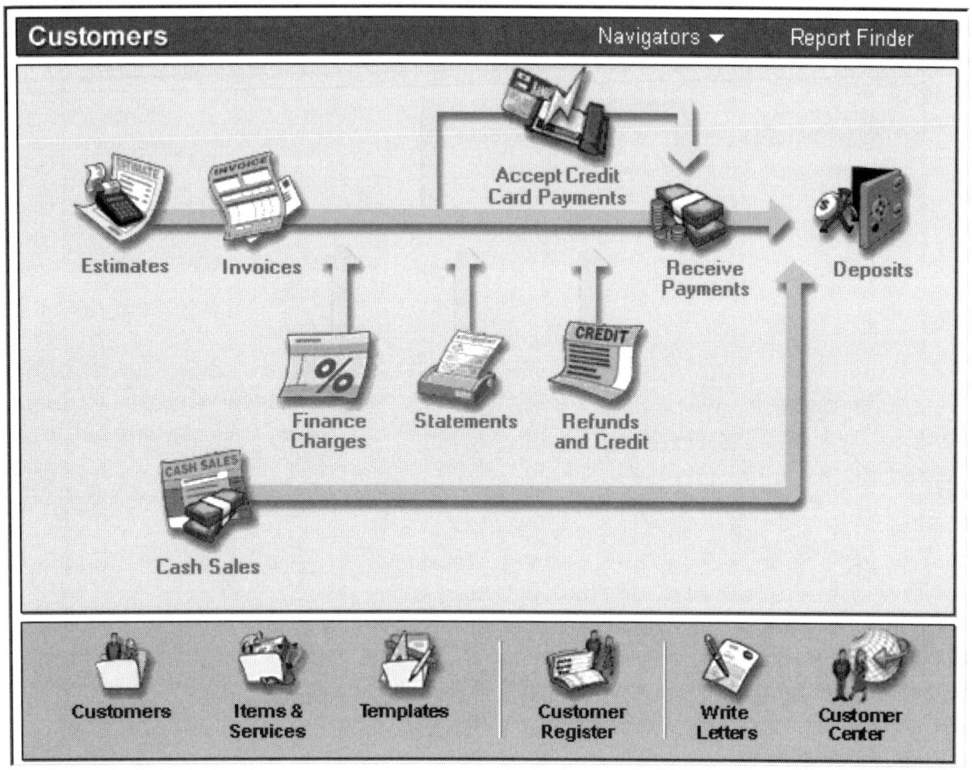

EXHIBIT I.1 Customers Navigator window.

related information such as customer:job fields, buttons, and checkboxes are understandable and located on the transaction windows to which they apply. If you customize your forms, the printed version may be laid out differently than the on-screen version and may include, or exclude, certain fields.

The Autofill feature also simplifies your accounting tasks. For example, when you begin typing the name of a vendor in an invoice, the system will match the keyed information with the vendors in the Vendor List. As you continue to type, QuickBooks Pro continues to match vendors already in the system. As soon as you see a vendor match on the screen you select it by pressing the tab key. The system accepts the vendor and, using information from your most recent transaction with that vendor fills in the current window. You then may choose to accept the information as presented by the system, or edit the information as necessary. This saves keystrokes and reduces the possibility of data-entry errors.

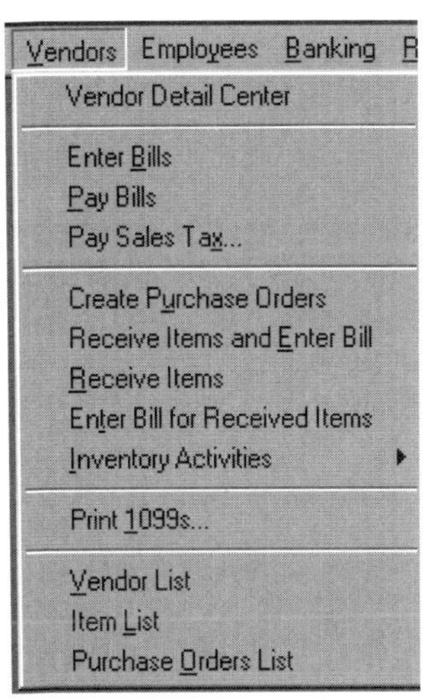

EXHIBIT I.2

Vendors pull-down menu.

System help features

Help is an important feature for any software program. QuickBooks Pro offers three, specific-task, onscreen help features:

- The Help with This Window feature is accessed by pressing the F1 key.
- The Help Index allows you to search for help by typing in key words. When you are in the Help window, highlight and click on the relevant words to see definitions, further explanations, examples, or related topics.

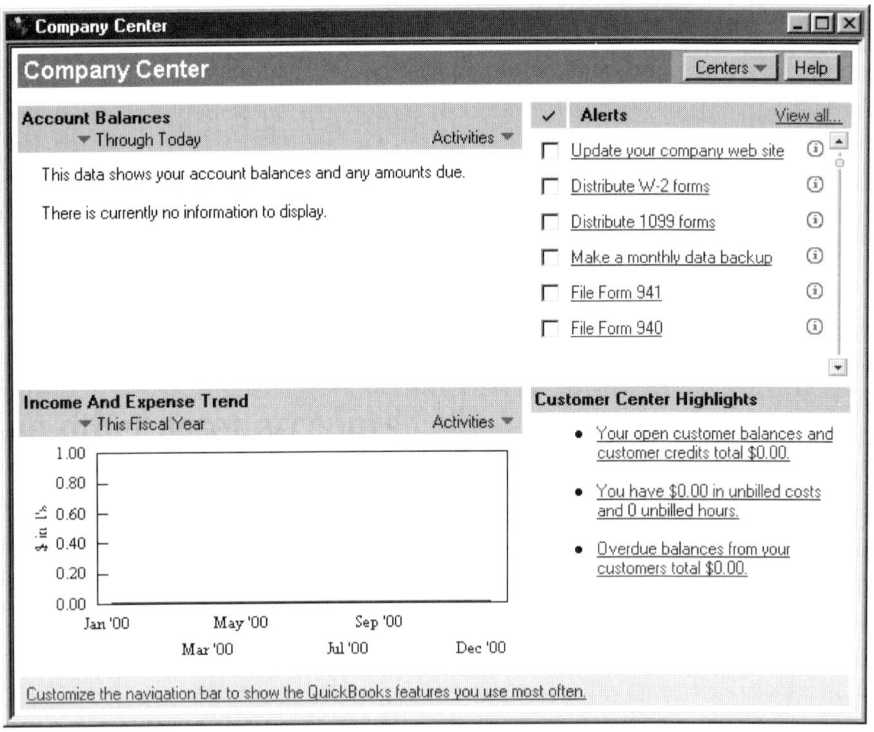

EXHIBIT I.3 Company center window.

- The How do I? Help menus are accessed from activity windows. For example, if you are writing a check, you can click on the How do I? Help button. A drop-down menu will appear from which you can select help for a variety of activities most often associated with writing a check. Highlight and click on a topic to access the correct Help window in the index.

Reliable customer support

Intuit, the company that makes QuickBooks Pro, has an extensive technical support operation that is responsive to your software questions. Keep in mind that no company in the technology field has a perfect track record and that response times for phone support can vary depending on the time of day, time of year, and if new versions have just been released.

Installing QuickBooks Pro

Before you install QuickBooks Pro, carefully read the installation requirements in the User's Guide, especially if you will be installing the software on a networked system. The minimum system requirements are:

- 32 mg of RAM, with 64 mg preferred
- 85 mg of free memory
- Online Internet access if you are processing and entering payroll-related information

Remember that hardware and software can be difficult to master and troubleshoot. Find a good technology-solutions provider to help you with your hardware and, if applicable, your network setup. This provider also can assist you with your QuickBooks Pro software installation, but probably won't provide QuickBooks Pro support.

QuickBooks Pro Install Checklist

Current version of QuickBooks: ______________________

Installed version of QuickBooks: ______________________

Internet connection: ______________________

- Run QuickBooks Install
- Keep prior preferences if previous file exists
- Create 'QB' or other folder for QuickBooks data files
- Datafile conversion if upgrading from earlier version
- Register
- Internet connection setup
- Run updates for software (registration code may be needed)
- Turn off auto update—Admin file
- Turn on share updates—Admin file
- Repeat for multiple users
- Shared drives mapped
- QuickBooks is pointing to the correct file.
- Printing works to all printers and trays
- Set Windows to make file extensions visible

Computer(s) running QuickBooks:

Notes:

Virus protection and backup routine:

EXHIBIT I.4 Installation checklist.

I have developed the checklist in Exhibit I.4 to help my clients who are installing QuickBooks Pro for the first time. While the checklist will not replace professional technical assistance, it should help ensure that you are aware of many of the steps required to successfully install QuickBooks Pro. Consider it a head start for your installation!

As in highway construction—drive slowly. Be prepared for delays and surface obstructions. It's a good idea to allow for unexpected delays related to hardware, software, or other unknowns. Professional technical assistance will increase your odds of moving through the technical part of the installation process smoothly.

Overall, I have found QuickBooks Pro to be a well developed, debugged, updated, and supported software package. With some time and effort, you can build an integrated accounting system using QuickBooks Pro that will provide you with up-to-date financial information you can use to make informed management decisions. Because of its large user base, you also are more likely to find trainers, consultants, and other users who are familiar with QuickBooks Pro if you need help accomplishing your financial management goals.

SECTION ONE

THE CONSULTATION PHASE

Remodelers and custom builders are in a high-risk, high-service business. While job income usually is fixed within the parameters of a job quote, cost gremlins, such as materials prices, weather problems, and labor overruns often eat away at a project's profits.

The first step to designing a successful financial management system is to understand how you run your business. Just as your clients know they need a kitchen, only by asking questions about the frequency with which they cook, the types of food they cook, their entertaining requirements, the number of cooks, their preferences for appliances, and the available space can you design a kitchen that they will love.

So it is with your accounting system. In the following chapters, you will be asked to define and clarify your vision of what you want for your company and your financial information requirements. You also will be asked to make choices about accounting measurement tools that have progressive levels of detail and complexity.

My goal is to assist you in making the correct choices. So I'll ask the questions and give you some information about what is available. You decide the level of detail you need.

CHAPTER 1

Incorporating controls and the control loop

Controls are the sum total of processes and procedures that a company owner or manager uses to make sure that the

- **Right things do** happen, and the
- **Wrong things don't** happen.

Stop for a moment and visualize what the "right things" might look like for your company. They might include:

- Generating sales revenue appropriate to the size of the company you wish to establish
- Defining and accurately estimating job costs
- Knowing which jobs to charge extra for, or to avoid entirely
- Assigning an appropriate gross profit to all of your jobs
- Training employees and trade contractors so they know their jobs and can accomplish their tasks effectively, efficiently, and with minimal supervision
- Managing a smoothly run office
- Reviewing operations reports on a regular basis to help avoid problems before they occur and to move swiftly to solve them if they do occur
- Being able to grant bonuses and provide regular vacations for you and for your employees

To increase the likelihood of producing each of these desirable outcomes, you will need to do some soul searching so you can honestly answer the following questions.

- What is your dream for your company?
- What is your company's mission?
- What are your financial goals?
- What are your service goals?
- What are your operations goals?
- What information will help you monitor your progress towards those goals?
- How will you know the right things are happening and the wrong things aren't?
- Are you willing to hear, face, and solve the wrong things when they occur?

What, you may be asking, do these questions have to do with accounting? The reality is that the production side of your business essentially is a goal setting, measurement, and achievement process. Would you purchase a truck full of construction tools and organize the tools in your truck if you didn't have a clear picture of what and when you are going to build? No, because you might spend too much money and effort on tools that aren't right for your purposes. The fact of the matter is it's nearly impossible to achieve a quality product without a vision, plans, measurements, and hard work.

This same principle applies to accounting. Because accounting is a *goal-setting and measurement system*, you must know *what* you're measuring, *why* you're measuring it, and even what *inaccurate* measurements look like. A good management accounting system is a set of tools and controls designed to help you create and measure your progress towards your company's financial, operating, and overall goals and vision. The clear vision, the detailed plan, the controls, and the measurement system all work together to dramatically increase your chances of success, which in turn enable you to achieve your personal, family, and other lifestyle goals.

So grab a cup of coffee, find a quiet place, and think about these questions. Then write down your answers. This is a good exercise for two reasons. The first is that you need to answer these questions, at least periodically, to keep your business on course. In addition, these answers will help determine how you will build your financial management system. Set aside some time now to focus on your financial and business needs and goals.

Implementing a control loop

A *control loop* is a management technique that can be used in both the financial and operating facets of your business. Use the control loop to increase your net profits and create dramatic changes in your business operations. Illustrated in Exhibit 1.1, the control loop is the four-step process of:

1. Establishing standards and goals
2. Measuring results
3. Comparing the results to your standards and goals
4. Taking corrective action to improve the results, or revising your standards or goals for the next round of operations

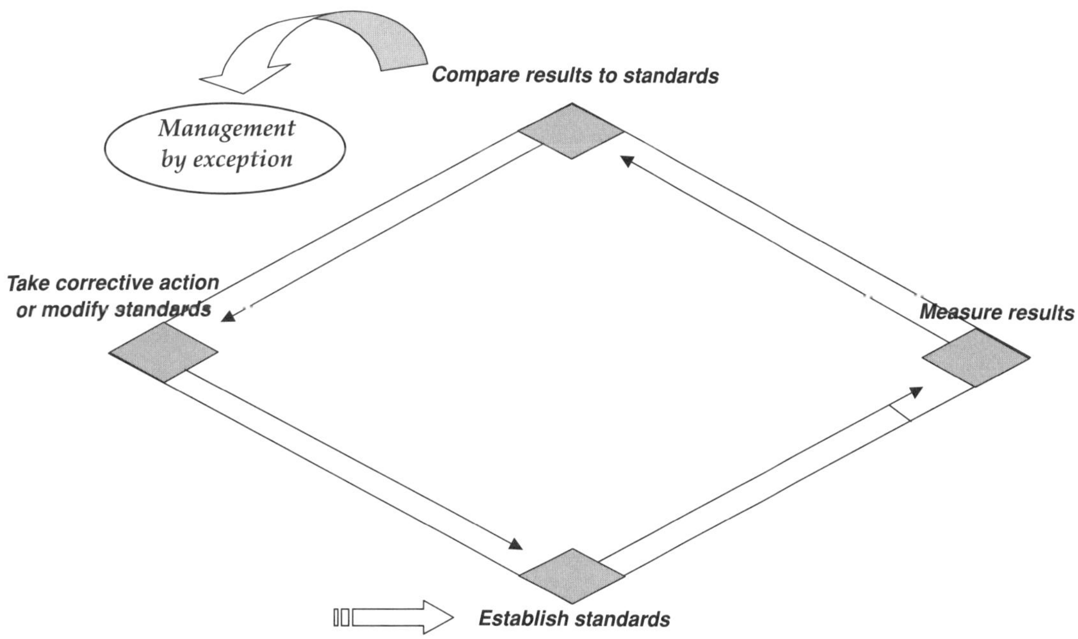

EXHIBIT 1.1 Control loop diagram.

The power of the control loop lies in completing each of the four steps, and then repeating them over and over again. Each time the loop is completed, both you and your employees become more knowledgeable about how to achieve specific financial and operating goals. You also will enhance the entire process each time the loop is completed, until you achieve a level of mastery, which I define as thoroughly understanding, using, and enhancing a process until you have upgraded it to a level of exceptional and individualized competence.

Completing a control loop doesn't have to be a strict 1-2-3-4 process. For example, when you build a home you:

- Establish standards by designing and planning the job
- Measure results by monitoring and measuring what's going on at the jobsite
- Compare those results against your plan
- Follow up with your employees or contractors when results don't go according to plan to discover why not, and how to fix the problem
- Adjust your plan if you overlook something

On a micro-basis, when your customer requests changes, you go through the same type of process by creating a change order, and making sure that it's properly completed. The following steps provide a conceptual overview of each process in the control loop.

Step 1. Create standards and expectations

As a first step to financial success:

- Create an annual budget for your company
- Prepare individual job estimates
- Document these financial goals in your accounting system

It's always easier to achieve a goal when you know exactly what the goal is. The process of creating your annual budget can help you quantify and clarify your company's goals and employee performance expectations for the upcoming year as the year progresses. Similarly, creating individual job estimates as the year progresses can help you quantify and clarify time, cost, and income expectations for each job, which, in turn, will help you achieve your company's annual goals.

Although the enhancement of operating procedures generally is beyond the scope of this book, the control loop concept is readily applicable to this side of your business as well. Examples of operational standards include:

- Company policies
- Standard procedures
- Organization charts
- Job descriptions
- Required number of sales calls
- Minimum or maximum size of job to be accepted
- Types of jobs to be accepted
- Trade contractor agreements
- Quality standards
- Process checklists

Step 2. Measure results

When you establish standards such as a goal, vision, or procedure, you also should establish a procedure to measure the results. If results are never measured, the original standard likely will fall into disrepair or oblivion. For example, many builders require trade contractors to remove discarded materials when they complete their work. As the business owner, you need to establish periodic checks to determine that your trade contractors are complying with this standard.

Consequences are a component of the system that measures results. What are the consequences if your trade contractor fails to comply with your standard and does not remove his or her discarded materials? By checking on your trade contractors and enforcing the consequences, you measure who complies with your standards.

Often, the very act of measurement has a positive influence on the outcome. When an activity or process is measured, employees perceive that it's important to management. They also realize that their work is being observed, which typically improves the quality or quantity of work. So be careful about what you measure, you just may improve the results!

Measure financial results

Your accounting system is a measurement tool. Just as it's wise to "measure twice, cut once," paying close attention to your financial measurements will help you determine your most effective and efficient subsequent business moves.

The accounting system I will help you design will house the actual financial transactions or financial measurements of your business. You will enter financial data into your accounting system and classify each transaction by telling the accounting system what activity is taking place. If the transaction is directly related to a job, you will tell the system which job, and, if you are measuring processes within a job, which process. QuickBooks Pro 2000 then will tabulate and summarize the results of your actual operations on request, thereby yielding financial measurements in the form of reports.

Measure operating results

Systematic measurement methods also are important to the operations side of your business. For example, you may want to know how many jobs have been completed without defects, or if the targeted number of sales calls and follow-ups were made, if established procedures were followed and checklists were completed properly.

The best way to create these measurement systems is to ask yourself or your staff which processes and activities are most important to achieving success. Then ask and determine how someone could confirm objectively, whether the standards were or were not achieved. The final step is to establish the required methods, such as spot checks or special reports, to periodically measure progress.

Step 3. Compare results to standards

At regularly scheduled intervals you will need to review results and compare them to your standards so you can take corrective action in a timely manner. Determine the actual reporting and reviewing frequency based on your company's "time needed to fix it before it blows up" circumstances. For example, how soon would you like to know that your roofing materials haven't been ordered or paid for yet?

Many companies monitor financial information on an as needed basis, whether hourly, daily, weekly, or monthly. Consider a company as large as General Motors. It would be hard to visualize their managers gathering around a table to review initial operations reports three or four months after a new line of cars rolls off the assembly line. A more likely scenario would show them reviewing the numbers on a day-to-day basis to see where they could modify operations, shift workers, reassign inventory, or enhance equipment to either smooth out problems or achieve or exceed projections.

This step requires a concentrated effort on your part. (Get the cup of coffee and find the quiet space.) Set aside time to review the results, compare the results against targets, and perhaps face specific unpleasant realities. This is the step that has the greatest potential for sidestepping disasters. When you compare results to standards you learn what's really happening in your business or on the job. This is the reality check!

Compare financial results to standards

I recommend you monitor your company's year-to-date actual vs. budget reports on at least a quarterly, if not a monthly basis. For individual jobs, review job-costs-to-date actual vs. estimate reports on a weekly or monthly basis. Examples of these reports are illustrated in chapter 4.

Although I don't cover operations reports in this book, a report that shows actual results, planned results, and the differences by volume and/or percent will be very useful to you. For example, you could build a spreadsheet that shows actual sales calls by salesperson, planned sales calls by salesperson, and the difference, whether favorable or unfavorable.

Manage by exception

The phrase "management by exception" means that after you've studied your actual vs. budget, or your actual job costs vs. estimates variance reports, you should expend the greatest effort on understanding, managing, and controlling areas that show the most significant variances or exceptions. Knowing where variances or differences exist helps you focus your attention on areas where your intervention will have the most impact. For example, if your roofing materials estimate is within 0.5 percent of what you expected and if the dollar variance from plan is minimal, you may not bother to investigate further. If your roofing labor is over or under by 12 percent and the dollar variance is more than minimal, however, you most likely should spend more time managing this exception area.

When you find variances, first double-check the underlying data. A mistake may have been made during the input of the income or expense item. Even if you find and correct the data, I recommend you review the revised set of reports because the favorable or unfavorable variances may merely have shifted to another area or job.

Take time to investigate and understand both positive variances (when income is higher than projected, or expense is lower than anticipated) and negative variances (when income is lower than projected, or expense is higher than anticipated). While positive variances generally are good news, an unusually large positive variance also could indicate errors or omissions in the accounting process, or that a vital operational process has not taken place or may have been missed altogether.

Step 4. Take corrective action or revise standards

After you're certain that your data is accurate, take immediate action to correct the situation. If corrective action is no longer possible, then view the information as an opportunity for improvement. A negative experience becomes an opportunity to:

- Learn more about a specific process for future reference
- Clarify roles and responsibilities with employees
- Enhance procedures
- Revise inaccurate or unrealistic standards
- Change your estimating procedures or formulas
- Reallocate or reassign staff

It's extremely important to understand why the variance occurred. Focus your investigative efforts on continually improving operations rather than affixing blame. You'll find it far more productive in the long run to understand what went wrong and why. Then you can create a system, procedure, or technique to help you keep the situation from recurring. Employees also will be more likely to help identify and fix problems if they are rewarded for understanding the cause of the problem and for their contribution towards the long-term fix.

Remembering lessons learned

When I've made a mistake or a bad decision, I find it helpful to keep a log of lessons learned. My log also contains the estimated cost of the lesson, and what I want to do differently next time. I find it useful to review this log on a regular basis—not to beat myself up, but to remind myself of the things I want to change and to help me identify my vulnerabilities. By regularly reviewing my log, I am able to fortify my company and myself so that I am less likely to repeat these specific lessons. It's a human tendency to want to spend our time where we're the most skilled and the most comfortable. Just like your blind spot when you're driving, however, you'll need to continually double-check these problem areas to see that you're covered.

CHAPTER 2

Factoring in the details

Let's look at some factors that will influence your decision regarding the degree of complexity you need to build into your company's accounting system and what you ultimately will want from your system.

The Factor column in Exhibit 2.1 shows some of the structural variables common to remodelers and custom home builders. The Continuum columns indicate the complexity of the accounting system you will need, given your company's current or anticipated situation.

	Continuum →	
Factor	***Simple (lower level of detail)***	***Complex (higher level of detail)***
Uses for/users of information:		
Tax information needs	Sole proprietor, some LLCs	Partnerships, corporations, some LLCs
Need to improve/change/grow business	No need to change	High need to improve/change/grow
Owner's willingness to use reports	Owner doesn't use acct'g reports	Owner uses reports to control business
Needs and number of users of information	Owner only	Lenders, stockholders, secured parties, internal managers, estimators, etc.
Financial risk:		
Life-stage of business	Start-up	Well-established
Asset investment	Insignificant	Significant
Dollars at risk on jobs	Cost-plus contracts	Fixed-price contracts
Number of jobs	Few	Many
Size of jobs	Small	Large
Diversity of jobs	All jobs similar	Wide variety of jobs
Complexity of jobs	Low complexity	High complexity
Profit margins/overall long-term profitability	Excellent margins/profitability	Low or negative margins/profitability
Owner involvement:		
Owner involvement in onsite work	Owner consistently onsite	Owner does little onsite work
Owner involvement in estimating	Owner does all estimating	Others do estimates
Owner involvement in purchasing	Owner does all purchasing	Others do purchasing
Number of layers of management	One	More than one
Number of employees	None	Many
Number trade contractors	None	Many
Accounting support:		
Accounting support	Little or no assistance	Excellent support
Accounting expertise	Little or no expertise	Excellent expertise
Accounting support turnover	High turnover	Low turnover

EXHIBIT 2.1 Factors affecting the level of detail required in an accounting system.

A more advanced and complex remodeling or custom home building company will need more detailed financial information to effectively control its operations. Let's examine some of the concepts in Exhibit 2.1 more closely.

Tailoring the information

Depending on the type of business you own or operate, you will have different uses for financial information. For example, if you are a mid- or large-size company, various employees may require different financial reports in order to effectively meet their responsibilities. Let's look at some of the different types and possible uses of information.

Annual tax return information

If you are a sole proprietor or a single-owner limited liability company (LLC) filing a Form 1040 Schedule C, you need to report only annual income and expense. In this case you probably will want to produce at least a simple balance sheet even though the tax return does not require it. If your primary goal is getting the Schedule C completed, a very simple financial management system may be acceptable, especially if you are a one-person operation with no employees and few trade contractors.

Both S and C corporations, partnerships, and certain other LLCs require balance sheets in addition to more detailed income and expense information. You therefore will find it necessary to develop a more advanced accounting system for these types of entities.

Specific financial measurements

If your business is *not* changing or growing then your need for specific financial measurements may be relatively low. If you already know which jobs are profitable and which aren't, which jobs to take and which to turn down, which of your lead carpenters deserve bonuses for outstanding job performance, and whether kitchen jobs have a higher profit margin than additions, then a simple accounting system designed primarily for tax preparation purposes probably will meet your needs.

On the other hand, if you suspect that you could enhance your control over your business by tracking these kinds of measurements, or if you want to make changes and track your profits and job costs, you probably should elect to utilize job-cost and variance reporting.

Accommodate users

If you, as the owner, don't want to review the accounting reports for management purposes, you probably don't need a detailed system because your report requirements are few. If, however, you want to review accounting reports on a regular basis (and I recommend you do), or if several managers want detailed information on a regular basis, you will need a more detailed system. Talk with all potential users within your company to find out the information they need and then build the details into your system.

In accounting, a good rule of thumb to follow is: The more people who need to use the information, the more detail you will need to design into your system. This is because different people need to access and use the information in unique ways.

If you need to provide information about your company operations to other interested parties, such as your tax advisor, bankers, stockholders, or partners, you will want to design a system capable of providing them what they need. Each of these interested parties probably will be looking for a slightly different version of the information, so create your system's structural "building blocks" in a way that will let you provide the appropriate information in a report format customized to each person's needs. Just as in remodeling or custom home building, a design that meets the needs of one occupant (user) will be simpler to create than if there are four occupants (users) using the same space (data).

Minimizing financial risk

A thoughtfully designed accounting system, accurate summary and detail-level reports, timely reviews, and immediate action are essential if you have a large investment to protect and/or potential risk to minimize. In short, the more you have at risk, the more important financial control becomes.

As the pace of work escalates, the complexity of operations increases and it becomes more difficult to keep all of the details under control. Loss of control usually equates to loss of money through errors, mishaps, or cost overruns—even if the loss is not immediately apparent. With more at risk, it becomes increasingly important to have a financial system that is robust and detailed enough to provide you with warning signs in time to avert losses before they occur.

Gauging the degree of owner involvement

If you are directly involved in the moment-to-moment operations of your business, it is less likely that you will need detailed reports about those activities. You most likely are in touch with reality and are able to course-correct as you proceed. As you add trade contractors, employees, and layers of management to your organization, however, you probably won't know what they are doing all of the time and won't be able to personally correct problems before they cost you money, time, or your reputation. For small businesses, one of the highest risk categories is the cost associated with other's inaction, such as wasted time, mistakes, or miscommunications.

A strong system of *accountability controls* is increasingly necessary as you deal with more people. Clear descriptions of who is responsible for what, and how each person's effectiveness will be *objectively* and *financially measured* against plans and goals is one example of this system. These accountability controls are supported and enhanced by a strong, accurate accounting, reporting, and feedback system, such as the control loop discussed in chapter 1.

Your ability to create and maintain an effective accounting and reporting system will be in direct proportion to your ability to attract and retain an employee with the right accounting expertise and mindset. Chapter 5 includes my tips for selecting, training, and retaining accounting help. If the complexity of your system requires that you hire accounting help, don't undervalue your inhouse accountant or bookkeeper. Financial record keeping is not a simple, clerical matter.

Relating job costing to system complexity

The job-cost level you require is a primary factor in your decision regarding the timing of your income and expenses. The matrix in Exhibit 2.2 plots the timing of income and expense (from simplest to most complex) against the level of job costing (from simplest to most complex) to help you determine the most appropriate accounting system for your business.

The nine boxes created by the matrix each represent a potential position of complexity. Each X indicates a basic type of management information report that you should expect to use on a regular basis from within that position. (You should be able to obtain a variety of other reports as well—these are just the basics.)

Timing of income and expense

The Timing of Income and Expense column represents a continuum from lower system complexity to higher system complexity as it relates to income and expense recognition. (A full discussion of the pros and cons of cash vs. accrual accounting is included in chapter 3.) Generally, management accountants prefer accrual or percentage of completion for management information or book purposes because it provides a more accurate measurement of your actual income-

Timing of income and expense	*Level of job costing:* None	Simple costs	Process costs
CASH BASIS			
Balance Sheet (no receivables or payables)	X	X	X
Income Statement (cash received & paid)	X	X	X
Budget & variance repots-company	X	X	X
Estimates & variance reports-by job		X	X
ACCRUAL BASIS		A	B
Balance Sheet (includes receivables & payables)	X	X	X
Income Statement (income earned & expense incurred)	X	X	X
Budget & variance reports-company	X	X	X
Estimates & variance reports-by job		X	X
PERCENTAGE OF COMPLETION			
Balance Sheet (incl receivables & payables by % done)	X	X	X
Income Statement (inc earned & exp incurred by % done)	X	X	X
Budget & variance reports-company	X	X	X
Estimates & variance reports-by job		X	X

Lower → *Higher* (level of job costing, left to right; timing of income and expense, *Lower* at top to *Higher* at bottom)

A	B
Direct job cost by Labor, Trades Materials, & Other	Direct job costs by Labor, Trades Materials, & Other -for each process

EXHIBIT 2.2 Factor matrix.

generating activities and related expenses than does cash-basis accounting, which only measures cash flow.

Level of job costing

Job costing means that in addition to company-wide income and expense reports, you track your income and direct expenses by job. Although you may choose not to use extensive job-cost reports, I recommend that you implement one of the following levels of job costing.

- **Simple job costing** records transactions by indicating the job and account, such as labor, trade contractors, materials, and so on. You then can obtain reports that summarize job costs by these accounts.
- **Process job costing** records transactions by job and account, plus it indicates the process that generated the cost, such as site preparation or framing. You then can obtain reports that summarize transactions by job process as well as by account.

If you also create and enter your job estimates for either level of job costing into your accounting system, you can obtain variance reports (see chapter 4) that summarize your income and costs and compare them to your estimates.

A and B positions

The positions labeled A and B in Exhibit 2.2 utilize accrual and job-cost features. They represent the complexity levels I typically find are most useful for the majority of remodelers, developers, and custom builders. You may, of course, initially choose to design your system in one of the cash-basis or no-job-cost positions, but I anticipate that as your business grows you will want to move your day-to-day system toward the positions labeled A and B.

Therefore, in the coming chapters, I will focus on helping you build the foundation and the framing for these two types of accounting systems.

Making some final decisions

After reviewing the decision-assisting information presented in this chapter, determine what reports you would like to get from your accounting system. Then decide whether you want to get there all at once, or if you prefer to work your way there in incremental steps. In summary, my recommendations are as follows.

- Unless you are willing to invest your own time and energy, and are committed to finding the right accounting assistant, don't attempt to start with the most complex system. Developing a system that is more complex than you need is a waste of energy, and has the potential to create confusion and frustration. Unless you plan to really use the additional level of detail that a more complex system will give you, or if don't have the resources to expend to create such a system, I recommend that you implement the simplest acceptable system.
- If you really need the information a more complex system can give you, and are willing to invest in it, develop it, support it, and use it wisely then you will find that it can pay for itself many times over. In that case, select either the A or B position in the matrix in Exhibit 2.2.
- If you choose to implement your new system in incremental steps (from simplest to more complex), think of it as a multistage remodeling or building job. Get the vision, the underlying knowledge, the plan, and the tools in place to ensure that you don't make any major errors or have to undo or redo previous work along the way. Build your system in the order presented in the upcoming chapters. Make sure you understand each piece of the structure and can work with it before you move to the next layer.

CHAPTER 3

Understanding cash and accrual accounting

"It's all about timing." Ever heard that said before? Well, that's the essence of the difference between *cash-basis* and *accrual-basis* accounting. The primary difference is *when* you recognize income and expense.

Why should it matter when you recognize either? If a job is profitable in the long run or a loss in the long run, the outcome is the same, isn't it? Yes, it is. You end up with the same net profit on that job before taxes either way. But what if you need to prepare financial reports on a particular date and need to see how each job is progressing against estimates? Your reporting periods usually don't coincide with the beginnings and endings of your jobs. How do you represent the beginnings, middles, and endings of projects with money earned, bills incurred, payments received, and so forth as of the particular date you prepare your financial report?

In cash-basis accounting the trade accounts receivable and trade accounts payable accounts do not exist. All the funds you receive related to services performed in the past or present or that will be performed in the future are considered current income. All payments made by the company for operating expenses also are considered current expenses.

NOTE **Even in cash-basis accounting, transactions related to fixed assets and their related depreciation, loan proceeds or payments, and equity accounts are handled in a manner that matches expenses incurred against income earned.**

In accrual-basis accounting the theory is that you want to recognize all of the income you have actually *earned* through your business activities within a given timeframe. After you recognize that income by creating invoices (trade accounts receivable) or recording entries for "income earned, but not yet invoiced" (other receivables) on your books, you then make sure all of the expenses related to earning that income are shown on your books. Any costs incurred that are not yet entered in your financial records then are entered into accounts payable or other liability accounts so that those expenses are included in the financial results.

To see how results can differ drastically within a given timeframe between cash-basis and accrual-basis accounting methods, please review Exhibit 3.1. Assume that both of these remodelers performed the same amount of work for the month, have the same beginning balances in their checking accounts, receive the same cash inflow, pay the same payroll, pay exactly the same bills, and receive exactly the same deposits and payments from their customers. The cash-basis remodeler in this scenario shows a *loss* of $11,607 for the month, while the accrual-basis remodeler shows income of $9,370—a difference of $20,977!

Activity or transaction	Cash basis				Accrual basis				
	Checking bal + or (-)	(Limited) Payables	Income +	Expense (-)	Checking bal	Receivables +/(-)	Payables +/(-)	Income +	Expense (-)
Beginning balances	$12,000				$12,000	14,879	6,542		
Contracted client drops off $1,500 deposit for future job-deposited to checking	1,500		1,500		1,500		1,500		
Invoices are prepared for customer work performed this month (cash basis must do this outside QBP)						42,425		42,425	
Current month's direct cost bills are entered into payables (cash basis must do this outside QBP)							19,091		(19,091)
Current month's overhead bills are entered into payables (cash basis must do this outside QBP)							6,364		(6,364)
Payroll paid for month	(6,156)	1,444		(7,600)	(6,156)		1,444		(7,600)
Some of prior month's remaining direct cost bills are paid, and some of current month direct cost bills are paid	(13,843)			(13,843)	(13,843)		(13,843)		
Some of prior month's remaining overhead bills are paid, and some of current month overhead bills are paid	(4,614)			(4,614)	(4,614)		(4,614)		
Customer payments are received	8,750		8,750		8,750	(8,750)			
Customer payments are received	4,200		4,200		4,200	(4,200)			
Ending acct balances	$1,837	1,444	14,450	(26,057)	$1,837	44,354	16,484	42,425	(33,055)
Net income for the month				($11,607)					$9,370

EXHIBIT 3.1 Cash basis results vs. accrual-basis results.

The differences occur because of the *timing*—when income is recognized and when expenses are recognized. The cash-basis remodeler has $27,975 ($42,425 – $14,450 = $27,975) less in income and $6,998 ($33,055 – $26,057 = $6,998) less in expense than the accrual-basis remodeler even though both of these remodelers worked just as hard and accomplished the same amount of actual work. Next month the cash-basis remodeler may have a sudden influx of cash, pay fewer bills, and show a $20,000 positive net income for the month. So over time they will both come out even.

At this point in time, however, which of these remodelers has better information about their company's performance for the month? Let's look at what this report tells each remodeler, supposing both companies estimate a gross profit margin for the year of 35 percent (income less 65 percent direct cost = 35 percent gross profit).

- The accrual-basis remodeler's direct costs of $26,691 ($7,600 for payroll + $19,091 for the current month's direct-cost bills) calculates out as 63 percent of gross income. That result sounds reasonable because the remodeler has estimated gross profit for the year at 35 percent.
- When the cash-basis remodeler looks at his or her direct costs of $21,443 ($7,600 for payroll + $13,843 for bills actually paid this month), it appears that direct costs are 148 percent of gross income.

NOTE When income and related expenses can't be matched, or the gross income is misstated, the resulting percentages aren't reliable or helpful.

Let's assume that each remodeler has done his or her accounting accurately and now has an income statement to review for the month's activities. What will the income statements tell them about their business?

- **Business viability.** If you were the cash-basis business owner showing the $11,607 loss, you might spend a few nights tossing and turning while you wonder about the viability of your business. If you were the accrual-basis business owner, you would be able to pinpoint how you're doing against budget projections. Your budgeting probably is fairly accurate, as you were able to predict how much work you could reasonably expect to perform each month and what proportion of income would be related to expenses based on historical information.
- **Budgeting.** As a cash-basis owner, your budgeting isn't as reliable because you can't predict actual cash flow as accurately and your variances from budget might be very dramatic based on your historical information. Because your actual income and costs don't represent actual work performed or the related expenses, they aren't good indicators of productivity or cost overruns. Your variance reports won't be as useful because the variances do not necessarily represent operational deviations from plan.
- **Cash flow.** Although cash flow should be carefully monitored by all businesses, cash-basis reports have a tendency to focus a business owner's attention on cash manipulation and other cash-flow issues. If you were the accrual-basis business owner, you might be more likely to focus on creating and maintaining consistently profitable operations. Profitable operations have a tendency to ultimately relieve cash-flow problems.
- **Problem recognition.** As the cash-basis business owner, you're constantly faced with "feast or famine" scenarios. So although you might be panic-stricken this month over an $11,607 loss, you may be more likely to overspend if next month's income statement shows a $20,000 profit. And because you're used to the wild cash ride, you might be less likely to go looking for an accounting problem or error. As the accrual-basis business owner, you probably will be suspicious of a $20,000 net income month and will go looking for accounting errors or other missing documents.

It's fairly obvious from this example that I recommend accrual-basis accounting for management reporting and information purposes. If the above examples haven't convinced you, here are some additional reasons to use accrual-basis accounting, and the activities you need to perform to achieve the results.

- **Timely invoices.** To see that income earned is shown correctly on your books, create timely invoices for your customers. For customers who aren't due to receive invoices, prepare journal entries to recognize the income you've actually earned within the accounting period, such as the month or year. This requires some extra monitoring and that you stay in touch with work performed (not a bad idea, all in all).
- **Income clarification.** Ensure deposits or draws received for work not yet performed are not included in income. Make a journal entry to move deposits or draws to "unearned income"—a liability account—because until you earn the funds you have received you technically owe those proceeds back to the customer.
- **Purchase orders.** If you choose to use the purchase order feature in QuickBooks Pro 2000, enter bills for payment using the job- and vendor-related information already entered in the purchase order. This actually saves entry and classification time and allows you to double-check the items you ordered and their cost against the vendor's bill.
- **Track payables and print checks.** Enter remaining bills into the payables system. This records the expense on your books, and lets you prepare a printout of your payables at any time. The printout shows who you owe, when the bill is due, and the discount due dates so you don't miss valuable prompt-payment discounts. After you select the bills to be paid, use your printer to print the checks.
- **Monitor expenses.** Compare estimated expenses against actual expenses to reassure yourself that all bills have been entered into accounts payable correctly.

Defining percentage-of-completion accounting

Percentage-of-completion accounting is another way of estimating how much income you've earned during a specific accounting period or life-to-date for a project. Accountants don't view it as an alternative accounting method, but rather as a computation method that describes how you might go about booking your earned income.

QuickBooks Pro can prepare percentage-of-completion invoices if you first enter the total detailed or summary estimated income for any given project into the system. You enter your estimate of the percentage of completion for each portion of the job and then QuickBooks Pro performs the calculations and creates the percentage-of-completion invoice for you. The software also will hold in memory the remaining amounts not yet invoiced until you are ready to prepare the next percentage-of-completion invoice for that job.

Reporting income for taxes

You can have accrual-basis accounting for your day-to-day or book accounting and still use cash-basis accounting for your taxes. This usually is a real relief for people who want to use accrual-basis accounting for business information and reports, but feel that they must use the same basis for both tax and book. Not true! This is one place where you can have your cake and eat it, too.

NOTE Once you've established yourself as a cash-basis or accrual-basis business with the IRS, changing your tax-basis accounting method can be a major undertaking, best attempted only with your CPA's full approval and involvement.

If you decide to set up an accrual-basis accounting system, and currently are on record with the IRS as a cash-basis business, you will need you to provide your accountant with a detailed list of your receivables and payables from the beginning to the end of the year. You also will need to provide the specific accounts to which each has been posted. Each of these reports can be created within QuickBooks Pro. Rest assured that having accrual-basis books and cash-basis tax returns does not involve keeping the proverbial "two sets of books!"

Using the information you provide, your accountant will prepare year-end adjusting entries on a separate spreadsheet in order to convert your accrual-basis reports into cash-basis results for your tax returns. He or she will:

1. Subtract your ending receivables from your income
2. Subtract your ending payables from your expense
3. Add your beginning receivables to this year's income
4. Add your beginning payables to this year's expense

You won't be able to anticipate your taxable net income without going through this conversion process. Your tax preparation and/or tax planning process may cost a bit more (ask for the estimate ahead of time), but you will be free to get accrual-basis management reports—reports that can help you understand and run your business much better and help you to sleep easier at night!

NOTE If your tax accountant isn't overly excited about the value of the extra work your new, accrual-basis accounting system requires, you might want to get a second opinion. Remember that the purpose of your accounting system is to provide you with the information you need to better manage your business. Easy tax forms preparation should be a secondary, albeit complementary, objective.

CHAPTER 4

Reviewing basic financial reports

When you work with new customers it only makes sense to find out within the first few minutes the type of remodeling job they have in mind—a kitchen, a bath, an addition? Do they have single or multiple uses for the remodeled space? Do they want the space to be basic and functional or are they looking for a more extensive custom design that will upgrade their entire living area? Is their budget the primary "driver," or is it secondary to getting exactly what they want? Are their time constraints two days, two weeks, or two months? Are they willing to have the work performed in stages to match their changing needs?

As a custom builder, you go through a similar process of defining your client's needs, designing potential solutions, and obtaining your client's buy-in and acceptance.

If you view the process of developing your accounting system in much the same way, you will see that your information wants, needs, and constraints influence your design and building processes. As in the construction industry, a sound fundamental design is the basis for all future enhancements. A good foundation will let you create a structure appropriate for the current situation, with the flexibility to add more detail as your company grows and your information needs change.

Understanding accounting terminology

There are three key accounting terms—double-entry bookkeeping, debits, and credits—that will help you understand your reports and where the information in the reports comes from.

Double-entry bookkeeping

Every transaction has two parts, hence the term *double-entry bookkeeping*. In accounting, each transaction is the equivalent of a two-sided coin. For example:

- When you create an invoice in your accounting system for products delivered or services rendered, two things happen—your income account increases and your accounts receivable account increases.
- When you receive a payment from the customer you invoiced, your cash account increases and your accounts receivable decreases. Income is not affected because you already recognized the income when you entered the invoice as a receivable.
- When you enter a bill as a payable in your accounting system two things happen—your expense account increases and your accounts payable increases.
- When you later pay the bill, your cash decreases and your accounts payable decreases. The expense account doesn't increase because the expense was recognized when you originally entered the bill.

Each account is a mini-system within the accounting equation:

Assets = Liabilities + Equity

Each account is increased and/or decreased through the use of debits and credits.

Debits and credits

In manual accounting systems, accounting records are organized in books called ledgers. The term "ledger," therefore, is one you probably will encounter in accounting discussions. For example, the main book of account pages is referred to as the "general ledger," while the other books that contain detailed records of customer or vendor accounts are referred to as "subledgers." You also may encounter the term "ledger account," which refers to an account that normally would be kept on the general ledger.

Each account in the ledger is assigned to a page. Each page or account in the ledger has columns to record the date of the transaction, the nature of the transaction, the amount of the transaction, with one column for increases and another for decreases, and a column that shows the resulting balance. The two columns used to record increases and decreases are referred to as debit and credit columns. The left column is referred to as the debit side of the account. The column to the right of the debit column is called the credit side.

So, in accounting the term *debit* does not necessarily mean a reduction. It simply means *left*, as in the left side of the account. In certain accounts, a debit can mean an increase.

Conversely, the accounting term *credit* does not necessarily mean on increase. It only means *right*, or the right column of the account. In some accounts, a debit can mean a decrease.

As you can see, this terminology can create confusion for people who use these two words in the normal day-to-day way. To help you understand the accounting usage of these terms, Exhibit 4.1 summarizes the affect of debits and credits on various types of accounts.

Yes, it probably seems totally and incredibly strange to think that a debit can represent an increase and that a credit can represent a decrease to an asset or expense account. But there it is! This is the way accounting has been done since its inception. Eventually, if you can let go of the debit = minus and credit = plus frame of reference, you'll find it's actually mathematically logical because it's based on the accounting equation and the position of each type of account within the equation.

Account	Increase	Decrease
Assets	Debit*	Credit
Liabilities	Credit*	Debit
Equity	Credit*	Debit
Income	Credit*	Debit
Expense	Debit*	Credit

Note: The * indicates this is also the normal status of the account. For example,assets typically exhibit a debit balance, liabilities typically exhibit a credit balance, and so on.

EXHIBIT 4.1 Summary of affect of debits and credits on Chart of Accounts accounts.

For example, your accounts receivable is an asset account with a normal debit balance. It is comprised of invoiced dollars sent to a number of customers who owe you money, each of which should normally show a debit balance. If one of your customers has a credit balance, it means that you owe him or her money or work, or it means that something has been erroneously posted to the account.

Another example would be an expense account that normally has a debit balance. If you see that it has a credit balance, you should probably investigate to see what's happening within the account to create the unusual credit balance. In most instances, a credit balance in an expense account arises from a refund of costs expensed in a prior period.

Fortunately, in QuickBooks Pro 2000, you won't need a nitty-gritty day-in, day-out familiarity with debits and credits. The forms you fill in and the individual systems within QuickBooks Pro will, in most cases, automatically post the debits and credits for you. But the accounting system is built on the debit-credit system, so you should have at least a rudimentary awareness of what the terms mean within different scenarios.

Reading basic reports

The types of reports you ultimately want will help determine the level of detail and complexity of the accounting system you build. Some reports are basic to all accounting processes, such as the balance sheet and income statement, while others are optional and useful only to certain companies, owners, individual managers, or specific industries, such as job-cost reports for remodelers or custom builders.

If you have just started your remodeling or custom home business, or if you are a sole proprietor, you may initially need only simple income statement and balance sheet reports. This may be acceptable if you work on one or two jobs per year. As you take on more or larger jobs and develop more extensive employee or trade contractor relationships, however, you should consider adding basic job-cost features and reports to your system. As your company grows in size and complexity, you probably also will want to add job-stage costing, whereby you track costs related to different stages of each job, such as site preparation, foundation, or roofing. I also encourage you to develop and use budgets and estimates regardless of the degree of sophistication you build into your system.

The five types of reports presented and discussed in this chapter are:

- Balance sheet
- Income statement
- Cost accounting by job
- Simple job-cost system—cost accounting variance for a specific job
- Detailed job-cost system—cost variance for a specific job

Each example presents a sample remodeling company's accounts. The reports progress from fairly basic to more detailed. Please remember that your company is unique. In the design section of this book you will determine which accounts you need and what level of detail is appropriate for your specific needs.

The first two reports—balance sheet and income statement—are considered essential financial accounting reports for all types of companies, regardless of industry. The third report—cost accounting by job—illustrates a simplified approach to job costing. Even though they each present very different kinds of information, all three of these reports reflect Step #2 (measure results) of the control loop discussed in chapter 1. The fourth report—cost variance for a specific job—illustrates a simplified approach to job costing that incorporates a comparison of standards. The fifth report—cost variance for a job-by-job stage—shows a more detailed approach to job costing. These two reports are very powerful because they actually incorporate Step #1 (create standards), Step #2 (measure actual results), and Step #3 (compare results to standards) from the control loop. Step #4 (take action or change standards) is up to you.

Balance sheet

The balance sheet, shown in Exhibit 4.2, often is referred to as a company's financial snapshot at a given point in time. The balance sheet reports the overall condition of your company to anyone who might have a financial interest in it, such as owners, creditors, or stockholders. If you were going to buy or sell your company on a particular day, you might think of the balance sheet as a "what you see is what you get" kind of measurement.

There are three major divisions within the balance sheet: dollars of *assets*, dollars of *liabilities*, and dollars of *equity*. These sections are the basic elements in what is commonly referred to as the *accounting equation* of:

ASSETS = LIABILITIES + EQUITY

or alternatively as

ASSETS – LIABILITIES = EQUITY

Each section is comprised of individual accounts that report the unique financial status of your company. You may want to think of the balance sheet sections as the "3 Os"—what you **O**wn, what you **O**we, and what's left **O**ver.

- **Asset accounts** show what you **own**, such as cash, receivables, or fixed assets. These assets generally are shown in order of their relative liquidity or how long it would take to convert the asset to cash, going from shortest to longest. The company in Exhibit 4.2 owns $44,691.62 in assets.
- **Liability accounts** list what you **owe**, such as trade accounts payable, payroll and related liabilities, other short-term liabilities, and longer-term liabilities such as notes or mortgages. Liabilities generally are listed in order of how quickly they must be paid, from the shortest to longest. This company's liabilities total $34,124.49.
- **Equity accounts** represent what's left **over**. This amount is comprised of contributions from business owner(s), plus what the business has earned during its life-to-date, less what has been distributed from the company to the owner(s). Account titles for a sole proprietor or partnership will be different than the titles for a corporation. The equity for this company totals $10,567.13.

As you review Exhibit 4.2 bear in mind that this is a very condensed, overview-level report. Each section has summarized a number of individual accounts in order to give a very succinct view of the basic elements of the balance sheet. Notice that the assets of $44,691.62 equal the liabilities of $34,129.49 plus the equity of $10,567.13.

In chapter 7, I'll lead you through the process of selecting balance sheet accounts customized to your company. In chapter 29, I'll discuss how to enhance the usefulness of your balance sheets.

Income statement

An income statement, shown in Exhibit 4.3, measures the results of operations for a limited period of time, such as the current month or year-to-date. Standard accounting practice dictates that you restart the measurement process for each income and expense account each year.

There are four basic sections within the income statement. Each section has accounts that show the detailed results of financial transactions, including:

- **Income** from operations, such as remodeling, building, sale of spec homes, design, or consulting
- **Direct costs** report the costs directly related to the production of income. To segregate these costs, use the QuickBooks Pro cost of goods sold (COGS) accounts. Examples of direct costs for remodelers and custom home builders include:
  - Labor
  - Labor burden, including payroll taxes, fringe benefits, paid time off, and training
  - Trade contractors

Dreamscape Remodeling, Inc.
Balance Sheet
As of December 31, 2000

	Dec 31, '00
ASSETS	
Current Assets	
Checking/Savings	
T.100 · Checking account	944.41
T.110 · Savings account	4,881.89
T.119 · Cash on hand	200.00
Total Checking/Savings	6,026.30
Accounts Receivable	
T.120 · Accounts receivable, trade	26,848.00
Total Accounts Receivable	26,848.00
Other Current Assets	
T.130 · Other current assets	2,500.00
Total Other Current Assets	2,500.00
Total Current Assets	35,374.30
Fixed Assets	
T.173 · Office FF&E, net book value	642.38
T.174 · Computers/software-net book val	1,433.68
T.175 · Vehicles-net book value	7,876.51
Total Fixed Assets	9,952.57
TOTAL ASSETS	45,326.87
LIABILITIES & EQUITY	
Liabilities	
Current Liabilities	
Accounts Payable	
2000 · Accounts payable, trade	12,055.00
Total Accounts Payable	12,055.00
Credit Cards	
T.210 · Company credit cards	3,994.60
T.215 · Shareholder cr cards	8,373.29
Total Credit Cards	12,367.89
Other Current Liabilities	
T.250 · Line(s) of credit	8,891.01
Total Other Current Liabilities	8,891.01
Total Current Liabilities	33,313.90
Total Liabilities	33,313.90
Equity	
3100 · Capital stock - 500 shares	500.00
3900 · Retained Earnings	11,512.97
Total Equity	12,012.97
TOTAL LIABILITIES & EQUITY	45,326.87

EXHIBIT 4.2 Condensed balance sheet.

  - Materials
  - Other costs that can be assigned to jobs, such as permits, direct financing, and sales costs associated with specific sales
  - Costs that cannot be assigned to specific jobs, but are related to the production of income, such as field cell phones, uniforms, or repair and depreciation on construction vehicles or equipment
- **Overhead expenses** (labeled Expense accounts by QuickBooks Pro) include accounts in which you record the costs of keeping your business running, such as general marketing, administrative salaries, occupancy or facilities expense, office supplies and equipment, and general communications costs.
- **Other income and expense** records nonoperating income and expenses such as interest or financing costs not related to a specific job, the gain or loss on the sale of fixed assets, or income taxes.

The income statement shows subtotals for each of these sections, along with subtotals of the profit after each type of expense has been subtracted. Each of the profit subtotals has a specific title. Net income, often referred to as the bottom line, is the final result of all income accounts minus all expense accounts.

TIP Accountants and business owners often compute percentages of costs in relationship to income to establish a handy measurement. For example, the cost of work performed ($113,526.04) is 70.5 percent of the total income ($160,932.00), and the gross profit of $47,405.96 is 29.5 percent of the total income. Once established, these percentages can be used to compare your company's measurements with other similar companies, and for various planning and estimating purposes.

In chapters 8 and 9, respectively, I'll guide you through the selection of income and expense accounts that you can tailor to your company. In chapter 29, I'll show you how you can use QuickBooks Pro to make your income statements more valuable.

Dreamscape Remodeling, Inc.
Income Statement
January through March 2001

	Jan - Mar '01
Ordinary Income/Expense	
Income	
T.450 · Design & consulting services	7,600.00
T.460 · Construction services	153,332.00
Total Income	160,932.00
Cost of Goods Sold	
T.529 · Direct employment cost	55,068.27
T.550 · Subcontractors	19,163.45
T.560 · Building materials & supplies	31,781.17
T.570 · Job-specific misc costs	8,692.50
T.580 · Direct expense-not job specific	323.06
Total COGS	115,028.45
Gross Profit	45,903.55
Expense	
T.710 · Sales & marketing expenses	5,324.58
T.800 · Admin employment costs	21,647.00
T.820 · Occupancy expense	3,071.00
T.830 · Office expenses	3,774.05
T.840 · Computer expense	24.30
T.850 · General & admin expense	4,791.64
T.860 · Vehicle expenses	2,805.32
T.870 · Meals, travel & entertainment	1,104.46
T.890 · Taxes & licenses	668.00
Total Expense	43,210.35
Net Ordinary Income	2,693.20
Other Income/Expense	
Other Income	
9100 · Interest income	50.79
Total Other Income	50.79
Other Expense	
T.920 · Interest	710.17
T.923 · Other financing expense	587.98
Total Other Expense	1,298.15
Net Other Income	(1,247.36)
Net Income	**1,445.84**

EXHIBIT 4.3 Sample income statement.

Dreamscape Remodeling, Inc.
Summary: Cost Accounting by Job Report
January through March 2001

	Act. Cost	Act. Revenue	($) Diff.
Buchanan, Bob & Barbara			
8103.Kitchen remodel	12,025.50	18,953.00	6,927.50
9021.Porch conversion	1,530.00	2,200.00	670.00
Total Buchanan, Bob & Barbara	13,555.50	21,153.00	7,597.50
Fellowes, Barbara			
9001.Bathroom remodel	8,862.51	18,250.00	9,387.49
9004.Handyman	781.76	895.00	113.24
Total Fellowes, Barbara	9,644.27	19,145.00	9,500.73
Fields, Bruce & Elaine			
9003.Kitchen	18,587.51	24,750.00	6,162.49
9011.Addition	29,300.00	40,000.00	10,700.00
Total Fields, Bruce & Elaine	47,887.51	64,750.00	16,862.49
Pets R Fun			
9006.Addition	18,090.00	25,400.00	7,310.00
Total Pets R Fun	18,090.00	25,400.00	7,310.00
Winworth, Charles & Marge			
9002.Handyman	734.05	1,437.00	702.95
9008.Kitchen	20,124.55	32,500.00	12,375.45
Total Winworth, Charles & Marge	20,858.60	33,937.00	13,078.40
zx.Jobs-non specific	3,813.66		(3,813.66)
TOTAL	**113,849.54**	**164,385.00**	**50,535.46**

EXHIBIT 4.4 Sample simple job-cost report.

Cost accounting by job report

A very simple job-cost report, shown in Exhibit 4.4, shows the summarized income and expense totals for each job. In the left column, you see customers, followed by specific jobs. The columns show you the total actual costs and revenue. The grand total at the bottom of the report shows actual revenue of $164,385, actual cost of $113,849.54 and a difference, or gross profit, of $50,535.46. This type of report can give you a quick, at-a-glance overview of the income and expense shown on your books for specific jobs during the current period.

Simple job-cost variance report

Job or project variance reports are designed to show you actual income and expense, estimated income and expense, and the difference or variance. Thevariance report presents the information an owner needs to complete Step #3 (compare results to standards) in the control loop process. This report, shown in Exhibit 4.5, shows the detail for only one job and includes job estimates as well as variances from those estimates. Please review and note the following features of a job-variance report.

Dreamscape Remodeling, Inc.
Job Estimates vs. Actuals Detail - Fields, Bruce & Elaine:
January through March 2000

	Est. Cost	Act. Cost	($) Diff.	(%) Diff.	Est. Revenue	Act. Revenue	($) Diff.	(%) Diff.
Service								
00. INCOME								
450.Design					2,200.00	2,000.00	(200.00)	(9.09%)
460.Rem-res'l					44,800.00	38,000.00	(6,800.00)	(15.18%)
Total 00. INCOME					47,000.00	40,000.00	(7,000.00)	(14.89%)
01. DIR COSTS								
530.Labor	10,000.00	10,348.00	348.00	3.48%				
532.Lbr brdn	2,500.00	2,587.00	87.00	3.48%				
550.Subs	4,300.00	5,874.00	1,574.00	36.61%				
560.Materials	7,500.00	8,283.00	783.00	10.44%				
570.Misc cost	3,000.00	2,208.00	(792.00)	(26.4%)				
Total 01. DIR COSTS	27,300.00	29,300.00	2,000.00	7.33%				
Total Service	27,300.00	29,300.00	2,000.00	7.33%	47,000.00	40,000.00	(7,000.00)	(14.89%)
TOTAL	27,300.00	29,300.00	2,000.00	7.33%	47,000.00	40,000.00	(7,000.00)	(14.89%)

EXHIBIT 4.5 Sample job-cost variance report.

- **Cost** estimates, actual project-to-date costs, and variance computations are shown in columns 1 through 4. This example shows an unfavorable project-to-date cost variance.
- **Income** estimate totals, actual project-to-date, and variance computations are shown in columns 5 through 8. The project-to-date income also is unfavorable.
- **Project-to-date variances** are computed by subtracting estimates from actual. The project-to-date variance in this example therefore currently stands at an unfavorable $7,000.
- The most important variance, of course, is the **sum of the variances**.

Although a job-cost variance report usually shows the *estimated total* income and expense for each job, it reports only the *actual* income and expense recorded as of the reporting date. Therefore, it is important to consider *how much* of your total income and cost has been recorded in the accounting system as of the report date. Depending on the progress of the job and/or how income and expense are entered on your books, the reports can show drastically different results from day-to-day or week-to-week.

Therefore, projecting into the future, the company owner is fairly certain that the remaining $7,000 of income will be collected. The owner also anticipates that $2,500 in costs still are not recorded on the books, and that there is a materials credit of $700 not shown on the books, which the owner says still needs to be negotiated with the supplier. This information creates a net additional expense of $1,800. Taking into account these factors, the owner's ending variance then will be an unfavorable total of $3,800.

The resulting percentages are computed as follows.

- The owner's original estimated revenue of $47,000 had a direct estimated cost of $27,300 or 58 percent, leaving a gross profit of $19,700 or a 42-percent gross margin.
- The anticipated cost over-run is $3,800 which reduces the owner's estimated total profit by 19.3 percent or $3,800/$19,700.
- The owner's anticipated gross margin therefore has been reduced to 33.8 percent ($15,900/$47,000)—from $19,700 to $15,900.

The payoff represents Step #4 (take corrective action or revise standards) in the control loop process. The company owner compares the actual anticipated costs against the original estimate to discover the cause of the variances. In this particular example, because the variance occurred due to a combination of estimating errors and cost over-runs, the owner:

- Corrects the estimating process so that the next estimate can be more accurate
- Discusses the reasons for the overruns with employees, trade contractors, and suppliers so similar situations can be minimized or avoided in the future
- Vows to implement a better change order process so that future customer-generated changes can be minimized to avoid cost overruns or converted to income so that target percentages can be reached

I recommend you utilize *at least* this level of job-cost information if your business is beyond the initial start-up stage.

Job costs by job stage variance report

A job-cost variance report by job stage is shown in Exhibit 4.6. Like the job-variance report in Exhibit 4.5, the job cost by job stage variance report presents the information an owner needs to complete Step #3 (compare results to standards) of the control loop. The difference from the previous report is that this variance report also shows detailed job costs and variances by the job stage or process for which the cost was incurred. Note that labor, trade contractors, materials, and other costs are recorded within each required job stage.

Both job-stage variance reports basically are the same except for the level of detail. If you are concerned about creating too much detail, please remember that:

- Although there are many possible job stages, few jobs utilize all of the potential stages. For example, one job may require 14 stages, while another needs only 3.
- The job stages can be developed to your personal preferences or they can be based on your current estimating system. I have developed an extensive detailed job-cost list for remodelers, custom home builders, and developers that is referenced in chapter 8.

This job-cost tracking report can provide great benefits if you need it and will use it, but it's also important to realize there are added costs involved in tracking, entering, and balancing this level of detail accurately in your accounting system.

On the other hand, by identifying income earned for each job and the related costs of employment labor, trade contractors, materials, and other costs, for each stage of a job, you gain immediate access to information regarding the job's profitability *as it progresses.* After the job is finalized you'll know whether it was a winner and met your profit projections or a loser, resulting in a loss or less profit than anticipated. If it was a winner, you'll want to know what you did right and do it again. If not, you'll want to pinpoint the reasons why so that you can build safeguards into future jobs. You may well find that this level of detail pays for itself many times over.

Dreamscape Remodeling, Inc.

Job Estimates vs. Actuals Detail for J801 Hickory

All dates

	Est. Cost	Act. Cost	($) Diff.	(%) Diff.	Est. Revenue	Act. Revenue	($) Diff.	(%) Diff.
Service								
1.Invoiced								
Change orders	0.00	0.00	0.00	0.0%	3,300.00	3,372.92	72.92	2.21%
Invoice-A	0.00	0.00	0.00	0.0%	21,500.00	21,454.00	(46.00)	(0.21%)
Invoice-B	0.00	0.00	0.00	0.0%	20,000.00	20,000.00	0.00	0.0%
Invoice-C	0.00	0.00	0.00	0.0%	20,000.00	20,000.00	0.00	0.0%
Invoice-D	0.00	0.00	0.00	0.0%	10,000.00	10,070.41	70.41	0.7%
Invoice-Final	0.00	0.00	0.00	0.0%	4,900.00	4,900.00	0.00	0.0%
Total 1.Invoiced	0.00	0.00	0.00	0.0%	79,700.00	79,797.33	97.33	0.12%
405.Est	0.00	128.00	128.00	100.0%	0.00	0.00	0.00	0.0%
410.Design								
a410.Design	0.00	224.00	224.00	100.0%	0.00	0.00	0.00	0.0%
b411.Prof des	1,454.00	1,454.00	0.00	0.0%	0.00	0.00	0.00	0.0%
Total 410.Design	1,454.00	1,678.00	224.00	15.41%	0.00	0.00	0.00	0.0%
415.Prod P&R	197.00	80.75	(116.25)	(59.01%)	0.00	0.00	0.00	0.0%
420.Demo								
a420.Demo	2,333.00	851.38	(1,481.62)	(63.51%)	0.00	0.00	0.00	0.0%
Total 420.Demo	2,333.00	851.38	(1,481.62)	(63.51%)	0.00	0.00	0.00	0.0%
450.Masonry								
b450.Masonry	2,150.00	2,150.00	0.00	0.0%	0.00	0.00	0.00	0.0%
Total 450.Masonry	2,150.00	2,150.00	0.00	0.0%	0.00	0.00	0.00	0.0%
460.Framing								
a460.Framing	5,976.00	5,677.94	(298.06)	(4.99%)	0.00	0.00	0.00	0.0%
c460.Framing	6,050.00	6,756.44	706.44	11.68%	0.00	0.00	0.00	0.0%
Total 460.Framing	12,026.00	12,434.38	408.38	3.4%	0.00	0.00	0.00	0.0%
470.Roofing								
a470.Roofing	292.00	162.00	(130.00)	(44.52%)	0.00	0.00	0.00	0.0%
b470.Roofing	2,250.00	2,620.00	370.00	16.44%	0.00	0.00	0.00	0.0%
c470.Roofing	0.00	156.99	156.99	100.0%	0.00	0.00	0.00	0.0%
Total 470.Roofing	2,542.00	2,938.99	396.99	15.62%	0.00	0.00	0.00	0.0%
480.Siding								
a480.Siding	1,458.00	1,126.75	(331.25)	(22.72%)	0.00	0.00	0.00	0.0%
c480.Siding	820.00	1,025.06	205.06	25.01%	0.00	0.00	0.00	0.0%
Total 480.Siding	2,278.00	2,151.81	(126.19)	(5.54%)	0.00	0.00	0.00	0.0%
490.Windows								
a.490.Windows	612.00	391.00	(221.00)	(36.11%)	0.00	0.00	0.00	0.0%
c490.Windows	2,890.00	2,995.09	105.09	3.64%	0.00	0.00	0.00	0.0%
Total 490.Windows	3,502.00	3,386.09	(115.91)	(3.31%)	0.00	0.00	0.00	0.0%
500.Doors								
a500.Doors	1,020.00	364.00	(656.00)	(64.31%)	0.00	0.00	0.00	0.0%
c500.Doors	1,370.00	2,121.21	751.21	54.83%	0.00	0.00	0.00	0.0%
Total 500.Doors	2,390.00	2,485.21	95.21	3.98%	0.00	0.00	0.00	0.0%
520.Plumbing								
a520.Plumbing	197.00	73.00	(124.00)	(62.94%)	0.00	0.00	0.00	0.0%
b520.Plumbing	1,730.00	1,730.00	0.00	0.0%	0.00	0.00	0.00	0.0%
Total 520.Plumbing	1,927.00	1,803.00	(124.00)	(6.44%)	0.00	0.00	0.00	0.0%

EXHIBIT 4.6 Sample job-cost variance report by job stage.

	Est. Cost	Act. Cost	($) Diff.	(%) Diff.	Est. Revenue	Act. Revenue	($) Diff.	(%) Diff.
530.Electric								
a530.Electric	444.00	317.50	(126.50)	(28.49%)	0.00	0.00	0.00	0.0%
b530.Electric	3,247.00	3,327.00	80.00	2.46%	0.00	0.00	0.00	0.0%
c530.Electric	1,050.00	1,272.37	222.37	21.18%	0.00	0.00	0.00	0.0%
Total 530.Electric	4,741.00	4,916.87	175.87	3.71%	0.00	0.00	0.00	0.0%
540.HVAC								
a540.HVAC	394.00	135.00	(259.00)	(65.74%)	0.00	0.00	0.00	0.0%
b540.HVAC	2,270.00	2,270.00	0.00	0.0%	0.00	0.00	0.00	0.0%
Total 540.HVAC	2,664.00	2,405.00	(259.00)	(9.72%)	0.00	0.00	0.00	0.0%
550-Insulate								
a550.Insulate	98.00	31.00	(67.00)	(68.37%)	0.00	0.00	0.00	0.0%
b550.Insulate	1,000.00	967.00	(33.00)	(3.3%)	0.00	0.00	0.00	0.0%
Total 550-Insulate	1,098.00	998.00	(100.00)	(9.11%)	0.00	0.00	0.00	0.0%
560.Drywall								
a560.Drywall	584.00	103.50	(480.50)	(82.28%)	0.00	0.00	0.00	0.0%
b560.Drywall	3,620.00	4,205.00	585.00	16.16%	0.00	0.00	0.00	0.0%
Total 560.Drywall	4,204.00	4,308.50	104.50	2.49%	0.00	0.00	0.00	0.0%
570.Pnt/Fin								
a570.Pnt/Fin	394.00	2,451.63	2,057.63	522.24%	0.00	0.00	0.00	0.0%
b570.Pnt/Fin	3,225.00	384.00	(2,841.00)	(88.09%)	0.00	0.00	0.00	0.0%
c570.Pnt/Fin	0.00	847.60	847.60	100.0%	0.00	0.00	0.00	0.0%
Total 570.Pnt/Fin	3,619.00	3,683.23	64.23	1.78%	0.00	0.00	0.00	0.0%
580.I trm/mw								
a580.I trm/mw	1,486.00	602.00	(884.00)	(59.49%)	0.00	0.00	0.00	0.0%
c580.I trm/mw	700.00	805.42	105.42	15.06%	0.00	0.00	0.00	0.0%
Total 580.I trm/mw	2,186.00	1,407.42	(778.58)	(35.62%)	0.00	0.00	0.00	0.0%
590.Cab/van								
a590.Cab/van	292.00	189.38	(102.62)	(35.14%)	0.00	0.00	0.00	0.0%
c590.Cab/van	800.00	789.15	(10.85)	(1.36%)	0.00	0.00	0.00	0.0%
Total 590.Cab/van	1,092.00	978.53	(113.47)	(10.39%)	0.00	0.00	0.00	0.0%
600.Cntrtops								
a600.Cntrtops	98.00	0.00	(98.00)	(100.0%)	0.00	0.00	0.00	0.0%
b600.Cntrtops	200.00	265.00	65.00	32.5%	0.00	0.00	0.00	0.0%
Total 600.Cntrtops	298.00	265.00	(33.00)	(11.07%)	0.00	0.00	0.00	0.0%
620.Flr covr								
a620.Flr covr	197.00	72.00	(125.00)	(63.45%)	0.00	0.00	0.00	0.0%
b620.Flr covr	0.00	0.00	0.00	0.0%	0.00	0.00	0.00	0.0%
Total 620.Flr covr	197.00	72.00	(125.00)	(63.45%)	0.00	0.00	0.00	0.0%
650.Applianc								
a650.Applianc	145.00	0.00	(145.00)	(100.0%)	0.00	0.00	0.00	0.0%
Total 650.Applianc	145.00	0.00	(145.00)	(100.0%)	0.00	0.00	0.00	0.0%
660.Gttr/dwn								
a660.Gttr/dwn	98.00	9.00	(89.00)	(90.82%)	0.00	0.00	0.00	0.0%
b660.Gttr/dwn	735.00	735.00	0.00	0.0%	0.00	0.00	0.00	0.0%
Total 660.Gttr/dwn	833.00	744.00	(89.00)	(10.68%)	0.00	0.00	0.00	0.0%
680.Handling								
a680.Handling	684.00	469.13	(214.87)	(31.41%)	0.00	0.00	0.00	0.0%
Total 680.Handling	684.00	469.13	(214.87)	(31.41%)	0.00	0.00	0.00	0.0%

EXHIBIT 4.6 Sample job-cost variance report by job stage. *Continued*

Dreamscape Remodeling, Inc.

Job Estimates vs. Actuals Detail for J801 Hickory

All dates

	Est. Cost	Act. Cost	($) Diff.	(%) Diff.	Est. Revenue	Act. Revenue	($) Diff.	(%) Diff.
690.Punchlst								
a690.Punchlst	584.00	378.25	(205.75)	(35.23%)	0.00	0.00	0.00	0.0%
Total 690.Punchlst	584.00	378.25	(205.75)	(35.23%)	0.00	0.00	0.00	0.0%
700.Cleanup								
a700.Cleanup	1,206.00	1,218.39	12.39	1.03%	0.00	0.00	0.00	0.0%
c700.Cleanup	0.00	32.46	32.46	100.0%	0.00	0.00	0.00	0.0%
Total 700.Cleanup	1,206.00	1,250.85	44.85	3.72%	0.00	0.00	0.00	0.0%
710.Supervis								
a710.Supervis	1,380.00	516.50	(863.50)	(62.57%)	0.00	0.00	0.00	0.0%
Total 710.Supervis	1,380.00	516.50	(863.50)	(62.57%)	0.00	0.00	0.00	0.0%
720.Warranty								
a720.Warranty	0.00	36.00	36.00	100.0%	0.00	0.00	0.00	0.0%
Total 720.Warranty	0.00	36.00	36.00	100.0%	0.00	0.00	0.00	0.0%
740.Admin								
a740.Admin	0.00	45.00	45.00	100.0%	0.00	0.00	0.00	0.0%
Total 740.Admin	0.00	45.00	45.00	100.0%	0.00	0.00	0.00	0.0%
750.Gn'l cond								
a760	0.00	252.00	252.00	100.0%	0.00	0.00	0.00	0.0%
b752.Trash R/								
d752	500.00	685.00	185.00	37.0%	0.00	0.00	0.00	0.0%
Total b752.Trash R/	500.00	685.00	185.00	37.0%	0.00	0.00	0.00	0.0%
c760.Misc	0.00	249.27	249.27	100.0%	0.00	0.00	0.00	0.0%
d753.Rental	100.00	0.00	(100.00)	(100.0%)	0.00	0.00	0.00	0.0%
d754.Tools/S	100.00	43.20	(56.80)	(56.8%)	0.00	0.00	0.00	0.0%
d755.Permits	674.00	589.50	(84.50)	(12.54%)	0.00	0.00	0.00	0.0%
Total 750.Gn'l cond	1,374.00	1,818.97	444.97	32.39%	0.00	0.00	0.00	0.0%
Total Service	57,104.00	54,380.86	(2,723.14)	(4.77%)	79,700.00	79,797.33	97.33	0.12%
No item	0.00	4,952.97	4,952.97	100.0%	0.00	0.00	0.00	0.0%
TOTAL	**57,104.00**	**59,333.83**	**2,229.83**	**3.91%**	**79,700.00**	**79,797.33**	**97.33**	**0.12%**

EXHIBIT 4.6 Sample job-cost variance report by job stage. *Continued*

CHAPTER 5

Finding accounting help

Small business owners often undervalue inhouse accountants and bookkeepers. Don't make this mistake! Once you have made the initial investment to personally design, understand, and use your accounting system, I think you will understand just how valuable a properly trained and experienced accountant can be to your business.

After you design and build your accounting system you will be more qualified to recognize whether a candidate is qualified and trainable for the position. Don't take the easy way out and delegate the accounting work to your spouse, unless he or she is trained in accounting basics. Instead I offer the following tips for successfully hiring an inhouse accounting assistant.

- Select your accounting assistant carefully. I advise against hiring a part-time student to handle your accounting work, primarily because students are more likely to change jobs just when they are finally trained.
- Look for someone with some bookkeeping or accounting expertise. A background in the building industry would be a real plus. Explain to him or her that your system is unique to your company and that because it involves job costing it will be a challenging position.
- Look for someone with the right personality for the job. Someone who is very detail-oriented and thorough might be a good candidate.
- Depending on the level of information you want to extract from your system, plan to invest fairly heavily in training for both you and your accounting assistant. Recognize the extent of your learning curve.
- Plan to make your accounting assistant a valued member of your management team. Let him or her know that you will be working together to understand the results of the reports in order to develop better operating controls and profits for your company.
- Review your accounting assistant's activities quickly and regularly. Give reports the "sniff test." Given what you know about the facts, do the results you see on the reports seem logical? It is very easy to skip the review steps when you're busy fighting other fires that might seem more urgent. Never allow yourself to be lulled into a false sense of security by an employee who seems competent. Review your reports!
- Encourage your assistant to develop a good working relationship with your CPA.

Your accounting professional will be working in a truly exciting management position in your company. Hopefully you'll be rewarded with a win-win situation—a long-term employee who feels appreciated for his or her contribution to your company, and to your success!

SECTION TWO

The Design Phase

As we begin designing your financial management system, visualize your new job-cost system as having three dimensions:

- **Dimension 1**—You can view information at the *account* level for your company. For example, you can see how much you spent on trade contractors in total.
- **Dimension 2**—You can view information at the account level and see the income and direct expense assigned to each *job*. As an accountant, I like to think of it as a spreadsheet, with the accounts down the left side and the jobs shown in each column across the top.
- **Dimension 3**—In addition, you can add *job stages* to the mix. Envision your spreadsheet with layered sheets from front to back, with each sheet labeled with a job stage. This dimension lets you assign, for example, labor expense for a given job to each job stage, and then trade contractor costs, followed by materials expense, and finally, other costs.

Unfortunately, I haven't quite figured out a way to magically convert those three dimensions into the two-dimensional physical limitations imposed by a piece of paper, but QuickBooks Pro 2000 has come up with the next best thing. By creating, combining, and linking accounts, jobs, and items in a logical, considered fashion within the QuickBooks Pro database, you can record, store, and then extract the various views and dimensions of information that, in effect, create those three dimensions on one piece of paper.

The chapters in this section help you build the various structural lists that will support your new three-dimensional information system and the integrated subsystems.

Think of your blueprints for a custom home or an extensive remodeling job. Each phase is contained in a separate blueprint, but when all are combined, the home is a three-dimensional structure with integrated systems. So it is with the lists that comprise your accounting structure.

I like to think of the Chart of Accounts as the foundation of your financial management system. Each financial transaction you enter is stored within your accounts. The accounts in total are referred to as the general ledger. Your account names and balances create the structure, layout, and subtotals for your balance sheet and income statement, as well as all other account-based reports that you may choose to create.

In QuickBooks Pro I like to refer to the Item List and the Customer:Job List as the framing of the structure. When you enter an expense you categorize the entry by item and assign it to a customer or job. QuickBooks Pro then directs the entry to the account you have designated in the Chart of Accounts. The Customer:Job List enables you to track all financial transactions that apply to a specific customer or job. The Item List enables you to track and subtotal job-cost details in a way that makes sense to you.

Take some time to consider the design and layout of your lists because they will determine the order and subtotaling of your job-cost and other management reports. For example, if you want to track the profitability of your fixed-price jobs versus your time and materials jobs, or if you want to track the profitability of certain subdivisions, or of jobs that you do for certain builders, you can build those features into your Customer:Job List. As an alternative, you may want to create job types such as kitchens, baths, decks, and other, so that you can extract profitability data about each job type and compare results. This setup also enables you to prepare mailing labels for just your "deck" customers.

Once your lists are created, you then can design management reports that provide full or finite-limited details, or summarize the details in broad categories based on your title and sub-headings. Because, within established boundaries, you will be involved in designing your own system, you can use names that are logical to you and complement the way that you do business.

Within each list, your names should look and "act" consistent. A consistent methodology removes the guesswork, and helps everyone involved learn and retain the process. Changing your approach mid-stream is difficult, time-consuming, frustrating, and may produce inaccurate results. As you name the elements in these lists, keep in mind that they determine the structure of your reports. Organize the data in a way that you believe will be most meaningful to you when you need to analyze the data.

Once you have the basics of your lists outlined, check with the people who will be using your management reports. Ask if the information they need is contained in either the title level or the detail level. I have found that consulting users early in the process ensures that their needs are met. Don't work in a vacuum. Often other users will have a perspective that can have a positive affect on the lists for everyone involved.

Near the end of this section we will look at the preferences you will need to set in order to effectively use the powerful features available in the QuickBooks Pro software. These preferences are yet another way that you can customize QuickBooks Pro to your company and your management needs.

With your lists established, I'll give you some tips on how to enter the opening balances in your accounts to get your system up and running as quickly as possible.

Do not allow yourself to become overwhelmed!

Have faith that the simplest of remodeling or building jobs is far more difficult than setting up your accounting system. We'll do it together, one step at a time. I'll ask guiding questions and provide tips and ideas about how to use each of your accounts and lists. You'll have time to think about the accounts, items, names, and lists. Place a checkmark in the margin along with any notes you want to refer to when you actually create your lists in QuickBooks Pro.

TIP Please carefully review the information and tips that this book provides. I recently read a quote that said, "An expert is someone who's made more mistakes than anyone else." I know exactly what the author was talking about!

CHAPTER 6

Setting up your Chart of Accounts

The Chart of Accounts represents the first and most basic level of accounting. I refer to it as the *first dimension* of accounting information. The Chart of Accounts is comprised of five basic types of accounts:

- Assets
- Liabilities
- Equity
- Income
- Expenses

The first three types of accounts—assets, liabilities, and equity—appear on your balance sheet. The remaining two types of accounts—income and expenses—appear on your income statement. These two reports are the basic reports considered essential for financial reporting, regardless of the type or size of business you operate.

QuickBooks Pro 2000, however, subdivides each of these broad categories into more detailed account types so that reports are presented properly. QuickBooks Pro account types are shown in Exhibit 6.1.

The first five account types listed in Exhibit 6.1—bank, accounts receivable, other current asset, fixed asset, and other asset—are asset accounts. The next four types—accounts payable, credit card, other current liability, and long-term liability—are liability accounts. The equity section has only one type of account, listed simply as equity. QuickBooks Pro also has two different types of income accounts—income and other income—and three different types of expense accounts—cost of goods sold (COGS), expense, and other expense.

The order shown for these accounts also is the order in which the accounts will appear on your reports. For example, your bank accounts will come first, your accounts receivable account second, and so on for all accounts that are appropriate for any account-based report.

On a periodic basis, such as monthly, quarterly, and year-end, you should perform special cleanup and control processes and accounting adjustments, so that at the end of any of these accounting periods you have a set of books that reflect your company's overall financial health. You also can enter budgets for each account and review dollar or percentage variances from budget using actual year-to-date or specific time-frame figures.

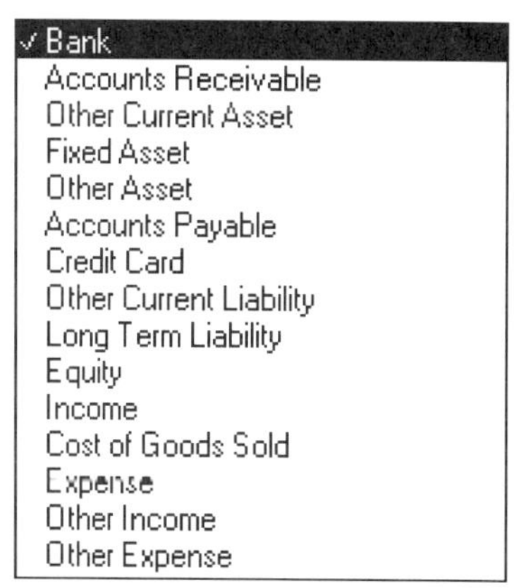

EXHIBIT 6.1 QuickBooks Pro account types.

Each account in the Chart of Accounts acts as a "bucket" in which financial information is held and summarized. For example:

- Your checking (asset) account holds the record of all of the checks you write, the deposits you make, and service charges your bank assesses. After each transaction you'll be able to see the balance remaining in your account.
- Your accounts payable (liability) account shows a record of increases (bills) and decreases (payments to vendors) and the balance payable after each of these transactions.

As you can see from the above examples, each transaction has two parts. This is referred to as double-entry bookkeeping (see chapter 4). Just remember that each transaction involves two accounts and is the equivalent of a two-sided coin.

Starting with a pencil copy

Since each account shows the results of various financial activities, you will want to select the accounts that correspond to your business. I recommend that you make a copy of the Chart of Accounts in appendix A to use as a "pencil" or working copy.

You definitely won't need to use all of the accounts from this list. Think of the account types and titles I've provided as a "hardware store of accounting buckets." Pick and choose the financial information buckets that are right for your company now and in the near future. You can come back later and add more accounts as your information needs change. Because I've included such an extensive list, chances are you'll be able to find exactly what you need. The accounts have been developed over many years of working with remodelers and custom home builders. Use it as an ongoing account naming and numbering reference.

Account names

If you do need to add accounts that aren't on the list, reference the potential accounts in appendix A to create similar accounts. For example, the list only includes one "Note payable, secured." If you have two different loans for vehicle purchases, add another account. Give the first one a descriptive title such as "Note payable, secured—'99 Ford pickup." Then add another note payable account titled, "Note payable, secured—'00 Chevy van."

Another frequently modified account name is the checking account. Let's assume that you have one full-service checking account at Standard Federal. Place a checkmark next to account S.09P.Checking. Note your bank name next to the account name because when you actually create this account you'll want to include the bank's name, initials, or other unique description in the account name. Have you made the checkmark? Terrific! You've just started to set up your Chart of Accounts.

TIP I advise against using your customer account number or your bank account number in the account name.

Account numbers

A number or a letter-number combination (see appendix A) precedes the name of each account in the Chart of Accounts. Since QuickBooks Pro does not require you to use a specific approach to account numbers, I have developed the following system to assist in your posting processes and to enhance the appearance of your final reports.

- A *four-digit number* indicates a detail-level account. These accounts are used to accumulate individual transactions, and the total of the transactions entered then is shown on detailed reports. For example, you would choose account 8291.Gas whenever you enter a gas bill or record a check written to the gas company. The total from this account also rolls up into a grouped total under the subhead.
- A *letter and a three-digit* number indicate that the account is a title account only. Transactions are not posted (entered) to these accounts. Instead, the account name appears on your

detailed, expanded reports as a title with the subtotal of the detailed accounts under that title. For example, 8291.Gas is located under S.8209.Utilities, which functions as the title account. Whenever a combination of title and subaccounts appear, the title account triggers reports to create a subtotal.

TIP

First-level title accounts in the Chart of Accounts are the only ones that show in a collapsed, summary report. When you create a collapsed, summary report the highest level title accounts show the sum of the detailed accounts within the title account.

NOTE

Although the letter at the beginning of the account acts as a signal to you that it is a title account, QuickBooks Pro will not prevent you from posting to this account. If you mistakenly post to it as you record your transaction, a separate line item will appear on the report with the name of the title account followed by "—Other." For example, if you posted a transaction to the title account "Utilities," it would appear as "Utilities—Other" on your report.

- A *letter, two-digit number, and another letter combination* (most commonly a "P"—signifying that you can post to the account) indicate an account that blends the two preceding numbering conventions. These accounts are designed to act as both title accounts and posting accounts. I developed this hybrid approach because these accounts typically are stand-alone accounts that do not need subaccounts. An example is S.93P.Loss on Sale of Assets. This is an account that does not have subaccounts, so it receives direct postings. Because you'll also want to see it as a title account on a collapsed report, the letter prefix keeps the appearance of the account name consistent with the other title accounts.

TIP

When entering transactions in QuickBooks Pro, the account pull-down lists are very narrow and do not fully display account names. Because some account names are very similar, I recommend you clarify the text names of certain accounts by inserting one of the following code letters before the name:

- D—Direct Labor
- S—Sales staff
- O—Owner
- A—Administrative staff

When it's necessary to utilize similar account titles for different purposes, these types of visual cues and codes can be useful for both accounting personnel and those who use the reports.

In chapters 7, 8, and 9 I will go through the individual accounts for the balance sheet—assets, liabilities, and equity—and income and expenses, respectively. I'll explain how each account is used. From these discussions you can identify and check-off the accounts you will need to use.

Creating an account

To create a new account, follow these steps.

1. From the **Lists** pull-down menu, select **Chart of Accounts**.
2. Click on the **Accounts** button and select **New** (see Exhibit 6.2) .
3. Enter information in the following fields, as applicable. For additional help, use the **How do I?** Help button.

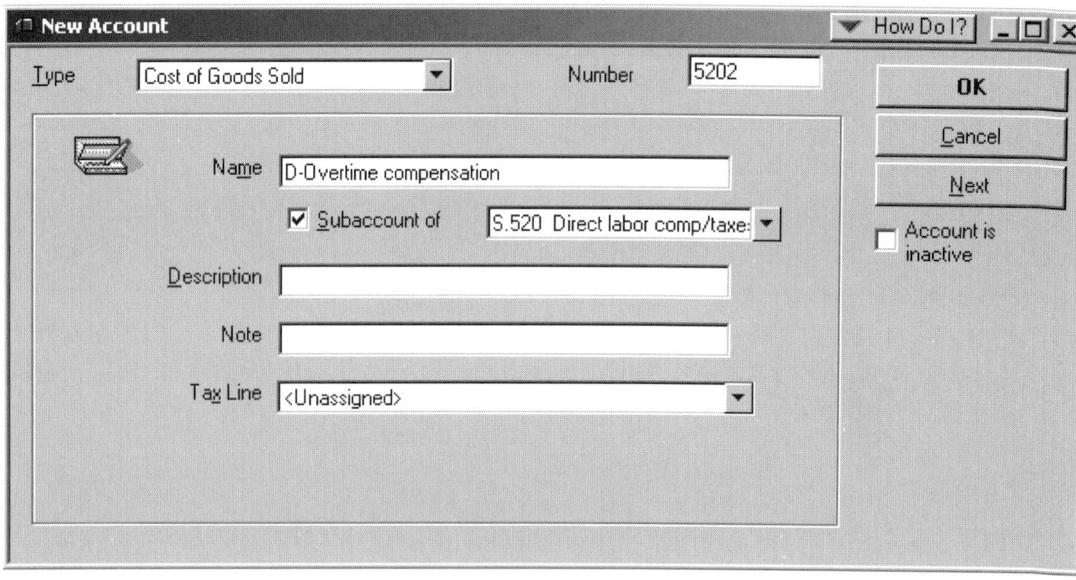

EXHIBIT 6.2 New Account window with subaccount.

- **Type**—Use the pull-down list to select the account type. Be sure to use the same account type as shown in the sample Chart of Accounts in appendix A.
- **Number**—Enter the account number as shown including letters and punctuation.
- **Name**—Enter the actual text for the account name. Abbreviations have been carefully selected to fit in the available field length, and the capitalization layout shown will give you a clean, professional looking report. Do not enter the account number in this field.
- **Subaccount**—If this is a subaccount of another account, check the box and then use the list to check-off the title account number under which the subaccount is located.
- **Description and Note**—These text fields, where they appear, can be used to enter brief information, such as loan terms and percentages.
- **Other selections you see below the description field**—These fields seldom are used by my clients.
- **Opening balance**—*Do not* type anything in this field (see chapter 14).

4. Click **OK** to create the account.

Editing an account

To edit an existing account follow these steps.

1. From the **Lists** pull-down menu, select **Chart of Accounts**.
2. Click on the **Accounts** button and select **Edit** (see Exhibit 6.3).
3. Edit the information in the applicable fields.
4. Click **OK** to edit the account.

Making an account inactive

You can delete an account from the Chart of Accounts only if it has not been used in any transactions. If it has been used in transactions, however, you can make the account inactive. Although inactive accounts are not displayed in your Chart of Accounts, financial activity related to that account still displays on your reports. To make an account inactive, follow these steps.

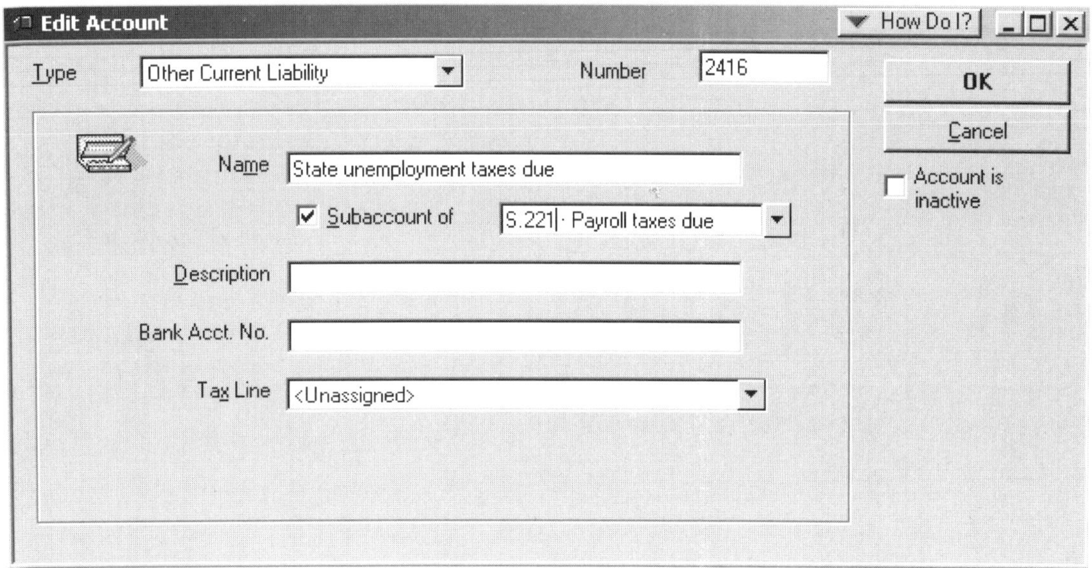

EXHIBIT 6.3 Edit Account window.

1. From the **Lists** pull-down menu, select **Chart of Accounts**.
2. Select the account you want to make inactive by clicking once with the left mouse button.
3. Click once with your right mouse button to access the submenu.
4. Select **Make Inactive** with the left mouse button (see Exhibit 6.4).

TIP If you want to see or re-activate inactive accounts, click in front of the Show All button at the bottom of your Chart of Accounts screen. When this button is checked, you can easily re-activate or inactivate accounts by positioning the mouse to the left of the account number and clicking once with the left mouse button when the cursor turns into an "X".

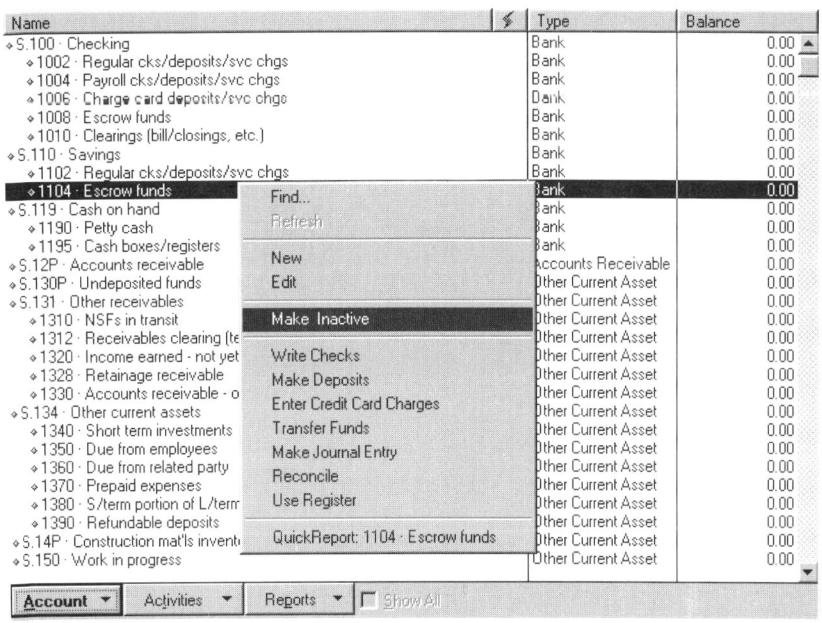

EXHIBIT 6.4 Make Inactive window.

CHAPTER 7

Entering asset, liability, and equity accounts

The balance sheet includes asset, liability, and equity accounts from the Chart of Accounts. This chapter looks at the individual accounts in the order in which they appear both in the Chart of Accounts and the balance sheet.

Setting up bank and current asset accounts

Assets generally are organized in order of liquidity, or the time needed to convert the asset to cash—from most liquid to least liquid. Exhibit 7.1 shows the first set of accounts related to banking, accounts receivable, and other current assets.

S.09P.Checking (your bank name) and S.100.Checking

As the names imply, checking accounts and their equivalents, such as money market, sweep, or other accounts considered immediately accessible via check or draft are included within the subaccounts of these title and posting accounts.

TIP QuickBooks Pro 2000 has a unique feature that can be used to track cash flowing into and out of your checking account for special activities or business functions. You can create separate subaccounts for each type of activity that you want to track, and the individual transactions will roll up into the title account. First add a subaccount called **1002.Operating funds.** Then add additional similar subaccounts for other functions, as desired. For example, my clients who are developers and custom home and spec home builders often like to use this feature to create subjob accounts for individual jobs. To do this you create a general business function subaccount as illustrated by the subaccount **1006.Job Funds** in Exhibit 7.1. Then you create sub-subaccounts for individual jobs, as illustrated by account **001.Job 001,** to track the inflow of funds such as client deposits and bank draws and the outflow of funds or bills paid. Except for credit card purchases, and payroll and payroll tax-related expenditures, this approach allows you to monitor net cash flow throughout the life of a specific project without opening a separate bank account. After the home-sale-closing process, transfer the remaining balance into your **1002.Operating funds** subaccount and inactivate the individual job-tracking subaccount.

Current Assets

Checking/Savings

- S.09P · Checking, your bank name
- S.100 · Checking
  - 1002 · Operating funds
  - 1004 · Payroll cks/deposits/svc chgs
  - 1006 · Job funds
    - 001 · Job 001
    - 002 · Job 002
    - 003 · Job 003
  - 1016 · Charge card deposits/svc chgs
  - 1018 · Escrow funds
  - 1020 · Clearings (bill/closings, etc.)
- S.10P · Savings, your bank name
- S.110 · Savings
  - 1102 · Regular cks/deposits/svc chgs
  - 1104 · Escrow funds
- S.119 · Cash on hand
  - 1190 · Petty cash
  - 1195 · Cash boxes/registers

Accounts Receivable

- S.12P · Accounts receivable

Other Current Assets

- S.12S · (Less) Allowance for bad debt
- S.131 · Other receivables
  - 1310 · NSFs in transit
  - 1312 · Receivables clearing (temp)
  - 1320 · Income earned- not yet invoiced
  - 1328 · Retainage receivable
  - 1330 · Accounts receivable - other
- S.134 · Other current assets
  - 1340 · Short term investments
  - 1345 · Refundable bond payments
  - 1350 · Due from employees
  - 1360 · Due from related party
  - 1370 · Prepaid expenses
  - 1380 · S/term portion of L/term assets
  - 1390 · Refundable deposits
- S.13P · Undeposited funds
- S.14P · Construction mat'ls inventory
- S.15P · Work in process
- S.16P · Lots/homes ready for sale

EXHIBIT 7.1 Income and current asset accounts.

TIP
Continued

You can use similar subaccounts to track escrow-like fund balances (1018. Escrow funds), charge card receipts and fees (**1016.Charge card deposits/svc chgs**), and payroll expenditures and the related Electronic Federal Tax Payment System (EFTPS) payments (**1004.Payroll cks/deposits/svc chgs**). In addition, you can set up accounts for special clearings, which are special entries requiring the use of QuickBooks Pro features available only through the use of a check (**1020.Clearings—bill/closings, etc.**), or funds designated for a special future purpose such as equipment replacement. Just label each of your cash subaccounts properly and use them for the deposits and payments specified by the subaccount title. Transfers between subaccounts are entirely proper as long as you pay close attention and understand that this type of in-and-out transfer has no affect on your total cash balance.

When you are ready to reconcile your bank account, choose the title account and it will present you with all the transactions in the subaccounts for reconciliation purposes. This approach can be adapted to any balance sheet account.

S.10P.Savings (your bank name) and S.110.Savings

Do you have a business savings account? More than one? A money-market account that acts and looks like a checking account? Have you closed any accounts during the current year that were open at the start of your accounting period? Check-off and note information about each of the accounts you need to create, such as bank name, description, and any subaccounts that you may wish to include.

S.119.Cash on hand

- **S.1190.Petty cash** is a frequently misused account name, term, and concept. I have seen the account used in a multitude of incorrect ways, so I'll share some of the common variations of ways to *not* use the account. Don't use your petty cash account to:
  - Accumulate small expenditures paid to employees or vendors
  - Write a series of checks to cash and post it to this account
  - Post transactions when the real account is unknown

Companies typically establish a petty cash fund because owners and/or employees encounter circumstances where they need to pay out a small amount of cash for day-to-day supplies or incidentals when it may not be convenient to pay with a company check or credit card. There are two ways to handle this situation.

1. If you or your employees pay for these types of items out of pocket, list them on an employee expense report and turn in the expense report for reimbursement on a monthly basis. When you enter the bill to repay yourself or your employee, itemize the expenses. This is the way I prefer to see these kinds of expenses handled. Each person keeps track of his or her reimbursable expenses and periodically turns them in for reimbursement. This method means that no one has to keep track of and balance a petty cash fund account, and your expense report receipts support your check.
2. Some owners still prefer to keep a physical petty cash fund on hand. A petty cash fund typically is somewhat difficult to control and track because it often "floats" between several employees. Unless it is tightly controlled by one individual, the fund often ends up short with unresolved questions about how, when, and where the money was spent.

TIP If you really want to keep a petty cash fund, here is how it should be handled. Keep your petty cash fund in an envelope, zippered bank deposit bag, or small cash box. Maintain a specific, predetermined balance, such as $200. At any given moment the total in the petty cash fund should be comprised of cash and receipts totaling $200. When the cash gets fairly low itemize the receipts to the appropriate accounts and write a corresponding check to "cash" to bring the total back to $200. Place the replacement cash in the petty cash bag or box so the process can begin again.

- **1195.Cash boxes/registers** is needed if you sell items for cash, and have a cash box or register. These funds also should maintain a specific, unchanging balance.

S.12P.Accounts receivable

These are special accounts that you will use if you have chosen to use accrual-basis accounting for your books. The account represents money owed to you for services you have completed. This account is linked directly to your invoicing and payments received processes.

S.12S.(Less)Allowance for bad debt

If you keep your books on an accrual basis and believe you have a debt you will not be able to collect, use this account to offset your receivables and simultaneously decrease the income you recognized with your invoice. Alternatively, I recommend that you directly reduce your receivables account and income with a credit memo coded to the appropriate receivable.

NOTE

You can only write-off bad debt expense against recognized income. If you use cash-basis accounting and/or are a cash-basis taxpayer you only recognize the cash you actually receive as income, therefore you don't have the option of a bad debt write-off. Bad debt is just money you haven't received and therefore haven't paid tax on!

S.131.Other receivables

- **1320.Income earned, not yet invoiced** is a special type of receivable accrual account. If the date you are using for financial measurement purposes does not coincide with the date you normally would invoice a client, use this account to record journal entries that recognize the income and receivable as of that date. Then reverse the entry on (as of) the day following the reporting period.

NOTE

This "temporary fix" allows you to continue with the regular QuickBooks Pro invoicing system, which assigns income to jobs and job processes within various job-cost reports. This account is used to record amounts not yet invoiced, but which sooner or later will be owed for services performed or products and other costs you have recognized as expense on your books, but have not yet invoiced to customers.

In accountant's terminology, this is your percentage-of-completion account. You also could view this account as "work completed and income earned that might be 'invoiceable' but, for whatever reason, you haven't yet invoiced it."

- **1328.Retainage receivable** typically is not used because amounts earned and owed already are included in **S.12P.Accounts receivable, trade**. It is possible that you, or your tax advisor, may wish to segregate and/or identify this unique type of receivable. If so, mark this account for setup.
- **1330.Accounts receivable, other** also is an accrual account. Use this account only if you have an unusual receivable not covered in any of the other due-from accounts discussed in the following **S.134.Other current assets** section.

S.134.Other current assets

- **1340.Short-term investments** is used to record certificates of deposit or other invested funds that are accessible quickly or available on demand.
- **1350.Due from employees** and **1360.Due from related party** is used for receivables from employees, such as an advance against pay, and from company owners or related parties. These accounts are only for short-term receivable situations.

NOTE

If you have employees you probably will need this account because it's almost inevitable that sooner or later you'll grant a pay advance to one or more employees. Because individual employee receivables can be tracked through the Payroll Items feature in QuickBooks Pro, it isn't necessary to set up individual accounts for each employee. I do recommend, however, that you keep each owner's receivable separate so the information is quickly available to your tax advisor.

- **1370.Prepaid expenses** is used to record expenses that you pay in advance, such as insurance. As time goes by and you use up your prepayment, this account is reduced and the expense is recognized. As a practical matter, however, most small businesses record insurance and other small, prepaid types of expenses as they are incurred. If you aren't sure whether you need this account or not, check with your tax advisor.
- **1380.Short-term portion of long-term assets** is an account your CPA may want you to use if you are submitting financial statements to a bank. The idea is to increase your short-term assets by the amount of long-term assets, usually loans receivable, within the next 12 months and to reduce your corresponding long-term assets by the same amount. The goal is to present an accurate picture of assets that are accessible within one year or less. In addition to meeting your banker's needs, the information can come in handy for internal purposes if you are trying to anticipate cash flow over the next 12 months.

NOTE This is an account your CPA may need if he or she prepares financial statements for you to present to outsiders. The related offset account is **1949.(Less) Short-term portion,** listed under the **S.192.Long-term notes/interest recble long-term** account.

- **1390.Refundable deposits** is used to record amounts that you have on deposit that will be refunded to you sometime in the future, such as security deposits and telephone or utility deposits. You will need this account if you are using accrual-basis accounting because refundable deposits aren't really an expense. They actually are a type of receivable because they represent money owed to you in the future. Cash-basis accounting requires you to expense these kinds of deposits immediately and then include them as income when they are refunded.

S.13P.Undeposited funds

This is a special type of temporary "holding account" for payments you have received, but not yet deposited. If you ever wait more than a day to make bank deposits, or ever deposit several checks at once, you will want to use this account. You will accumulate checks you receive and then, as part of the deposit process, select the checks to deposit. QuickBooks Pro will add them together and post them to your bank account in total. This will make your bank reconciliation process simpler because the deposit totals on your books will match the deposits shown on your bank statement.

S.14P.Construction materials inventory

You'll need this account only if you purchase and hold materials for resale. Most remodelers and builders buy specific materials for specific jobs, and general consumable supplies are charged to jobs or to the company as they are purchased. Occasionally you may get a special price on a particular item that you will be using in the near future and may want to use this account as a temporary holding account. If you believe you have a retail or inventory situation to address, consult your tax advisor before creating this account because inventory issues can be very complex.

S.15P.Work in progress (WIP)

These accounts are used to record the asset associated with a home purchased to remodel on a speculative basis, spec homes being built, or land being developed into lots for sale (or to hold and ultimately build on). As you incur costs on speculative projects you are, in effect, building an asset. Although you have expended funds, you haven't earned any matching income, so the amounts you have spent are considered WIP, which is classified as an "other current asset" assuming you intend to sell it within the near future.

NOTE

For a variety of reasons I recommend that spec home builders and developers initially record these costs in cost of goods sold accounts (COGS) (see chapter 8). Because of the way QuickBooks Pro is set up, I suggest that you assign all of your day-to-day in-progress costs to COGS accounts so they appear on the income statement instead of the balance sheet. (The balance sheet in QuickBooks Pro is not designed to present job costs in a columnar fashion, whereas the income statement is.)

Then, when you need to prepare a financial report that shows your correct, current, year-to-date income and expenses, you temporarily move the costs out of the COGS accounts and into the WIP accounts using a journal entry. On the day following the reporting period or the day you prepare the financial report, you then prepare another journal entry to reverse the process and move the costs back into the COGS accounts, where they await their matching "income" entry. When the gross income arrives via a closing, the net income is correct for the job. If the home doesn't sell within the next recording period, repeat the process.

By periodically moving costs into and back out of these accounts through period-end adjustments and reversals you get better and clearer profit and loss job-cost reports, and your in-house accountant will be able to perform job-cost proofing and balancing procedures. You also will be able to see accurate company results on a periodic basis and still meet your banker's and tax advisor's needs.

S.16P.Lots/homes ready for sale

This account is nearly identical to the preceding WIP account, but should be used to hold the costs related to homes ready for sale. The accounting process is the same.

Selecting fixed-asset accounts

These accounts represent assets you own with a useful life greater than one year. For instance, if you have office furniture and equipment, own shop tools, have purchased a vehicle for your business, or own construction equipment, these fixed assets are recorded *at their cost to you* on your books.

As your assets' useful or taxable lives pass by, their original cost is periodically recognized in a special kind of expense called *depreciation*. For leasehold improvements and non-tangible assets that are expensed over time, the term *amortization* is used in place of depreciation. Even though the terminology is different, the amortization concept is nearly identical to that of depreciation.

"Wait a minute," you might be saying to yourself. "That filing cabinet cost me something. Why shouldn't I recognize the related expense when I buy it?" The answer is that because you receive *benefit* or income-producing capacity from the asset over time, you should record the expense related to that benefit over time as well. The idea is that you match expenses to benefits using the accumulated depreciation (or amortization) accounts. Each year, as you recognize the appropriate part of your asset expense, your *depreciation expense* accounts increase, and your accumulated depreciation accounts increase.

As the accumulated depreciation increases, the resulting combined value of the asset, minus the accumulated depreciation, decreases. The result becomes the *net book value* of your assets or the portion of your asset not yet recognized as expense.

So, instead of recognizing expense at the time of your purchase, you increase an asset account. Simultaneously, using traditional double-entry bookkeeping, you increase another account called accumulated depreciation (or amortization). This account is an offset to the original at-cost fixed-asset account. When you look at a detailed balance sheet, you can see:

- The original costs of your assets
- How much depreciation (or amortization) has accumulated on the assets
- The current net book value, which is the original cost of your assets minus accumulated depreciation or amortization

Although you could create just one accumulated depreciation account to cover all of your asset types, I prefer to create a separate accumulated depreciation account for each title account in the fixed-assets section so that I can see the net book value of each group. The fixed-asset accounts in Exhibit 7.2 show a (Less)Accumulated depreciation (or amortization) account for each title account.

Fixed Assets

S.170 · Land & improvements
- 1700 · Land, at cost
- 1705 · Land improvements, at cost
- 1805 · (Less) Accum depr-land imprvmts

S.171 · Buildings, net book value
- 1710 · Buildings, at cost
- 1810 · (Less) Accum depr-buildings

S.172 · Leasehold improvements, NBV
- 1720 · Leasehold improvements, at cost
- 1820 · (Less) Accum amort-L/H imprvmts

S.173 · Office furn/fixt/equip, NBV
- 1730 · Furn/fixt/equip, at cost
- 1830 · (Less) Accum depr-FF&E

S.174 · Computers/sftware, net book val
- 1740 · Computers/software, at cost
- 1840 · (Less) Acc deprec & amort-C&S

S.175 · Vehicles, net book value
- 1750 · Vehicles, at cost
- 1755 · Construction vehicles, at cost
- 1850 · (Less) Accum deprec-vehicles
- 1855 · (Less) Acc deprec-cnstr equip't

S.176 · Construction equipment, NBV
- 1760 · Construction equipment, at cost
- 1860 · (Less) Accum depr-cnstr equip't

S.177 · Construction tools, NBV
- 1770 · Construction tools, at cost
- 1870 · (Less) Accum depr-constrn tools

S.178 · Organizational costs, NBV
- 1780 · Organizational costs
- 1880 · (Less) Accum amort-orgz'l costs

S.179 · Goodwill & NC covenants, NBV
- 1790 · G'will/non-compete cov, at cost
- 1890 · (Less) Accum amort-GW/NCC costs

EXHIBIT 7.2 Fixed asset accounts.

TIP If you have donated assets to your company, consult with your tax advisor to establish the correct values and accounting entries.

NOTE There are several reasons why you should talk with your tax advisor about depreciation and amortization calculations.

- There are a variety of tax implications involved in these computations.
- Your tax advisor will have special software to calculate the correct depreciation and amortization costs for the tax year.
- In certain cases, you may be able to depreciate an entire asset for tax purposes immediately.
- It is very important to keep your accounting records consistent with your tax return. When you receive your return from your tax advisor remember to ask for the depreciation and amortization detail, as well as any other accounting adjustments. Then post the results to your fixed-asset and depreciation (or amortization) accounts.
- If you ask, your tax advisor also will be able to provide you with the amount of depreciation and amortization you will be entitled to take during the next year, assuming there are no changes in your fixed-asset accounts. Then you can make entries to recognize estimated depreciation and amortization on prior-year assets as the year progresses, making your year-end depreciation entry less dramatic.

S.170. Land and improvements

Land is the only type of fixed asset *not* considered depreciable for accounting or tax purposes, so there is no offsetting accumulated depreciation cost for land. Certain land improvements, however, are depreciable.

S.171. Buildings, net book value

If you own your building, choose these accounts.

S.172. Leasehold improvements

If you rent you may have made improvements to your leased space. If the improvements have a useful life greater than one year and don't qualify as repairs or maintenance you will need these accounts.

S.173. Office furniture/fixtures/equipment

These accounts are used for office furniture, fixtures, and equipment (FF&E), such as filing cabinets, desks, calculators, chairs, copy machines, faxes, telephone systems, cameras, or lights. Nearly everyone needs these accounts.

S.174. Computers and software

It's usually helpful to keep your computers and long-term software purchases separated into these accounts. Please note that most software upgrades should be posted to an expense account rather than to this asset account because their useful life often is fairly brief.

S.175. Vehicles

If you have donated administrative vehicles to, or purchased vehicles for, the company, use these accounts.

S.176. Construction equipment

This account is for vehicles used for construction purposes, as well as for construction equipment that does not qualify as tools.

S.177. Construction tools

Put tools with a useful life greater than one year in this set of accounts. Ask your tax advisor if he or she also uses a specific price guideline, such as greater than $100, to differentiate between tools that are immediately expensed and those that become fixed assets. For example, a hammer has a useful life greater than one year so it technically is a fixed asset, but the price is insignificant so it normally would be expensed when purchased.

S.178. Organizational costs

If you incurred costs before you were officially open for business under your current entity type, you may have incurred costs to create your business such as:

- Legal fees
- Accounting fees
- Research costs
- Travel costs

It is important to keep track of these costs because, although they cannot be immediately expensed for tax purposes, they can be expensed or amortized over time with the correct tax election. Consult your tax advisor. If this tax election is not done correctly and on a timely basis you could lose the deduction. Select these accounts if you have these types of start-up costs.

S.179.Goodwill and covenants not to compete

If you purchased a business, you may have purchased *goodwill*, which is the price of the business in excess of assets minus liabilities. You also may have purchased the owner's agreement or covenant to not compete directly with you for a specific period of time and/or in a certain geographic region. If so, you'll need these accounts.

Choosing other asset accounts

Other assets are the least liquid type of assets. They include notes receivable, which by contract are received over time, and thus theoretically are less readily convertible to cash. This category also covers long-term investments and the cash value on life insurance policies because the stated purpose is to hold these types of assets over time.

S.190.Capitalized loan costs

You may need these two accounts if you recently obtained a loan with origination fees. Certain loan origination fees must be capitalized or turned into assets and then expensed (amortized) in equal installments or on a straight-line basis over the life of the loan. For example, if your loan origination costs were $1,200 for a 60-month loan, your entry to **1900.Capitalized loan costs** account would be $1,200 and your monthly entry to **1905.Accumulated amortization-loan costs** would be $20. Only some of the various loan-related fees should be amortized, so check with your tax advisor for specifics.

S.192.Long-term notes and interest receivable

Has your company issued a note to anyone, or have you sold property on a land contract? If so, select the correct accounts. It's often useful to see the interest receivable broken out from the principal, so I have included that account as well (see Exhibit 7.3). Account **1949.(Less) Short-term portion** allows you to move the next 12 months' worth of receivables, as listed in amortization tables, out of long-term status and into current assets if you wish. This is the offset to the account mentioned under other current assets.

Other Assets
- S.190 · Capitalized loan costs
  - 1900 · Capitalized loan costs
  - 1905 · (Less) Accum amort-loan costs
- S.192 · Long-term notes/interest recble
  - 1920 · Note receivable, borrower/descr
  - 1945 · Accrued interest receivable
  - 1949 · (Less) Short-term portion
- S.195 · Long-term investments
  - 1950 · Investments, at cost
  - 1960 · Cash value, co. life insurance
  - 1970 · Invstmnt in property to remodel

Accounts Payable
- S.20P · Accounts payable

Credit Cards
- S.210 · Company credit cards
  - 2100 · Co credit card name #1
  - 2101 · Co credit card name #2
- S.215 · Propr/prtnr/sh'holder cr cards
  - 2150 · CC name, owner of cc #1
  - 2151 · CC name, owner of cc #2
- S.217 · Business exp reimbursements due
  - 2170 · Reimbursements due to (owner)
  - 2171 · Reimbursements due (employee1)
  - 2172 · Reimbursements due (employee2)
- S.21P · Reimbursements due petty cash

EXHIBIT 7.3 Other assets and credit card accounts.

S.195.Long-term investments

If your company owns other long-term investments, such as stocks or investments in other companies, property held to develop or remodel more than a year in the future, or if your company owns a life insurance policy with a cash value, check-off the appropriate account(s).

S.20P.Accounts payable

If you have chosen to record your income and expenses on an accrual basis, this is the account where QuickBooks Pro records your bills. As each bill is entered, you also will recognize the related expense, asset acquisition, or liability reduction. This account links directly to the bill entry process, bill payment process, and the invoicing process for pass-through costs. If you are using a cash-basis approach, do not select this account.

S.210.Company credit cards and S.215.Shareholder credit cards

Credit card purchases are considered *cash* expenditures for tax purposes regardless of when they are actually paid. In some cases, owners pay off cards each month as the bill comes due. In other cases, owners do not pay the balance off immediately and incur finance charges that are assigned to credit card interest expense and included in the card balance. Regardless of whether you pay the balance in full each month, or whether you carry an ongoing balance, I advise you to create and use these accounts to record expenditures and payments as of the *actual* transaction dates.

If you have acquired one or more cards in the company's name, note the credit card name next to the accounts on your pencil copy.

NOTE

If you have personal credit cards designated as company-use-only cards, note the card name and type, and the shareholder's initials. I suggest segregating these cards as shown in Exhibit 7.3 because, under certain circumstances, your tax advisor may need to include your personal credit card accounts as your basis or personal investment in your company. If your personal cards were used exclusively for your business, all of the related interest and service charge expenses are deductible for tax purposes.

If you have placed both business and personal purchases on your personal credit card account, I suggest that you turn in your company purchases as an expense reimbursement request, just as you would if you worked for another company. Enter the amounts due as an accounts payable item or bill due to yourself. Pay yourself, and then use those funds to pay the credit card. This approach is appropriate because it becomes nearly impossible to segregate deductible interest and reconcile credit card statements in a mixed-purchases situation.

TIP

You may want to reorganize existing credit card balances so that certain cards contain only business expenditures. You then can deduct the related interest and service charges as described above.

Adding liability accounts

The second category of accounts in your general ledger and on your balance sheet is liability accounts. Liabilities generally are organized in the order of when they are due—from those due most currently to those due over a longer-term. Exhibit 7.4 shows the other current liability accounts from the Chart of Accounts included in appendix A.

S.221.Payroll taxes/withholdings due

Set up this account with the detailed subaccounts shown in Exhibit 7.4. Some Charts of Accounts consolidate all the different types of payroll taxes and withholdings payable into one account. Most accountants, however, prefer to use these detailed subaccounts because it is much easier to track down discrepancies when the accounts are segregated and clearly labeled. Also, since different taxes are due at different times, you can quickly determine outstanding amounts when you look at the balance sheet. Please note the following abbreviations:

- FIT = Federal Income Taxes
- MC = Medicare
- SIT = State Income Tax
- FUTA = Federal Unemployment Tax

Additional accounts can be established if your employees have other withholdings such as friend-of-the-court payments, savings account withholdings, simple IRA withholdings and match payable, or other retirement account withholdings payable to outside parties. (See chapter 8 for a discussion of the concepts related to payroll posting in the direct cost or COGS section.)

S.22P.Sales taxes payable

Select this account if you charge or collect sales or use taxes in your business, regardless of your cash or accrual status. When you charge sales tax to your customers it is not income or expense to you. You are acting as a tax collector for the state or federal government, so as it is charged and/or collected, the tax becomes a liability because that amount actually is due to the government). If you purchase materials or supplies for your company's use from out-of-state sources, you also may owe use tax to your home state. Each state and sometimes each county or locality has different rules regarding sales and use taxes. See your tax advisor for specifics about your particular situation.

Other Current Liabilities

- S.221 · Payroll taxes/withholdings due
  - 2210 · FIT withheld
  - 2211 · (Less) Adv earned income credit
  - 2212 · FICA/MC withheld & matched
  - 2214 · SIT withheld
  - 2216 · State unemployment taxes due
  - 2218 · State disability insurance due
  - 2220 · FUTA due
  - 2222 · Local taxes withheld
  - 2230 · 401K w/held & co. contribution
  - 2232 · SIMPLE IRA w/held & co contrib
  - 2240 · Friend of the Court withholding
  - 2242 · Garnishments payable
  - 2245 · Other withholdings payable
  - 2250 · Other contributions payable
  - S.221 · Payroll taxes/withholdings due - Other
  - S.22P. · Sales tax payable
- S.230 · Other taxes payable
  - 2320 · Real estate taxes payable
  - 2375 · Personal property taxes payable
  - 2380 · State taxes on earnings payable
  - 2385 · Fed corporate inc taxes payable
  - 2390 · Other taxes due
  - 2395 · Tax-related interest/penalties
  - 2410 · Compensation payable
  - 2420 · Client deposits (unearned inc)
  - 2422 · Work billed - not yet performed
  - 2430 · Retainage payable
  - 2450 · Accounts payable, other
- S.250 · Line(s) of credit
  - 2500 · Bank #1-line of credit
  - 2502 · Bank #2-line of credit
- S.260 · Due to related parties
  - 2600 · Due to affiliates
  - 2610 · Due to employees
  - 2620 · Due to owner(s)
  - 2630 · Due to other related parties
  - 2690 · Dividends payable to owner(s)
- S.27P · Other current liabilities
- S.28P · Plus S/Term portion-L/T liab

EXHIBIT 7.4 Other current liability accounts.

S.230.Other taxes payable

- **2320.Real estate taxes due** can be used if you own real estate and use accrual-basis accounting.
- **2375.Personal property taxes payable** is used only if your city or district charges tax on your production facilities or equipment and you use accrual-basis accounting. Alternatively, I prefer to include either of these two payables in the accounts payable listing.
- **2380.State taxes on earnings payable** is applicable to accrual-basis businesses with state taxes due based on company earnings.
- **2385.Federal corporate income taxes payable** is used primarily by accrual-basis C Corporations with earned income and taxes payable at year-end. The account is not applicable to sole proprietors, partners, S Corporations, and most LLCs because the company passes the income directly through to the owners' personal tax returns when taxes are paid on an individual level.
- **2410.Compensation payable** is an accrual account that you may need to use periodically. On a practical basis, most small-business compensation is not expensed until the date it actually is paid. From an accrual standpoint, however, at year-end you may need to enter outstanding gross payroll expense if you have employees. For example, if payday is once every two weeks, and at year-end you owe eight days' worth of payroll, recognize the expense in the current year by posting a journal entry and a follow-up entry to this account. Use this account for any type of compensation payable such as direct labor, administrative, sales, and commission compensation payable. Post each type of expense at pre-tax in gross cost, to the correct expense account.

NOTE

Because company payroll taxes are not due until the payroll actually is paid, it technically is not necessary to record company payroll taxes as an accrued amount. You'll probably want to coordinate with your tax advisor on this issue however.

- **2420.Client deposits (unearned income)** is the accrual-basis account you will need to reflect deposits you have received from clients before the related income is earned. This is a liability account because it really represents an amount you owe to your client—either in work to be performed or as a cash refund if you don't perform the work. For example, near the end of a quarter, your receivable records show that you have a client with a credit balance because you received a $3,000 deposit for an upcoming job, but you haven't performed any work (earned any income) on the job yet. Prepare a journal entry to move the amount into this account and out of accounts receivable. On the first day of the following quarter, reverse the journal entry.
- **2422.Work billed, not yet performed** is very similar to the account listed above. For example, you may invoice your clients because the terms of your contract say they are to submit payments on certain dates or before certain parts of the job are started. The invoice creates a receivable, but income has not yet been earned, so this is an appropriate account for these amounts.
- **2430.Retainage payable** represents amounts due to service providers that are being withheld until certain conditions have been met. Unless there is a specific reason to list them separately, these amounts often are already included with the accounts payable detail. Set the bill due date to the estimated future payment date.
- **2450.Accounts payable, other** is an accrual-basis account where you record any remaining payables not covered above or under the following current liabilities.

S.250.Line(s) of credit

A line of credit is designed to be a short-term loan that you can draw on for *temporary* cash shortages. If you have one or more of these short-term accounts, include the bank's name or a description of the loan in the account name so that it is easy to spot in your Chart of Accounts. Depending on your loan agreement, you may pay interest charges as they are incurred or you may add accrued interest due to the loan balance. In either case, match the balance in this account to the amount shown on your bank's monthly loan statements.

S.260.Due to related parties

This type of liability is separated from other liabilities because *who* is owed is useful information to various users of financial statements. Generally speaking, a *related party* is anyone who has a direct interest in the business, such as an owner, or someone who has a direct family, financial, or legal relationship with the company. Your accountant can offer more guidance if you have questions. You may want to create more subaccounts if the company owes money to more than one individual within a particular account. For example, if the company owes money to three different related parties, you may want to create three subaccounts under this title account.

NOTE If the liability is long-term in nature, create the account under the **S.280.Notes/leases/interest** due account rather than in the short-term liabilities section.

Picking out long-term liability accounts

Long-term liabilities typically are arranged according to the nature of the contract. Each land contract, secured note wherein you pledge collateral to the lending institution, unsecured note, mortgage, and capitalized lease should be listed separately under these long-term liability accounts, which are shown in Exhibit 7.5.

S.28P.Plus short-term portion of long-term liabilities

The idea behind this account is to show the current portion, or the portion due within the next 12 months, of your long-term debt. To create a balance in this account, use a journal entry to subtract the current portion of long-term debt from the long-term debt accounts and show the amount here. The offset account is **S.29^.(Less)S/Term portion-L/T liab.** You'll use this account primarily for your banker, but you also may find it useful to help predict annual cash-flow needs.

S.280.Notes/leases/interest due

If you have more than one type of note, lease, or loan create subaccounts for each, such as Note payable, '99 Ford truck. The notes payable account usually includes only the remaining principal due as verified by the loan statement or loan amortization schedules.

Long-Term Liabilities

S.28P. · Plus short-term portion of long-term liabilities
S.280 · Notes/leases/interest payable
2800 · Land contract payable-prin
2810 · Note payable, land-principal
2815 · Note payable, development loan
2820 · Note payable, construction-prin
2840 · Note payable, secured-principal
2850 · Note payable, unsecured-prin
2860 · Mortgage payable-principal
2880 · Capitalized lease-principal
2890 · Accrued int due-notes/leases
S.29^ · (Less) S/Term portion-L/T liab

EXHIBIT 7.5 Long-term liability accounts.

In this example, if the final income for the year was $24,488 and the partners had an agreement to split income equally, on the first day of the new year you would make a journal entry to move their contributions and draws or distributions into their individual Capital prior-years account, which would result in the balances previously computed. Then you also would move half of the preceding year's income into each of their individual Capital prior-years accounts. Capital balances as of the first day of the new fiscal year would be as shown in Exhibit 7.9.

Follow the same process for a sole proprietor, but 100 percent of the income would move to the proprietor's Prior years' capital account. You may wish to ask your tax advisor for assistance with the journal entry.

S.310.Capital and outstanding shares

This section applies if you have incorporated your business. When you incorporated your business you requested an authorization for a certain number of shares. Most companies are authorized for far more shares than they actually purchase at the inception of their company.

- **3100.Capital stock-### shares** is used to record only the shares you actually purchased from your company when it was incorporated. Enter the number of shares in the "###" part of the title. The account balance will be the amount that you paid for the shares. This account likely will maintain a constant balance, unless you raise capital through the sale of stock to yourself or others. You may show stock with a par or set value established as part of your incorporation process or you may have no-par stock that does not have a set value. This will depend in part on your state's incorporation procedures.
- **3110.Capital in excess of par value** is used if your stock has an established par value and you have sold stock to yourself or others at a value higher than the actual par value. The account balance represents the amount not entered into the capital stock account. Alternatively, if you invested additional capital in your company without purchasing additional shares, and if the funds invested were not set up as a loan. The account balance will be the contributed amount.
- **3120.Treasury stock** is used only if your company repurchased stock from shareholders and the stock has *not* been formally retired or closed against existing stock by an action of the board of directors. The balance normally will be a debit (negative), as it represents a reduction of outstanding capital stock. When the stock is retired this balance can be closed with a journal entry into your capital stock account.

Capital, Bob	
Prior years' capital, Bob	$33,744 ($11,500 + 12,244 income)
Plus current-year contributions, Bob	0
(Less) Current-year draws (or distributions), Bob	0
Capital, Don	
Prior years' capital, Don	$24,244 ($12,000 + 12,244 income)
Plus current-year contributions, Don	0
(Less) Current-year draws (or distributions), Don	0

EXHIBIT 7.9 Capital balances for owners.

S.35P.Opening balance equity (temporary)

This is a temporary account you will use to enter your beginning balances. See chapter 14 for an explanation about how this account is used.

S.36P.Retained earnings

This account shows the life-to-date results of the company's earnings less amounts paid out to owners. In QuickBooks Pro, current-year earnings are summarized on the balance sheet in the net income line throughout the year. At the beginning of the next year the net income is moved into this account.

NOTE I like to create separate accounts to segregate the distributions to shareholders from the rest of the retained earnings. This keeps the cumulative corporate earnings shown in the retained earnings account clearly visible. The balances in these distribution accounts should show up on your reports as a negative balance because they represent a reduction to equity. If you wish, you can move these account balances into the retained earnings account at any time.

S.37P.(Less) Current-year S Corp distributions and S.38P.(Less) Current-year C Corp dividends

These accounts are similar to the sole proprietor's or partners' draw accounts.

In Sub S Corporations owners are taxed at the personal level on the corporation's net income regardless of how much they actually have withdrawn from the business. The distribution occurs at the owner's discretion and is not a taxable event. Record the actual withdrawals or distributions in this account.

In C Corporations, the company is taxed at the corporate level on earnings before dividend distributions. The board of directors may choose to retain the company's net income, or may choose to distribute some or all of it in the form of dividends to shareowners. Each share of the business receives a certain dividend. Recipients are taxed on the dividends at the personal level. You must report dividends paid during the year to shareholders to the IRS on form 1099 DIV.

TIP You can create subaccounts for any of the preceding accounts for individuals. For individual shareholder's stock purchases, list individual owners in subaccounts, and list individual shareholders in subaccounts for distributions or dividends declared. That way you can keep the records for each shareholder very clear. When there is more than one owner, listing each separately lets them quickly and easily see their capital position within the company.

CHAPTER 8

Selecting income and cost of goods sold accounts

In chapter 7 we looked at the accounts associated with the balance sheet—accounts for assets, liabilities, and equity. Now let's begin to look at the various types of income and expense accounts that make up your income statement reports. In this chapter we will help you select your income and direct cost, or cost of goods sold (COGS), accounts. The remaining expense accounts are reviewed in chapter 9.

As with the accounts for the balance sheet, appendix A includes a comprehensive list of income and expense accounts, some of which will apply to your business. Using these account numbers, names, and groupings as the basis for your income statement accounts will save you extensive design time and allow you to take advantage of other remodelers', custom home builders', developers', and accountants' experience.

The logical grouping of accounts and the subtotaling of specific accounts by and within business functions can provide you with immediate and lasting benefits for current periods, prior periods, and budgets. If you group and subtotal accounts by business function your reports will immediately become more understandable and logical. So if your current reporting system is not organized in this fashion, you'll want to make these two changes your first order of priority because you'll receive outstanding "bang for the buck."

The QuickBooks Pro account types for income and direct costs that I use when working with remodelers, custom home builders, and developers are:

- Income
- Cost of goods sold (COGS)—Direct costs, which are costs directly related to the production of income
- Other income—Non-operational income, including interest income

In the Chart of Accounts, accounts have been assigned to the appropriate business function, and subaccounts created to produce useful subtotals within the business functions. These computer-generated subtotals save you from having to create subtotals with an adding machine tape, and also can be used to provide summarized or collapsed management reports.

Posting to income and COGS accounts

All postings to income and COGS accounts should "arrive" in these accounts via the QuickBooks Pro 2000 items in the Item List that are linked to these accounts. In the QuickBooks Pro way of doing things, you access the Item List from the transaction screens, such as purchase orders, bills, estimates, and invoices. Each of these forms includes an Items tab or an Items column that allows you to type in an item or highlight an item on the Item List (see chapter 10).

To ensure that all income or direct cost is assigned to a job, assign a customer or job name every time you use an item. (Here's that three dimensional concept at work.) In my sample Customer:Job List I have created several special jobs for costs that cannot be assigned to actual customer jobs, so that you can assign every item to an appropriate customer or job name.

Exercise diligence when you have transactions relating to income or COGS accounts. Even though QuickBooks Pro doesn't require you to use items for certain transactions or require you to assign a customer or job to a transaction, you will want to do so in order to make your job-cost reports as accurate as possible.

If you follow my recommendations regarding the use of the Item List you will be able to see your job-cost information in a format designed to show job processes and their corresponding subtotals, as well as actual vs. estimated costs for individual jobs. You also will be able to see individual and consolidated customer and job costs as well as your company's total COGS costs in the general ledger account order and income statement format.

Selecting income accounts

Income accounts are the "buckets" in which you place dollars that are earned by your business. On your pencil copy of the accounts in appendix A select the income groupings and accounts that apply to your business as you see it functioning now and in the near future. You probably won't need all of these accounts in the beginning. Use what you need now and in the near future. You can create a new grouping, a new account, or move an account into another grouping if you need to do so later.

Let's look at the specific accounts in appendix A and review their purposes. Exhibit 8.1 lists the first set of income accounts we will look at.

S.40P.Unclassified income

Use this account to hold income-like transactions that you don't know how to classify at the time you post the transaction. Although entries into this account must be classified into another account at a later date, this technique allows you to make sure the transaction is recorded promptly. This allows you the opportunity to gather more information and then come back at a later time and assign the transaction to the appropriate account. If you haven't made the correction by the time you review your reports, you will see the balance and know the amount still needs to be reclassified.

Income

- S.40P · Unclassified income
- S.41P · Design & consulting services
- S.42P · Construction services
- S.440 · Income from sold units
  - 4400 · Lot sales
  - 4410 · Home sales
  - 4490 · Add-ons
- S.45P · Misc. reimbursed expense
- S.46P · Other income
- S.47P · (Less) Discounts to clients
- S.48P · (Less) Refunds to clients
- S.49P · Income from/(to) unearned inc

EXHIBIT 8.1 Income accounts.

S.41P.Design and consulting services

I encourage my clients (especially my remodeling clients) to charge for these services. Even though most remodelers readily acknowledge that they primarily are a service business, I find they seldom charge separately for the time they spend delivering their valuable services.

TIP

If you currently do not charge separately for these services, you might consider a transition in which, for each job, you:

- Determine how many hours you anticipate spending consulting and/or providing design service
- Build a reasonable hourly return for your time into the price of the job

TIP
Continued

- Discuss your budgeted time with each client for each stage of the job, and then charge by the quarter hour on a regular basis for the excess required by your clients

If you set clear expectations for both you and your client at the outset of a project, you can limit what sometimes seems like an endless drain on your time. Value yourself. Value your value. Value your time. Don't be timid about your hourly rate; charge a rate that clearly reflects your value as an expert professional. If you're new to the business, charge a rate somewhat higher than a trade rate. You are the business owner and you should be reasonably compensated.

S.42P.Construction services

This account is used to record income for custom home building, remodeling, or other construction services rendered to clients. I usually consider income to be *services income* when it occurs as a result of direct interaction with clients. In QuickBooks Pro, services income, such as remodeling services or insurance restoration services, is tracked through the use of detailed items on the Item List.

S.440.Income from sold units

When builders or developers work in spec mode, the sale is recorded in one or more of the following sales subaccounts.

- **4400.Lot sales** is used to record the sale of a lot when the amount is clearly and objectively defined (as distinguished from the sale of the home or condo).
- **4410.Home (condo) sales** is used to record the gross sales price of the home or condo. This may or may not include the sale of the lot. For management accounting purposes, I suggest you record your sales (either with or without lots) in a consistent manner. If the gross sales price includes extras or add-ons, however, I would record that part of the sale in the next account.
- **4490.Add-ons (or extras)** is where I prefer to record the income from add-ons. Although the sales closing documents may fold the price of customer-selected extras or add-ons into the gross sales price, I recommend you separate out this income so that future sales can be compared against one another. Separating out this income also can help you quantify the additional income generated by customer requests.

S.45P.Miscellaneous reimbursed expenses

If you need to include reimbursed expenses in income for tax purposes (see your tax advisor), or if you simply want reimbursed expenses shown separately from the income you generate from operational activities, you may choose to place it here.

Alternatively, you may wish to position this account in the "other income" section so that your tax advisor can clearly identify the reimbursed expenses for tax purposes. In the "other income" section, reimbursed expenses won't inflate your gross sales total or distort your gross profit margin percentage for management reporting purposes.

S.46P.Other income (and services)

If you provide other services or generate other kinds of income as an inherent part of your construction business, you can include that income here. It may be better, however, to place accounts for income not directly related to your primary business, such as rental activities in **9150.Rental property income** under title account **S.910.Interest and other Income**. This posts your additional sources of income in a separate section at the very end of your profit and loss reports. Then your other income is segregated from your construction or remodeling income and won't interfere with your operating margin percentages.

S.47P.(Less) Discounts to clients and S.48P.(Less) Refunds to clients

I typically don't show either of these as separate line items. If you give a client a discount, credit, or refund, apply the reduction in income back against the applicable job. If the job has been completed for a long time, however, or if you want to show discounts or refunds separately, these are the accounts to use.

NOTE These accounts are not to be used for refunds, credits, or rebates to you from your vendors. When you receive a rebate from a vendor, use the cash inflow to offset the original expense and job. If the cash inflow doesn't relate to expense, such as a referral bonus or a prize, use account **9190.Income from other sources** under title account **S.910.Interest and other income.**

S.49P.Income from/(to) unearned income

Use this account to transfer income *out* of the income statement and *into* the unearned income balance sheet account at the end of an accounting period. Perform this transfer when you or your accountant determine that it is not yet appropriate for accounting or tax purposes to recognize income on certain jobs. You then will reverse the entry into the appropriate income accounts on the first day of the following accounting period so that you can always see the job's income to date in your profit and loss by job report.

Entering direct cost accounts

QuickBooks Pro will create subtotals and gross profit amounts if you place your direct costs into the COGS accounts. I recommend you place the following accounts in the direct costs section of your Chart of Accounts as your circumstances warrant. Exhibit 8.2 shows the first group of COGS accounts.

Cost of Goods Sold

- S.50P · Unclassified COGS
- S.510 · Construction costs
- S.520 · Direct labor comp/taxes/burden
  - 5200 · D-Regular compensation
  - 5202 · D-Overtime compensation
  - 5204 · D-Payroll taxes/specif benefits
  - 5206 · D-Allocated burden & NJS cost
- S.522 · Employee labor burden
  - 5220 · E-Workers' comp insurance
  - 5222 · E-Health insurance
  - 5224 · E-(Less) Employee-paid hlth ins
  - 5226 · E-Private disability insurance
  - 5228 · E-Life insurance
  - 5230 · E-Retirement benefits
  - 5232 · E-Employee assistance programs
  - 5234 · E-Gifts/promotion
  - 5236 · E-100% deductible meals & ent
  - 5238 · E-Recruiting and hiring costs
  - 5248 · E-(Less) burden allocation

EXHIBIT 8.2 Cost of goods sold (COGS) construction and direct labor accounts.

S.50P.Unclassified COGS

Use this account the same way you would use the **S.40P.Unclassified income** account described previously under income accounts. Use this account if you believe that the mystery transaction relates to an undetermined COGS expense, but don't yet know which account is correct.

S.510.Construction costs

If you are a developer or builder of spec homes, create this title account to differentiate direct construction costs from land or lot costs and interest, land development, and infrastructure.

S.520.Direct labor compensation/taxes/burden

Use the series of accounts under this title account to record the costs associated with employees who work primarily on your *jobs*. If you have no employees, don't choose any of the following accounts. You might want to review the concepts, however, for future reference.

> **TIP** The title account number uses S- to designate that it is used as a summary sales or title account. Within certain account *text names*, I use D as a prefix to designate that the account is a direct labor account. This is designed to be both a memory and visual posting aid.

5200.D-Regular compensation and 5202.D-Overtime compensation

In a program such as QuickBooks Pro I try to keep the number of compensation-related general ledger accounts to a minimum. I do, however, believe that overtime pay should be monitored continually, and since profit and loss reports are a good vehicle for owners to use for review and investigation, I recommend creating specific general ledger accounts for overtime costs in all compensation categories.

You then can track very detailed information on time and costs by job for direct labor, supervisory time, and production management through the payroll subsystem using payroll items (see chapter 24).

In addition, payroll items can track costs for other non-job-specific (NJS) payroll time and costs, such as paid bonuses or profit sharing, paid time off, training time, and other NJS duties including company errands, staff meetings, and timekeeping.

5204.D-Payroll taxes and specific benefits

This account holds the cost of the *company's* payroll taxes including Social Security, Medicare, federal unemployment taxes, and state unemployment taxes.

> **NOTE** If you link your payroll tax items to this account, QuickBooks Pro will compute, post, track, and assign payroll tax costs to specific jobs based on timesheet input. In addition, by using special design and setup procedures you can reassign or allocate certain company-paid benefits from their original accounts into this account on a per-job-cost basis. I have done this for costs, such as retirement benefits, worker's compensation, union costs, and health, life, or disability insurance.

If you want to use this option, set your QuickBooks Pro file preferences accordingly (see chapter 13). Then the costs you see in this account will have been assigned to specific jobs or to the NJS direct labor job based on timesheet input.

5206.D-Allocated burden and NJS cost

Create this account when you choose to move on to the more complex process of assigning the following labor burden costs to specific jobs.

> **TIP** When you first begin job costing you may want to measure and monitor burden and NJS costs for six months to a year before you begin to allocate costs within your system.

S.522.Employee labor burden

I will use the term labor burden to refer to the remaining costs associated with labor. For example, in addition to the costs previously listed for direct labor you probably pay for some of the following direct labor costs.

- **5234.E-Gifts and promotional items** is used for items such as shirts, caps, or jackets with your company logo.
- **5236.E-100-percent deductible meals and entertainment** is used to record company-sponsored meals and entertainment such as holiday parties or paid onsite or on-the-job overtime meals. (See your tax advisor for details.)
- **5238.E-Recruiting and hiring costs** is used to record fees from newspaper ads or employment agencies.
- Payroll mandated insurances such as **5220.E-Worker's compensation insurance**
- Company-paid insurances such as **5222.E-Health insurance**, **5228.E-Life insurance**, and **5226.E-Private disability insurance**

NOTE Because some of these accounts apply to employees other than direct labor employees, I use an E- prefix for each of these account names rather than the D- used for the other direct labor accounts.

S.53P.Trade contractors

This is an account used almost universally by remodelers, custom home builders, and developers (See Exhibit 8.3.)

NOTE I'd like to offer a note of caution to those of you not familiar with IRS rules relating to trade contractor relationships. See your tax advisor for guidelines defining the difference between employees and trade contractors. If the IRS, your state, or an attorney representing a "former trade contractor turned employee" determines that you should have been treating a trade contractor as an employee, you can be liable for exceptionally high fines and penalties. Understand the rules and proceed cautiously.

S.54P.Building materials and supplies

Use this account for materials and supplies that you can assign to specific jobs. Alternatively, some builders who purchase and hold inventory may choose to use this account as a temporary holding account for purchases until the year-end inventory is adjusted to its actual value. You will use another account, listed later, for small tools and general building supplies that would be difficult or impractical to assign to specific jobs.

S.55P.Other job-specific costs

Use this account for costs that can be assigned to jobs, but are less significant and don't fall into any of the three preceding categories. I create a number of specific items in the Item List that link to this account. This allows you to track specific costs without creating an over-abundance of general ledger accounts. Typical job-cost items that link to this account include:

- Permits
- Blueprints and copies
- Site utilities

- Rental of tools or equipment for a job
- Travel and auto expense
- Trash removal
- Job promotional costs such as flowers, gift baskets, or meals out
- Association (subdivision) dues
- Builders' risk insurance
- Other miscellaneous job-specific construction costs

Exhibit 8.3 shows the next set of income accounts.

S.560.NJS direct costs

Use this account to subtotal direct costs that can't be attached to a particular job, but are directly related to job production. Subaccounts under this title account include:

- **5600.General liability insurance** for the portion of your insurance related to job liability protection
- **5602.Small tools and miscellaneous job supplies** for costs that can't be assigned to any specific job, such as saw blades, hammers, tool belts, sandpaper, and so on
- **5604.Construction equipment/vehicles-gas/oil**
- **5606.Construction equipment/vehicles-licenses/insurance**
- **5608.Construction equipment/vehicles-repair/maintenance**
- **5650.Uniforms and laundry** for field personnel
- **5660.Cell phones-field** for communications costs for field personnel
- **5670.Depreciation on construction equipment, tools, vehicles** for the depreciation cost of construction-related fixed assets
- **5690.(Less) D-NJS burden allocation** to move costs or potential costs from one location in the accounting system to another (Use this account when you *remove* or offset the NJS costs out of the job called NJS and to assign the costs to individual jobs using QuickBooks Pro payroll items.)

S.53P · Trade Contractors
S.54P · Building materials & supplies
S.55P · Other job-specific costs
S.560 · NJS (Non-job specific) dir cost
    5600 · General liability insurance
    5602 · Small tools & misc job supplies
    5604 · Cnstr equipt/vehicles-gas/oil
    5606 · Cnstr equipt/vehicles-lic/ins
    5608 · Cnstr eqpt/vehicles-repair/mntc
    5650 · Uniforms & laundry
    5660 · Cell phones-field
    5670 · Depr, cnstrn equipt/tools/vehic
    5690 · (Less) D-NJS burden alloc
S.57P · Cnstr costs from/(to) WIP

EXHIBIT 8.3 Non-job specific cost of goods sold (COGS) accounts.

TIP You might want to accumulate and measure total NJS costs for a period of time before you begin using QuickBooks Pro to allocate these costs to jobs.

S.57P.Construction costs from/(to) WIP (work in process)

This account is used to transfer construction costs *out* of the income statement and *into* the work in process (WIP) account on the balance sheet for jobs still in process at the end of an accounting period. You or your accountant will perform this transfer when you determine that, for accounting or tax purposes, it is not yet appropriate to recognize costs on certain jobs as of the end of a specific accounting period. You then reverse the entry to the appropriate cost accounts on the first day of the following period so that (except for the last day of the accounting period) you always see the job's costs-to-date in your profit and loss by job report.

Use the next section of the COGS accounts to track interest and fees directly related to land purchase, development, and construction loans; to allocate interest and fees out to units or jobs; and to move costs for units not yet completed to and from the WIP account as needed. Exhibit 8.4 shows the accounts that will be discussed below.

S.620 · Direct interest costs
- 6200 · Interest & fees-land prch loans
- 6202 · Interest & fees-developmt loan
- 6204 · Interest & fees-construction ln
- 6206 · Interest allocation to units
- 6208 · Interest expense from/(to) WIP

S.650 · Developmt/infrastr costs
- 6506 · Land development costs
- 6508 · RE taxes/utilities/maint costs
- 6560 · Dev cost allocation to units
- 6590 · Developmt costs from/(to) WIP

S.670 · Land costs
- 6700 · Land/lot purchases
- 6710 · Land cost allocation to units
- 6790 · Land costs from/(to) asset acct

S.680 · Gen/admin/warranty costs
- 6800 · G&A costs from expense
- 6810 · Gen & admin allocation to units
- 6860 · Warranty/repair alloc to units
- 6890 · G&A & warranty from/(to) WIP

EXHIBIT 8.4 Balance of COGS accounts.

S.620.Direct interest costs

If you can identify the specific job for the following types of loans, assign the cost to that job. If you are a developer, assign the actual costs to the interest and fees job. To assign interest and fees costs to individual units, see the discussion under **6206.Interest allocation to units**.

Subaccounts under this title account include:

- **6200.Interest and fees—land purchase loans** is used to accumulate interest and fees incurred for this type of loan.
- **6202.Interest and fees—development loans (developers only)** is the account where interest and fees information incurred for development loans is stored.
- **6204.Interest and fees—construction loans** is used to accumulate interest and fees for construction loans.
- **6206.Interest allocation to units (developers only)** is used to allocate interest to individual units on a reasonable basis, based on beginning estimates and costs incurred for the development as a whole.

NOTE

From a cost-accounting standpoint, certain costs incurred in the process of creating and holding a development, such as interest, pose various issues. Not all costs can be assigned logically or directly to units. Even if they could, they don't provide especially useful information for management and/or decision-making purposes. For example:

- How can interest for land, development, and construction loans be assigned to individual units when the interest paid doesn't have a direct, identifiable relationship? (Developers usually face this concern because loan repayment arrangements don't match up nicely to individual units.)
- Even if you can match interest to each unit, should the first unit sold show minimal interest cost and the last unit sold be loaded-up with a disproportionate share of interest expense because it happened to be the final unit sold?
- When there are a large number of units is there a good cost-benefit reason to go to the effort and expense of posting interest to each unit?

From a cost management and control standpoint, the developer looks at the development as one large project with an overall projected interest cost—which unit is sold first, and which unit is sold last, is relatively unimportant. What *is* important to the development as a whole are the sizes, interest rates, and timeframes of the loans, which depend for the most part on how quickly units are sold. Even so, for owners to monitor the project's unit-basis costs, and from an accounting and tax standpoint, the interest still must be assigned to individual units. So how can we resolve this mix of differing needs?

I use the following approach. At the beginning of a project, allocate a reasonable amount of interest to the individual units based on the beginning estimates using **6560.Development cost allocation to units**.

Allocating interest in this way lets you view your job costs in a profit and loss by job format. The allocation will appear as an offset or cost subtraction from the interest and fees job, and as an addition to the cost of individual units. Anticipated interest cost therefore is added to all jobs even before it actually is incurred.

NOTE Remember that an allocation just moves costs, or potential costs, from one location in the accounting system to another. It has no immediate affect on your total profit and loss when it is recorded.

Discuss the actual amount to allocate to each unit with your tax advisor. In some cases the anticipated costs may be allocated equally to each unit. In other cases it may be more appropriate to allocate costs based on relative sales price or some other well-documented approach.

If, at some point, it becomes obvious that the initial estimate was substantially incorrect, make a correcting allocation to assign cost or a reduction in cost to the current and remaining units.

As previously discussed, when *actual* costs are incurred use the correct item to assign them to the appropriate interest and fees account and job. Also enter your month-by-month projected loan and fee costs for each type of loan into QuickBooks Pro as budget amounts.

Then, on a monthly basis, review your life-to-date actual costs vs. budgeted costs to monitor whether you actually have spent more or less than anticipated on interest to date. (The results could affect your pricing decisions!)

TIP For better reports, enter your budget amounts at the account level only. Do not add a job or class when entering the budget.

6208.Interest expense from/(to) WIP is used to transfer the allocated interest and fees for jobs still in process *out* of the income statement and *into* the asset WIP account at the end of an accounting period. You or your accountant will perform this transfer when you determine that it is not appropriate for accounting or tax purposes to recognize costs on certain jobs as of the end of a specific accounting period. You then reverse the entry into the appropriate cost accounts on the first day of the following period so that (except for the last day of the accounting period) you always see the job's costs to date in your profit and loss by job report.

S.650.Development/infrastructure costs (developers only)

Use this section of the COGS accounts to track various costs for the development and infrastructure of land, developments, and lots, and to move costs for units not yet completed to and from the WIP account as needed. Conceptually, these accounts should be used in a fashion nearly identical to the preceding interest and fees accounts, except that actual costs should be assigned to the special land and development job.

- **6506.Land development costs** includes costs such as:
  - Design and engineering costs
  - Professional fees related to developing or presenting the development

- Architectural fees to design homes or common buildings for the development
- Surveying
- Excavating
- Utilities for the development
- Paving and sidewalks
- Lighting
- Landscaping
- Leisure and other common area costs

I have not created separate accounts in the Chart of Accounts for these detailed expenses, but you can create detailed items in your Item List and link them to this account.

- **6508.Real estate taxes/utilities/maintenance costs** are costs incurred while you still hold properties awaiting sale such as taxes on individual units, interim utility costs, and non-warranty maintenance and repair costs on individual units or within the development.
- **6560.Development cost allocation to units** is used to allocate estimated total development costs to each unit using a method identical to the allocation of estimated interest. Create an entry excluding land, interest, loan fees, and warranty work. The results appear as an allocated cost for each unit and as an allocated cost offset in the **6506.Land and development costs** account.

TIP

If you enter monthly budgets for accounts **6200, 6202, 6204, 6506, 6508, 6860,** and total estimates by item for your land and development and general and administrative (G&A)/warranty jobs, you can prepare actual-to-date vs. budget-to-date reports, and actual to-date vs. estimate reports. These reports will allow you to monitor actual costs against your to-date budgets and estimated total development costs. (There's that Step #3 of the control loop showing up again!)

- **6590.Development costs from/(to) WIP** is used in a manner identical to that described for the **6208.Interest expense from/(to) WIP** account.

S.670.Land costs

Use this group of COGS accounts to track various costs for the acquisition of land and lots, and to move costs for units not yet completed to and from the WIP account as needed.

- **6700.Land/lot purchases** is used to record purchases of land that you will be using for home building or development. If the purchase will be used for one or two homes, assign the costs directly to the job. If the purchase is for a development, assign the acquisition costs to this account.
- **6710.Land cost allocation to units (developers only)** is used in a manner identical to that described in the preceding development costs section. At the beginning of the project, allocate estimated total land costs to individual units.
- **6790.Land costs from/(to) asset account** is handled in a manner identical to that described in the **6208.Interest expense from/(to) WIP** account.

S.680.Gen/admin/warranty costs

Use this grouping of COGS accounts to move various costs incurred by the G&A side of your business to the direct costs section of the income statement. These accounts also are used to allocate estimated G&A costs to various jobs or units.

- **6800.G&A costs from expense** is used primarily by developers to move their G&A costs into the COGS section so they can allocate these costs out to individual jobs. If you choose to do this, select this account as well as the following account.

- **6810.Gen and admin allocation to units** is used in a manner identical to that of **S.650.-Development/infrastructure costs**. Allocate estimated total G&A costs to individual units at the beginning of the project. Create your budgets and estimates in the same way so that you will see those results in your comparative reports as well. See the note under **S.620.Direct interest costs** regarding adjustments if your initial allocations were incorrect.
- **6860.Warranty/repair costs allocated to units** is used primarily by spec home builders and developers to allocate estimated total warranty and repair costs to individual units at the beginning of a project. Create your budgets and estimates in the same way so that you will see those results in your comparative reports as well. See the note under **S.620.Direct interest costs** regarding adjustments if your initial allocations were incorrect.

NOTE Post actual warranty and repair costs from the G&A and warranty job to the **S.520.Direct labor, S.522.Employee labor burden, S.53P.Trade contractors, S.54P.Building materials,** and **S.55P.Other job-specific** direct cost accounts by using warranty items from the Item List. Then when you review a profit and loss by job report for the G&A and warranty job you will be able to monitor actual costs to date less the amount allocated to units in total.

- **6890.G&A and warranty from/(to) WIP** is handled in a manner identical to that described in the **6208.Interest expense from/(to) WIP** account.

NOTE For specific procedures related to special developers' allocations, see chapter 34.

Choosing other income accounts

At the end of the Chart of Accounts are income accounts that are related more to the financial status of your business than to the operations side of your business, such as:

- Finance charges to clients
- Other business ventures, such as income from partnerships and joint ventures or rental property income
- Occasional, relatively unusual non-operations-related activities, such as a gain on the sale of fixed assets

If you keep the types of accounts shown in Exhibit 8.5 segregated from operations, your reports will reflect the unadulterated results of your remodeling, building, or development operations in the net ordinary income subtotal.

Other Income

S.910 · Interest and other income
- 9100 · Interest income, banks/invstmts
- 9110 · Finance charge income, clients
- 9140 · Net inc-prtnrships/joint vent
- 9150 · Rental property income
- 9160 · Gain on sale of assets
- 9190 · Income from other sources

EXHIBIT 8.5 Other income accounts.

S.910.Interest and other income

These subaccounts commonly are found under this title account.

- **9100.Interest income, banks/investments** is used to record income you earn on bank balances or investments.

TIP

Earnings tip: If you have sporadic, large cash inflows from deposits or draws and then pay bills as they are due, you may be carrying a relatively large balance in your non-interest-bearing checking account. You might want to ask your bank about a "sweep account." Because banks legally cannot pay interest on checking accounts, sweep accounts are separate interest-bearing investment accounts tied to your checking account. Each night the bank "sweeps" the funds out of your checking account into this interest-bearing account where they stay until your in-clearing checks create a need for a certain amount of cash. The funds then are swept back into your checking account to meet the needs of the in-clearing checks. You legally have earned interest on those funds in the interim, and continue to earn funds on the amount remaining in the investment account. The bank has performed a cash management service for you, for which they charge a fee.

Investigate this option carefully before enrolling. Sweep investment accounts do not carry the same guarantees as typical bank funds. The bank also will charge fees for this cash management service. Ask for an analysis of your historical data to find out whether, assuming a similar future cash-flow scenario, you would come out ahead using this type of product. If your current bank doesn't give you satisfactory answers, check with other banks to see what types of plans they offer.

- **9110.Finance charge income, clients** is where you record finance charges charged overdue clients. QuickBooks Pro has a feature you can use to selectively charge finance charges to overdue clients. Consider implementing this feature and using this account if you have problems with overdue invoices. Remember that if you had their cash in hand you could use it to either pay off some of your interest-bearing debt or you might be earning interest on the money through investments. Use this feature to motivate overdue clients to pay.
- **9140.Net income-partnerships/joint ventures** should be used to record only actual income earned from partnerships and joint ventures.

TIP

If your company is involved in and loans money to, or pays bills on behalf of, a partnership or joint venture (P/JV) company, record the expenditures as though they were a note related-party receivable account. When the funds are repaid, record the deposit as a reduction to the note related-party receivable account. When the companyP/JV earns money, record the income in the **9140.Net income-P/JV** account and as an increase in the amount due from the related party. When you receive the money, increase your bank account and reduce the related-party receivable account.

You, or one of your partners, also will need to keep a separate set of accounting records for any P/JV in which you are involved. Consider creating a separate company in QuickBooks Pro to record the P/JV's activities.

- **9150.Rental property income** can be used in two situations.
  - **Situation #1.** If you have income from rental properties that you own personally, I strongly recommend setting up a separate rental company file to handle all of the income, receivables, expenses, fixed assets, and depreciation. If your remodeling or building company performs some of the services or provides materials to the rental properties, treat your properties just like another customer by creating invoices to them

for the amounts you choose to charge. To keep your building company ratios in line, fair market value would be best, but never invoice less than cost.

- **Situation #2.** If you have income from rental properties that legally are owned by your remodeling or building company, I suggest setting up a customer called "zy.Rentals." (When alphabetically sorted, this entry falls to the bottom of the list.) For that customer, create a job for each property, and a subjob for each renter within each property. Use this account to record rental income by subjob. If you need to use receivables, I suggest you set up a separate accounts receivable-rentals account.

- **9160.Gain on sale of assets** is first used to record the initial deposit of proceeds from the sale of a fixed asset. When you sell a fixed asset, the ultimate object is to remove the fixed asset at cost and its related accumulated depreciation from your books, increase cash by the amount you received, and post the remaining amount to either the **9160.Gain on sale of fixed assets** or to the **S.93P.Loss on sale of asset** account. Because of current-period depreciation issues, it may be difficult for you to locate the accumulated depreciation amount for a specific fixed asset. Record the deposit of the proceeds in this account and let your tax advisor know all of the specifics about the sale so he or she can prepare the adjusting entries. Because none of the entries involve QuickBooks Pro subsystems, a traditional journal entry is entirely appropriate.
- **9190.Income from other sources** can be modified to fit your needs.

CHAPTER 9

Choosing expense accounts

Categorizing expenses by business function provides you with management information on direct, general and administrative (G&A), and facilities costs from a control and decision-making perspective. Some of the business functions I use to categorize expenses are:

- Sales and marketing costs
- Administrative costs
- Occupancy costs
- Office expenses
- Non-operational expenses such as:
  - Interest income
  - Interest expense
  - Taxes

Within each of these categories are associated accounts. For example, rent, janitorial services, property and contents insurance, and utilities are all costs associated with the office occupancy function.

Some accounts may appear to be similar types of expense. If they are used to accomplish separate business objectives, however, I assign them accordingly. For example, rather than grouping various insurance expense accounts together under a title labeled insurance, I post the health insurance with the company-wide labor burden direct costs. I group workers' compensation insurance, general liability insurance, and insurance on construction vehicles with direct costs. I list insurance on general-use vehicles with the G&A function, and include the building insurance with the occupancy costs. Ultimately, it is more useful to see your expense totals by business function than it is to know the total of your insurance costs.

QuickBooks Pro 2000 has two account types that you can use to create accounts for non-direct costs: *expense* accounts and *other expense* accounts. You'll notice that I've placed accounts that are commonly referred to as overhead accounts in QuickBooks Pro's expense accounts, and that only a limited number of remaining costs are assigned to the other expense accounts.

I like to think of overhead costs as the expenses you incur simply by being open for business. The costs are not related to production activities, although if production levels change dramatically these costs can be affected.

Overhead costs usually increase or decrease in "chunks" or "lumps." When operations and management stay consistent in size, spending habits, and philosophy, these costs also have a tendency to stay fairly constant or "fixed." When the business grows, however, overhead costs generally increase in some way. Unfortunately, the timing of the expenditures often precedes the corresponding increase in income.

Growth requires policies and procedures, documentation, information, facilities, machinery, human resources, and technology—in short, capital investment. As the business owner, your challenge is to understand how these costs should act, from a timing and amount standpoint, within your current environment and in relation to your current plans, projections, and cash flow. You will be selecting accounts from this section to reflect your current situation, and based on your need to understand, monitor, and control overhead costs to achieve your cost-benefit objectives.

Organizing expense accounts

I find it helpful to organize the accounts and subtotals in a logical sequence within each business function. I recommend placing the accounts that typically contain the higher dollar amounts before those with lower dollar amounts. For example, rent costs precede utilities costs within the occupancy grouping because rent typically accounts for a higher dollar amount than an individual utility.

Other times it might make more sense to order the accounts in a relationship sequence, or in a sequence that facilitates the sorting, subtotaling, and viewing of accounts and reports. For example, subtotaled groupings of "Professional services and fees" follow individual accounts within G&A expenses because the report is easier to read when the accounts are organized this way.

I've also found that fairly detailed accounts, such as those used for communications and utilities, are helpful for review and budgeting purposes because they allow you to analyze and predict one type of expense at a time. For example, suppose you change long distance phone carriers, hoping to decrease your phone charges. As you examine these costs under the subtotal for **S.838. Communications**, you realize your costs under **8380.Telephone** decreased $2,250, but your costs under **8382.Cell phone** increased $3,600. After a little research you find that the cost increase is because your two lead carpenters used cell phones to report in twice a day resulting in costs of $1,800 each or $150 per month. If the long distance and cellular phone expenses had been combined into one account, you would have seen only an increase of $1,350 and might not have followed through with corrective action. Or, if you had wondered how each of the different costs are acting, you would need to take several more steps to segregate the cost changes between these two types of phone cost.

Although many business owners shudder at the word "detail," this is an example of when having the detail immediately at hand can be extremely helpful. The ability to examine and then clarify the individual elements driving your financial results makes them easier to control and predict for the future. Your ability to predict your costs, in turn, will drive your pricing and, ultimately, your company sales, production goals, and financial results.

Selecting expense accounts

Many of the following accounts have little or no explanation, as the account title is self-explanatory. I have included notes where I think they would be helpful. Expense accounts associated with marketing, business development, and advertising are shown in Exhibit 9.1.

S.70P.Unclassified expense

Use this account in much the same way you would use **S.40P.Unclassified income**, which was described in chapter 8. Use this account if you believe that the transaction in question relates to a non-direct, undetermined expense.

S.710.Marketing and sales expenses

S.711.Marketing and business development (subaccount also used as a title)

> **TIP** If you wish to track employee compensation costs related to specific marketing activities or events, create subjobs under the **7.Sales/Mktg** job (see chapter 11). When your employees spend time on those activities, have them charge their time to that job so you can track the related cost.

- **7100.Events** is used to track all costs related to special events like home tours or building shows.
- **7110.Newsletters** is used to track costs such as development or copy fees, printing, mailing, and postage.
- **7120.Donations** is used to track donations separately so that your tax advisor can handle each donation as it relates to your company's organizational type. Post the tax-deductible portion of donations to charitable organizations to this account.

> **NOTE** Organizations generally will state if a portion of your donation is non-deductible. Post the non-deductible portion to **9300.N/D dues, penalties, gifts, contributions** under title account **S.930.Nondeductible expense.**

If a portion of a fund-raising ticket is for a meal or entertainment, post that part of the cost to **8720.Meals/entertainment-50% deductible** under title account **S.870.Travel and meals/entertainment**.

Post contributions to individuals or political organizations to **9300.N/D dues, penalties, gifts, contributions** under title account **S.930.Nondeductible expense**.

Expense
- S.70P · Unclassified expense
- S.710 · Marketing & sales expenses
- S.711 · Marketing & business developmt
    - 7100 · Events (shows, tours, etc.)
    - 7110 · Newsletters
    - 7120 · Donations
    - 7130 · Sponsorships
    - 7140 · Giveaways
    - 7150 · Referral expense
    - 7160 · Business promotion
    - 7170 · Gifts - ded'bl @ $25/person
- S.720 · Advertising expenses
    - 7200 · Business cards/brochures
    - 7210 · Yellow pages
    - 7220 · Print-nwsppr/mag/guides, etc.
    - 7230 · Specialty publications
    - 7240 · Radio/television
    - 7250 · Site signs
    - 7260 · Website
    - 7290 · Model home expense

EXHIBIT 9.1 Marketing and sales expense accounts.

- **7130.Sponsorships** is used for costs related to a company-sponsored softball or bowling team, for example.
- **7140.Giveaways** includes expenses for low-priced items such as pens, paper, caps, letter-openers, chip clips, and mugs with your company name and logo that are given to customers and potential clients to promote your company.
- **7150.Referral expense** is used to record the expense associated with revenue sharing or reimbursing referral sources. These payments may be subject to 1099 reporting, so check with your tax advisor to find out the form your referral expense should take.
- **7160.Business promotion** is the final resting place for a variety of promotional expenses that don't fit nicely into other categories, such as a promotional banner or a sympathy gift for a business associate.
- **7170.Gifts-deductible at $25 per person** is where you would record holiday gifts up to $25 per person per year for business associates. Costs for gifts exceeding that limit should be split out to **9300.N/D dues, penalties, gifts, contributions** under title account **S.930.Nondeductible expense**.

S.720.Advertising expenses (subaccount also used as a title)

- **7200.Business cards/brochures**
- **7210.Yellow pages** is used to record costs associated with your *Yellow Pages* advertising. Separate the expense from your telephone bill so you can see what you are spending on this type of advertising.

- **7220.Print-newspapers/magazines/guides, etc.** is used to record promotions and advertising that you put in print for general or target audiences.
- **7230.Specialty publications** is used to record costs incurred if you contract with a company to create a special promotional piece, such as a glossy magazine-type publication that features your business, or another type of specialty publication.
- **7240.Radio/television**
- **7250.Site signs** can be assigned here, or if the costs are high, your tax advisor may want to reclassify them as fixed assets.
- **7260.Website** is used to record the costs for Website development, maintenance, upgrades, and so on.
- **7290.Model home expense** can be used to record operating expenses associated with model homes if you chose to classify these expenses as a sales advertising expense instead of a direct cost. Because almost all model home operating costs are deductible in the period incurred, except for fixed-asset types of expenditures such as furnishings, I have chosen to include the operating costs here.

Employee- and owner-related expense accounts are shown in Exhibit 9.2. Subaccount names under these title accounts are designated by S- for sales-related, E- for employee-related, and O- for owner-related.

S.730.Sales-comp/taxes/burden

These accounts are used in nearly the same way as the direct labor accounts discussed in chapter 8. If you have sales staff on your payroll, or if you believe you will assign costs for other employees to sales-related activities, select these accounts.

- **7300.S-Regular compensation**
- **7302.S-Overtime compensation**
- **7304.S-Payroll taxes/specific benefits**
- **7306.S-Allocated burden**

S.800.Administrative employment costs

S.802.Administrative comp/taxes/burden

These accounts are used in nearly the same way as the direct labor accounts. If you have administrative staff on your payroll, select these accounts.

- **8020.A-Regular compensation**
- **8022.A-Overtime compensation**
- **8024.A-Payroll taxes/specific benefits**
- **8026.A-Allocated burden**

S.804.Owner's comp and benefits

The first three of these accounts are used almost exactly as the direct labor accounts. The other accounts are very similar to the expenses you see in the employee labor burden group, but they have been created specifically to record costs for owner-related benefits due primarily to tax reporting needs. It generally is easier to record the information as you go, rather than split it out at tax-planning time or at year-end when everyone seems to be under excessive time pressures. Your tax advisor may well send blessings your way for having this information readily available!

S.730 · Sales-comp/taxes/burden
    7300 · S-Regular compensation
    7302 · S-Overtime compensation
    7304 · S-Payroll taxes/specif benefits
    7306 · S-Allocated burden
S.800 · Admin employment costs
S.802 · Admin comp/taxes/burden
    8020 · A-Regular compensation
    8022 · A-Overtime compensation
    8024 · A-Payroll taxes/specif benefits
    8026 · A-Allocated burden
S.804 · Owner's comp & benefits
    8040 · O-Regular compensation
    8044 · O-Payroll taxes/specif benefits
    8046 · O-Allocated burden
    8050 · O-Health insurance
    8051 · O-(Less) Owner-paid hlth ins
    8052 · O-Life insurance
    8054 · O-Private disability insurance
    8056 · O-Retirement benefits
S.81P · Subcontracted admin labor

EXHIBIT 9.2 Employment/owner costs.

- **8040.O-Regular compensation**
- **8044.O-Payroll taxes/specific benefits**
- **8046.O-Allocated burden**
- **8050.O-Health insurance** (Your health care premium notice should allow you to determine the amount of health insurance cost being paid for the owner.)

> TIP
>
> **Health insurance tax tip:** If your company is a sole proprietorship or a sub-S corporation and your spouse is an employee, you may be able to deduct 100 percent of your health care premiums if your spouse is the primary insured and you are the dependent on the policy. Currently, sole proprietors and sub-S corporation owners are not able to directly deduct 100 percent of their health care premiums, whereas premiums for employees are 100-percent deductible. Check with your tax advisor.

- **8051.O-(Less) Owner-paid health insurance** (The payroll system will record any health insurance co-pays deducted from the owner's pay.)
- **8052.O-Life insurance**
- **8054.O-Private disability insurance**
- **8056.O-Retirement benefits**

S.81P.Subcontracted admin labor

If you subcontract for administrative office help, post the costs to this account.

The next category of expenses relates to occupancy and office expenses. These expense accounts are shown in Exhibit 9.3.

S.820.Occupancy expense

- **8210.Rent/lease on office space**
- **8220.Property/liability insurance** includes insurance costs for property damage and for on-premises liability insurance. Assign general liability insurance for your production activities to the correct item and account in the cost of goods sold (COGS) accounts (see chapter 8).
- **8230.Repairs and maintenance, office space** generally includes costs that are "return to useable condition," ongoing maintenance, relatively low in cost, or short-lived in nature, such as costs to repair a leaky faucet, replace a broken window, or an air conditioner maintenance contract. Assign larger or longer-term expenditures such as creating new managerial offices in the warehouse, rewiring the office, or installing built-in cabinets, to the correct fixed-asset account (see chapter 7). Keep in mind, however, that fixed-asset costs related to buildings or leases are written off over a much longer period of time, so repairs and maintenance expenditures create a more immediate tax benefit. In this situation, your tax advisor may help you with the timing and a strategy that will be to your best advantage.
- **8240.Janitorial/cleaning services**
- **8250.Personal property taxes** is used to record taxes paid on personal property, such as office furniture, owned by your business. Assign taxes on real estate that you own or rent to the following account.
- **8260.Real estate taxes**
- **8270.Depreciation, land improvements** is used to record deprecation expense related to assets in **S.170.Land improvements, at cost**, a fixed-asset account.
- **8275.Depreciation, building** is used to record depreciation for **S.171.Building, at cost**, a fixed-asset account.
- **8280.Amortization, leasehold improvements** is used to record the related depreciation arising from **S.172.Leasehold improvements, at cost**, a fixed-asset account.

S.820 · Occupancy expense
- 8210 · Rent/lease on office space
- 8220 · Property/liability insurance
- 8230 · Repairs & maint, office space
- 8240 · Janitorial/cleaning services
- 8250 · Personal property taxes
- 8260 · Real estate taxes
- 8270 · Deprec, land improvements
- 8275 · Deprec, buildings
- 8280 · Amort, leasehold improvements

S.829 · Utilities
- 8291 · Gas
- 8292 · Electric
- 8293 · Water
- 8294 · Trash removal
- 8295 · Security system
- S.829 · Utilities - Other

S.830 · Office expenses
- 8310 · Office supplies
- 8320 · Printing & copying
- 8330 · Postage & delivery
- 8340 · Bank service charges
- 8370 · Misc. office expense

S.838 · Communications
- 8380 · Telephone
- 8382 · Cellular/digital phone services
- 8384 · Pagers
- 8386 · Answering services/voice mail
- 8388 · On-line services

S.839 · Office equipment, general
- 8390 · Equipment rental/lease
- 8392 · Repairs & maint, office equipmt
- 8398 · Deprec, furn/fixt/equip't

S.840 · Computer expense
- 8410 · Computer supplies
- 8420 · Computer rental/lease
- 8430 · Computer repairs/maintenance
- 8440 · Computer software support
- 8490 · Software amortization
- 8495 · Computer depreciation

EXHIBIT 9.3 Occupancy and office expense accounts.

S.829.Utilities

- **8291.Gas**
- **8292.Electric**
- **8293.Water**
- **8294.Trash removal**
- **8295.Security system**

S.830.Office expenses

- **8310.Office supplies** is used to record non-fixed-asset types of office supplies. If you don't wish to track computer supplies separately, computer supplies, low-cost software, and periodic software upgrades can be posted to this account as well. Post shop supplies to the correct COGS account (see chapter 8). Consider any sales tax or shipping charges part of the cost of the item purchased.
- **8320.Printing and copying** is used to record costs for general printing and copying for the office. Charge printing and copying expenses related to specific jobs to the job. Assign advertising or promotional printing and copying to the correct account within the **S.710.Marketing and sales** title account.
- **8330.Postage and delivery** is used to record general postage and delivery costs for typical office operations. Charge postage and delivery expenses related to specific jobs to the jobs. Assign postage and delivery for advertising or promotional mailings to the correct account within the **S.710. Marketing and sales** title account.
- **8340.Bank service charges** typically is used to record the charges shown on your bank statement for regular processing. Assign fees for checks or other banking supplies to **8310.Office supplies**.

TIP If you accept credit cards, you may want to create a separate account called **8345.Credit card acceptance fees** to record the credit card handling fees charged by your credit card service providers.

- **8370.Misc. office expense** is used only rarely to record very small amounts that do not fit in any other account. Make every effort to assign each transaction to an account with a properly descriptive title.

S.838.Communications

- **8380.Telephone**
- **8382.Cellular/digital phone services** is used for office-related cellular or digital phone services. The COGS section (see chapter 8) also has an account for cell phone costs that are field related.
- **8384.Pagers**

- **8386.Answering services**
- **8388.On-line services**

S.839.Office equipment, general

- **8390.Office equipment rental/lease** is used to record postage meter rental, office copier rental, and so on.
- **8392.Repairs and maintenance, office equipment**
- **8398.Depreciation, furniture, fixtures and equipment** is used to record the depreciation expense for office furniture, fixtures, and equipment only. If computers are included in **8310.Office supplies, furniture, fixtures, and equipment, at cost**, rather than in **1740.Computers/software, at cost,** then their depreciation cost must be included here as well.

S840.Computer expense

Some of my clients prefer to ignore the following accounts and just use office accounts to record their computer-related costs. Others clients like to select some or all of the following accounts. Choose the ones that work for you and ignore the rest.

- **8410.Computer supplies**
- **8420.Computer rental/lease**
- **8430.Computer repairs/maintenance**
- **8440.Computer software support**
- **8490.Software amortization**
- **8495.Computer depreciation**

S.850.General and administrative expense

The expense categories under this title account do not relate to a specific job. The subaccounts and detailed accounts are shown in Exhibit 9.4.

- **8510.Dues and memberships** is used to record general business memberships and dues to organizations such as your local chamber of commerce, home builders association, and other trade associations. Sometimes a portion of the dues is not tax-deductible. If this is the case, the organization usually will tell you on the face of the bill the portion that is not deductible. Split that amount out to **9300.N/D dues/penalties/gifts/contributions** under the **S.930.Non-deductible expense** title account.
- **8530.Books and subscriptions**
- **8540.Training/education fees and materials** includes registration fees paid for seminars, conferences, conventions, or other training, and any related materials needed for the course. Assign other costs

S.850 · General & admin expense
- 8510 · Dues & memberships
- 8530 · Books & subscriptions
- 8540 · Training/education fees/mat'ls
- 8550 · Amort of organization cost
- 8555 · Amort of goodwill & NC covenant
- 8560 · Bad debts & collection expense

S.858 · Professional and service fees
- 8580 · Accounting
- 8582 · Payroll services
- 8584 · Legal services
- 8586 · Consulting services
- 8590 · Other professional fees

S.860 · Vehicle expenses
- 8600 · Mileage reimbursement
- 8610 · Vehicle lease/rental
- 8620 · Vehicle gas & oil
- 8630 · Vehicle licenses & insurance
- 8640 · Vehicle repairs & maintenance
- 8690 · Depreciation, vehicles

S.870 · Travel & meals/entertainment
- 8700 · Travel expense
- 8710 · Lodging expense
- 8720 · Meals/entertainment-50% ded'ble

S.890 · Other taxes & licenses
- 8920 · License fees
- 8930 · State taxes (misc non-income)
- 8990 · Other taxes

S.89P · (Less) G&A allocation to COGS

EXHIBIT 9.4 General and administrative (G&A) expense accounts.

such as travel, lodging, meals or entertainment to the specific accounts under title account **S.870.Travel and meals/entertainment**.

- **8550.Amortization of organization cost** is used to record the related amortization in **1780.Organizational costs**, a fixed-asset account.
- **8555.Amortization of goodwill and NC covenant** (covenant not to compete) is used to record the related amortization in **1790.Goodwill/non-compete covenant**, a fixed-asset account.
- **8560.Bad debts and collection expense** is used to write-off income recognized in a prior tax period and still on your books as a receivable, which you now believe—based on your collection attempts, or on outside verification such as a bankruptcy notice—is not collectible.

NOTE

If the amount is minimal, it may be more efficient to issue a credit memo during the current period to clear out the receivable and reduce your current-period income. If the amount is considerable, you probably want to discuss your options, documentation, requirements, timing, and the tax affect on you and the debtor with your tax advisor, and possibly with your attorney. If you incur costs related to collection efforts, assign them to this account.

S.858.Professional and service fees

- **8580.Accounting**
- **8582.Payroll services**
- **8584.Legal services**
- **8586.Consulting services**
- **8590.Other professional fees**

S.860.Vehicle expenses

These accounts generally relate to expenses for the owner's vehicle or to other vehicles used for administrative, sales, or another purpose not related directly to production activities. The best overall advice I can offer without getting into complicated if-then tax explanations is to record everything in the following accounts as accurately as you can. Then let your tax advisor review the resulting detail, ask the questions, collect any remaining data needed, crunch the numbers, and suggest journal entries to keep you aligned with your tax return.

- **8600.Mileage reimbursement** generally is used to record payments to employees, owners, or trade contractors who use their own vehicle for company business.

TIP

Tax Tip: As the owner of the company, if you use your own vehicle be sure to document all of your business mileage (think of all those little errands and all of those sales calls) and have the company reimburse you for that mileage. The mileage adds up quickly, and it is deductible. Even if you think your cash flow can't handle mileage payments to you, keep track of your mileage anyway. Then you can submit the amount to your tax advisor at year-end so the cost can be deducted from your company's income.

- **8610.Vehicle lease/rental** is used to accumulate costs related to the lease or rental of company vehicles. Your tax advisor may adjust the amounts you post to this account to reflect totals allowable for tax purposes.

- **8620.Vehicle gas and oil** records expenditures on company vehicles. You may wish to create an account for each vehicle.
- **8630.Vehicle licenses and insurance** records expenditures on company vehicles. You may wish to create an account for each vehicle.
- **8690.Vehicle repairs and maintenance** records expenditures on company vehicles. You may wish to create an account for each vehicle.
- **8690.Depreciation, vehicles** is used to record the related depreciation in **1750.Vehicles, at cost**, a fixed-asset account under the **S.175.Vehicles, net book value** title account.

S.870.Travel and meals/entertainment

- **8700.Travel expense** is used to record costs for air, rail, coach, taxis, car rentals, parking, tolls, and transportation and baggage tips.
- **8710.Lodging expense** is used to record only the costs related to your lodging, which most likely will include some telephone expense as well. Receipts are required for deductibility purposes. Assign in-room meals, snacks, liquid refreshments, and in-room movies to the **8720.Meals/entertainment-50% deductible** account.
- **8720.Meals/entertainment-50% deductible** is used to record all business meals and entertainment costs that do not qualify for the **5236.E-100% deductible meals and entertainment** COGS account under the **S.522.Employee labor burden** subtitle account. Check with your tax advisor about deductibility requirements and be sure to properly document all expenses showing the amount, time, date, and place of the expenditure, purpose of the business discussion or nature of the business benefit, and the parties involved. Receipts are not required for expenses under $75. Regardless of amount, all costs also must meet the IRS' criteria of being "ordinary and necessary."

S.890.Other taxes and licenses

- **8920.License fees** is where you post the cost of your builder's license and any other special licenses you may need to obtain for yourself, your employees, or for the business.
- **8930.State taxes (miscellaneous non-income)** is used to record fees paid to your state to keep your incorporation statute current or to keep your state sales tax license active.
- **8990.Other taxes**

S.89P.(Less) G&A allocation to COGS (primarily used by developers)

Please see the **6800.G&A costs from expense** account under the **S.680.Gen/admin/warranty costs** COGS title account in chapter 8 for a full explanation of how this account is used. This is an offset account that is used to move general and administrative costs accumulated in the overhead accounts into the COGS group so these costs can be allocated to jobs or units.

In QuickBooks Pro's profit and loss reports you will find a total for these expense categories at the conclusion of this list of accounts.

Entering other expenses

As with the "other income" accounts, these "other expense" accounts relate primarily to costs arising from your company's legal, tax, or financial status. Moving your financing-related expenses, unusual or non-operations-related expenses, and entity-related tax costs (or lack thereof) to this section can help keep you focused on operating results. The accounts in this section of the Chart of Accounts are shown in Exhibit 9.5.

S.920.Interest and other financing costs

- **9210.Interest, credit cards** is used for interest charged by your credit card company. In QuickBooks Pro these entries can be generated automatically as part of the credit card reconciliation process.
- **9212.Annual credit fees** is used to record annual credit card fees or recognized annual fees on a general business bank loan, such as a line of credit.
- **9214.Interest, loans (non-direct)** is used to record interest paid or accrued on general business loans from unrelated parties.
- **9216.Interest, stockholder loans** is used to separate interest paid on stockholder loans for tax-tracking purposes.
- **9280.Interest, late tax payments** is here to record the interest charged by government entities if you are late making payroll or other tax payments. (Hopefully you won't need it!)
- **9290.Discounts lost and late fees** is used to record the discount you weren't able to take advantage of because an invoice was paid late.

Other Expense

S.920 · Interest & other financing cost
  9210 · Interest, credit cards
  9212 · Annual credit fees
  9214 · Interest, loans (non-direct)
  9216 · Interest, stockholder loans
  9280 · Interest, late tax payments
  9290 · Discounts lost & late fees
S.930 · Non-deductible expense
  9300 · N/D dues/penalties/gifts/contrb
S.93P · Loss on sale of assets
S.950 · Federal & State taxes
  9500 · Federal income taxes
  9550 · State earnings taxes

EXHIBIT 9.5 Other expense accounts.

NOTE

For management purposes, I recommend implementing a company rule that all bills are entered net of the discount. For example, a $10,000 invoice with 2/10 net 20 terms—meaning if you pay within 10 days the amount due is $9,800; after 10 days the total $10,000 is due by the 20th day—would result in an entry of $9,800. If the invoice was not paid within 10 days, the $200 lost discount would be posted in the **9290.Discounts lost and late fees** account.

When you compute the effective rate of interest on discounts lost you will be shocked to find that the imputed rate of interest can reach as much as several hundred percent! For example, let's examine "the case of the lost-lost discount." In the above example, let's say your accounts payable person was busy and didn't get to pay the bill until the 13th day, so the full $10,000 was due and you lost the $200 discount.

The result is that you effectively paid $200 for the use of $9,800 for a three-day period. This calculates out to an interest rate of 2.0408 percent ($200/$9800 = 2.0408 percent) for a year. But you didn't use the money for a year. You only used it for 3/365ths of a year or approximately 1/122nd of a year. So you actually paid 248.3 percent for the use of that money. As you can see, this is an expensive form of alternative financing!

Some companies enter their bills at gross cost and post the discount to an income account when they take the discount and pay less than the gross. If this approach had been used in the preceding example, nothing would have shown up in either income or expense, and you, the owner, would have been blissfully ignorant of the lost discount. This scenario creates "the case of the lost-lost discount." Unfortunately, this is the way QuickBooks Pro's Discount Calculation feature works.

If this lost-lost discount scenario happens once or twice a year it might not be a big issue. But what if it happens repeatedly? Or what if cash-flow problems

NOTE
Continued

cause you to miss discounts regularly? How will you know how much this problem is costing you?

By requiring that all bills be entered net of the discount (in this case at $9,800), and that the lost discount be posted into the **9290.Discounts lost and late fees** account, you will be able to see the cost of the lost discounts and manage by exception. You'll also be able to take corrective action within the bill payment cycle or search for cheaper sources of financing. Based on to the preceding discussion I suggest that you avoid using QuickBooks Pro's Discount Calculation feature.

S.930.Nondeductible expense

9300.N/D (nondeductible) dues/penalties/gifts/contributions is used to record the portion of dues, gifts, and contributions that are not tax deductible. The other costs that are always nondeductible are penalty costs. For example, if you decide to reimburse an employee's parking ticket because he or she received it while on company business, post the cost to this account. If you missed a filing date on one of your payroll returns, use this account when you enter the "penalties" portion of the bill. The interest portion can be assigned to account **9280.Interest, late tax payments**.

S.93P.Loss on sale of assets

Please see the discussion in chapter 8 under **9160.Gain on sale of fixed assets**.

S.950.Federal and state taxes

- **9500.Federal income taxes** is used only if your company is subject to federal income tax.
- **9550.State income taxes** is used only if your company is subject to state income tax, or a similar tax on earnings.

NOTE

If your company pays any of your personal, non-payroll federal or state taxes on your behalf, that expenditure should be classified as a draw or distribution against your balance sheet equity account.

CHAPTER 10

Building an Item List

The Item List is a subsystem within QuickBooks Pro 2000 that I use to record detailed transactions to direct costs and income. Each item can link to one expense account for purchase transactions, and to one income account for sales transactions. For example, when you purchase an item called roofing materials, the cost would post to your Materials account, and if charged directly to a customer, would post to your Construction income account.

With the Item List, you can record information at a more detailed level than is generally practical using accounts in the Chart of Accounts. Since you name the detailed items, create the titles and subtitles, and establish the item groupings, the Item List can be tailored to meet the unique needs of your business (see Exhibit 10.1). If you use an estimating program, you may build your Item List to match the categories in your estimate. If you want to see your job costs shown by specific phases, you can build your Item List accordingly.

The Item List included in appendix B includes 73 title accounts, such as:

- Non-job-specific (NJS) costs
- Employee burden
- Exterior site preparation
- Framing
- Roofing
- Other direct construction costs

Because your Item List incorporates all your job details it probably will be the longest list on your system, even if you don't need all the items in the sample Item List provided in appendix B.

Linking items with the Chart of Accounts

The Item List allows you the flexibility of preparing traditional income statements acceptable to your tax advisor and banker, while designing estimated costs vs. actual costs reports that let you analyze each stage of a specific job.

Detail in the Item List can be created in a different "view" or format than the accounts in the Chart of Accounts. For example, the general ledger cost of goods sold (COGS) accounts traditionally summarize labor, trade contractors, materials, and other costs. Within the Item List, however, you can have items for site preparation materials, foundation materials, framing materials, roofing materials, drywall materials, and so on. These five materials items all would link to one COGS account, such as **S.54P.Building materials and supplies**, and one income account, such as **S.42.P.Construction services** in the Chart of Accounts.

You also could have cleaning supplies, hand tools, and disposable job supplies, all of which would link to **5602.Small tools and miscellaneous job supplies** under title account **S.560.NJS**

Item List

Name	Description	Type	Account	On Hand	Price
◆304b.Ductwork	Duct work-subs	Service	S.42P · Construction services		0.00
◆304c.Ductwork	Duct work-materials	Service	S.42P · Construction services		0.00
◆304d.Ductwork	Duct work-other costs	Service	S.42P · Construction services		0.00
◆304v.Ductwork	Duct work-variance from	Service	S.42P · Construction services		0.00
◆305a.Venting	Venting-D/L-pay	Service	S.42P · Construction services		0.00
◆305b.Venting	Venting-subs	Service	S.42P · Construction services		0.00
◆305c.Venting	Venting-materials	Service	S.42P · Construction services		0.00
◆305d.Venting	Venting-other costs	Service	S.42P · Construction services		0.00
◆305v.Venting	Venting-variance from P	Service	S.42P · Construction services		0.00
◆32.ROOFING	ROOFINGXXXXXXXXXX	Service	S.40P · Unclassified Income		0.00
◆320a.Roofing	Roofing-D/L-pay	Service	S.42P · Construction services		0.00
◆320b.Roofing	Roofing-subs	Service	S.42P · Construction services		0.00
◆320c.Roofing	Roofing-materials	Service	S.42P · Construction services		0.00
◆320d.Roofing	Roofing-other costs	Service	S.42P · Construction services		0.00
◆320v.Roofing	Roofing-variance from P	Service	S.42P · Construction services		0.00
◆322a.Shingles	Shingles-D/L-pay	Service	S.42P · Construction services		0.00
◆322b.Shingles	Shingles-subs	Service	S.42P · Construction services		0.00
◆322c.Shingles	Shingles-materials	Service	S.42P · Construction services		0.00
◆322d.Shingles	Shingles-other costs	Service	S.42P · Construction services		0.00
◆322v.Shingles	Shingles-variance from F	Service	S.42P · Construction services		0.00
◆323a.Tiles	Tiles-D/L-pay	Service	S.42P · Construction services		0.00
◆323b.Tiles	Tiles-subs	Service	S.42P · Construction services		0.00
◆323c.Tiles	Tiles-materials	Service	S.42P · Construction services		0.00
◆323d.Tiles	Tiles-other costs	Service	S.42P · Construction services		0.00
◆323v.Tiles	Tiles-variance from P.O.	Service	S.42P · Construction services		0.00
◆324a.Roof cov	Other roof coverings-D/l	Service	S.42P · Construction services		0.00
◆324b.Roof cov	Other roof coverings-sut	Service	S.42P · Construction services		0.00
◆324c.Roof cov	Other roof coverings-ma	Service	S.42P · Construction services		0.00
◆324d.Roof cov	Other roof coverings-oth	Service	S.42P · Construction services		0.00
◆324v.Roof cov	Other roof coverings-var	Service	S.42P · Construction services		0.00
◆325a.Ventilat	Ventilation-D/L-pay	Service	S.42P · Construction services		0.00
◆325b.Ventilat	Ventilation-subs	Service	S.42P · Construction services		0.00
◆325c.Ventilat	Ventilation-materials	Service	S.42P · Construction services		0.00
◆325d.Ventilat	Ventilation-other costs	Service	S.42P · Construction services		0.00
◆325v.Ventilat	Ventilation-variance from	Service	S.42P · Construction services		0.00
◆326a.Skylight	Skylights-D/L-pay	Service	S.42P · Construction services		0.00
◆326b.Skylight	Skylights-subs	Service	S.42P · Construction services		0.00
◆326c.Skylight	Skylights-materials	Service	S.42P · Construction services		0.00
◆326d.Skylight	Skylights-other costs	Service	S.42P · Construction services		0.00
◆326v.Skylight	Skylights-variance from I	Service	S.42P · Construction services		0.00
◆34.MASONRY	MASONRY & PREPXXXX	Service	S.40P · Unclassified Income		0.00
◆340a.Masonry	Masonry & prep-D/L-pay	Service	S.42P · Construction services		0.00
◆340b.Masonry	Masonry & prep-subs	Service	S.42P · Construction services		0.00
◆340c.Masonry	Masonry & prep-material:	Service	S.42P · Construction services		0.00
◆340d.Masonry	Masonry & prep-other cc	Service	S.42P · Construction services		0.00
◆340v.Masonry	Masonry & prep-variance	Service	S.42P · Construction services		0.00
◆341a.M walls	Masonry walls-D/L-pay	Service	S.42P · Construction services		0.00
◆341b.M walls	Masonry walls-subs	Service	S.42P · Construction services		0.00

EXHIBIT 10.1 Segment of Item List.

direct costs. In the latter example, the costs posted to 5602 would roll up under the S.560 title account on the income statement. On an item report, each would be listed as a line item, providing you with the details.

Because the estimate vs. actual job-cost reports are based on the Item List, if you want your detailed job-cost reports to agree with your account-based reports, it is critical that you *always* use an item for income or direct cost (COGS) transactions.

Accessing items

QuickBooks Pro *requires* you to use items when you create customer- or job-related transactions such as:

- Estimates
- Invoices

- Customer credit memos or refunds
- Cash sales
- Purchase orders
- Finance charges

You also can *choose* to use items for transactions that are not necessarily related to customer or job transactions by clicking on the Items tab when:

- Creating bills and credits
- Creating credit card charges and credits
- Writing checks

You *cannot* access the Item List when creating:

- Deposits
- Journal entries
- Budgets

Incorporating job-cost details

I use the Item List to create detailed job-cost information by:

- Creating a title item for each major phase of construction
- Organizing the items in an order that approximates the order in which most jobs are completed
- Creating subitems that also post to the appropriate general ledger accounts. For example, 14.Exterior site preparation is a title item that may have the following subitems:
  - **140a.Site preparation-direct labor** posts costs to a subaccount under the **S.520. Direct labor compensation/taxes/burden** subtitle account in the Chart of Accounts.
  - **140b.Site preparation-trade contractors** posts costs to the **S.53P.Trade contractors (trade contractors)** account in the Chart of Accounts.
  - **140c.Site preparation-materials** posts costs to the **S.54P.Building materials and supplies** account in the Chart of Accounts.
  - **140d.Site preparation-other** posts costs to the **S.55P.Other job-specific costs** account in the Chart of Accounts.
  - **140v.Site preparation-variance from purchase order** usually posts costs to the **S.53P.Trade contractors (trade contractors)** account in the Chart of Accounts.

TIP Set all title items to post to your unclassified income and expense accounts. This will alert you immediately to situations where the person doing the entry has chosen a title item rather than the correct detail item.

When you enter estimates, purchase orders, bills, checks, and so on, enter the information or transaction at the detailed item level, and then indicate the specific job to which the item is related. The results are saved at the item level, and also post to the correct general ledger accounts.

When you subsequently prepare a report for a specific job based on the Item List, the subitems roll up or subtotal into each of the title items. This lets you see the detail for each of the individual items, as well as a total for the title item. Reports from the Jobs and Time reports category that are exceptionally useful include:

- Job estimates vs. actual detail
- Job profitability summary
- Job profitability detail

Alternatively, when you prepare a standard general ledger report, such as a profit and loss report, you will see the correct amounts for direct labor, trade contractors, materials, and other job-specific costs. Because you also indicated the specific job when the transaction was entered, you can see your job income and costs in this format as well—either in the profit-and-loss-by-job spreadsheet-type report, or in a filtered report for just one job. In either case, you also can check the % of Income box in the Customize Reports window so that you can see the gross profit for each job.

Designating item types

Although QuickBooks Pro separates the Item List into 10 different types, most remodelers, developers, and custom home builders need only designate items as service type, unless they operate a retail store or maintain a true inventory. As you designate the type of item, keep in mind that QuickBooks Pro first organizes and totals item-based reports by type.

Service items

This is the type of item I recommend you use to create the reports described previously. Although materials and other costs technically are not services, if we designated the site preparation materials item as a non-inventory item, it wouldn't subtotal with the other site preparation items entered as service items. It would show on a different part of the report. For this reason, I recommend you use the service category almost exclusively for the items you create. This will ensure that you receive reports that are easy to read and use. Also, only service items can be selected on timesheets.

Non-inventory items

You could create a few special purpose items in this item type. For example, you may need to create an item for a purchase order so you can purchase a new computer. I would suggest inactivating these special purpose items after you use them for their intended purpose to avoid cluttering up your Item List. (It already will be long enough!)

Inventory items

Use inventory items only if you have a true inventory situation, such as retail sales, or if you buy quantities of specific items that you hold until you use them at a later time. When you purchase an inventory item, QuickBooks Pro increases your inventory (asset) dollars and quantity for that item. When you sell the item using an invoice or through a cash sale, QuickBooks Pro reduces your inventory and moves the *average* cost into your COGS account.

The invoice or cash sale form that you complete when you sell an inventory item also will record your income and your increased receivable or cash. There are several ways accountants record changes in inventory. The way that QuickBooks Pro records each purchase and sale of inventory may or may not be the best approach for your particular situation. If you have a true inventory situation, spend some time with your CPA to determine how you want to handle inventory in your accounting system.

Organizing your Item List

I recommend that you organize your Item List in the following order:

1. Income-only items, including typical income and cost-reimbursement items encountered during closing transactions
2. Simple cost items that almost mirror the accounts in the Chart of Accounts (Use these items when you don't want to track job-process details for small jobs. You will have some additional detailed items that will post to the **S.55P.Other job-specific costs** account.)

3. Non-job-specific (NJS) items that are direct costs but cannot be assigned to a specific job
4. Employee-burden items that match up to each of the accounts under **S.522.Employee labor burden**
5. Building- or remodeling-process items, including all of your estimating, remodeling, and building process titles with detailed subitems
6. Project administration time
7. Add-ons or extras
8. Allocation items for developers (see chapter 34)
9. Model home operations costs, if you have a model home
10. Land development costs (If you develop land, segregate those costs here.)

Link income-only items in #1 above only to your income account(s). These items are used to invoice fixed-price contracts and/or to record loan closings.

For all other items, click on the box that says, "This service is performed by a trade contractor, owner, or partner." This will let you set purchase transactions to post to COGS accounts (the left side of the window), and sales transactions to post to income accounts (the right side of the window).

When you use these items to record transactions, the actual costs for the specific job(s) to which they are linked will accumulate in the item "bucket." If you have cost-plus contracts and invoice these line items, the income will show up on your detailed job-cost reports on the same line as the expense, and on your account-based reports in the income account.

Naming your items

To manipulate the Item List into the order in which you wish to have it appear on your reports, you need to adopt a lettering-numbering system. The numbering system also will help efficiently record the type of work performed on timesheets and make data entry much faster.

In the past, I've found it necessary to change the order of, or add to, the major titles in items #1 through #7 above. So instead of using numbers, I use double letters with intervening gaps, such as aa, ad, aj, and so on for the first major title items. This avoids the need to go through the process of renumbering subitems to keep the numbers in line with their title items.

Then, I try to use a fairly stable numbering system, with gaps built in where possible, to track the more specific job processes, such as 140=Exterior site preparation.

You will want to keep item names very short because QuickBooks Pro aggregates the names together in certain reports. Therefore, I recommend the following letters as add-ons to the basic number:

- a = Direct labor
- b = Trade contractors
- c = Materials
- d = Other
- v = Variance from purchase order

I also like to use a "+" in the areas where it is appropriate to add a special item for allowances. Using these naming conventions, a final detailed name for an item will appear something like: 140b.SitePrep. You will know that the letter "b" means trade contractor because you will use the same coding throughout your whole system.

Using special items

There are several features incorporated into the Item List that will help you distinguish certain items and dictate their role in reports.

- **Other charge** item—I use this for the finance charge item. This places the finance charge totals at or near the bottom of item-driven reports.

- **Subtotal** item—Allows you to create a subtotal on whichever form you are using.
- **Group** item—A special feature that allows you to combine various items into a single posting. This feature is best used for product sales. See the QuickBooks Pro Help window or a QuickBooks Pro consultant for help if you need to group the sales of specific products.
- **Discount** item—Calculates a discount on amounts listed in your invoice or cash-sale form.
- **Payment** item—Used to accept a payment at the time an invoice is created.
- **Sales tax** items—Needed for each of the types of sales tax that you collect. If you need to report amounts of nontaxable sales, it probably will be helpful to create a sales tax item with 0% for each type of nontaxable sale, such as resellers, institutions, finance charges, and so on. These items will let you extract the sales total associated with each type of tax or nontaxable transaction from your QuickBooks Pro sales tax liability report.
- **Sales tax group** item—Designed for states with multiple sales taxes assigned to taxable sales. Using this allows you to choose a group of sales taxes rather than having to select each tax separately.

In addition, you can add up to five custom fields to the items on your Item List. You cannot set up different custom fields for each item so you must decide ahead of time which fields you wish to create. If you typically want specific information on your forms, such as material-related information like color, finish, and dimension-size, you can link these custom fields to items in your Item List. The custom fields can be included as columns on your customized form templates, serving as a reminder to fill in the details.

Creating new service items

Because I recommend you designate your items almost exclusively as service items, the process for creating a new income item follows.

Creating an income item

1. On the **Customers Navigator**, click on the **Items and Services** icon or from the **Lists** pull-down menu, select **Item List**
2. In the Item List window, click on the pull-down **Item** menu and select **New**.
3. In the Type field, select **Service**.
4. Enter information in the following fields, as applicable. Most of the information is self-explanatory. If you need help, click on the **How Do I?** Help button.

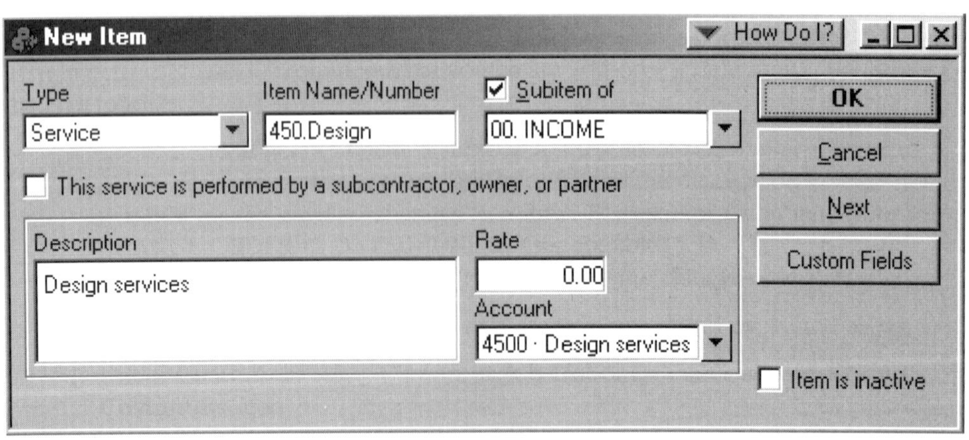

EXHIBIT 10.2 New Income Item window.

- **Item Name/Number**—Assign a name and/or number to the new item. Keep in mind that items appear in numerical order in the Item List.
- **Subitem of**—You can further classify or sort an item by making it a subitem.
- *Do not* place a checkmark in front of the statement: **This service is performed by a trade contractor, owner, or partner**.
- **Description**—Type a description for the item. The description will appear on invoices, credit memos, and estimates.
- **Rate**—Assign a rate or price for the item. This is the default rate you charge for this item. You can always change the rate in the Invoice window.
- **Account**—Select the income account to which you want to attribute the sale of this item.

5. When you have finished entering item information, click **Next** to continue adding items or **OK** to create the item and exit.

Creating a cost/cost-plus item

1. On the **Customers Navigator**, click on the **Items and Services** icon or from the **Lists** pull-down menu, select **Item List**.
2. In the Item List window, click on the pull-down **Item** menu and select **New**.
3. In the Type field, select **Service**.
4. Enter information in the following fields, as applicable. Most of the information is self-explanatory. If you need help, click on the **How Do I?** Help button.
   - **Item Name/Number**—Assign a name and/or number to the new item. Keep in mind that items appear in numerical order in the Item List.
   - **Subitem of**—You can further classify or sort an item by making it a subitem.
   - Place a checkmark in front of the statement: **This service is performed by a trade contractor, owner, or partner**.
   - **Purchase Description**—Type a description for the item that will appear on purchase orders, bills, credits and credit card charges.
   - **Cost**—Assign your estimated cost for the item, or leave it as 0.00. This is the default rate you are charged for this item. You can always change the cost later on your forms.
   - **Expense Account**—Select the account to which you want to attribute the purchase of this item (you will in almost all instances want this to be a direct cost account).
   - **Preferred Vendor**—Enter this if you generally order the item from one particular vendor. If so, their name will drop in as the default vendor if you enter this item on a purchase order or bill and another name has not already been entered.

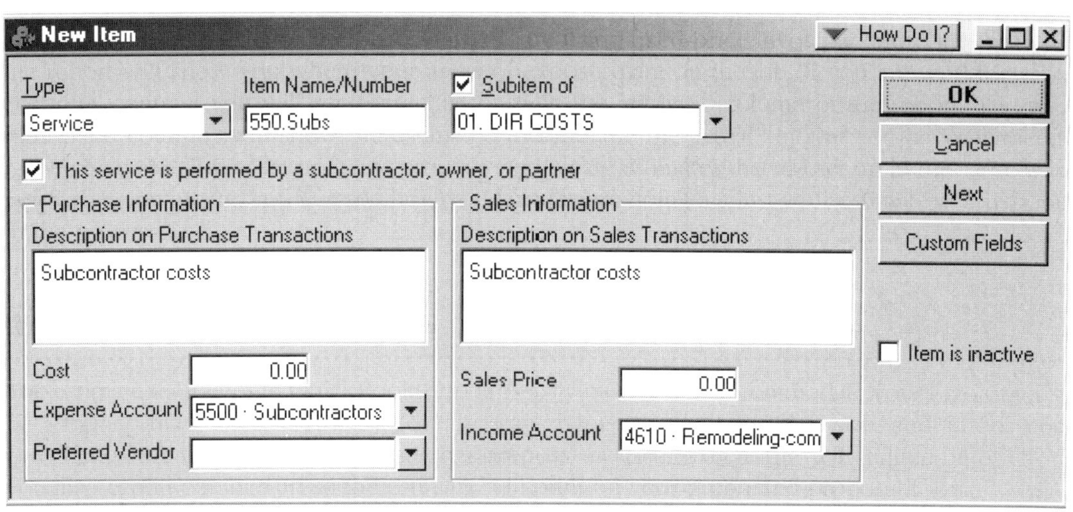

EXHIBIT 10.3 New Cost Item window.

- **Sales Description**—Type a description for the item that will appear on estimates, invoices, and credit memos.
- **Sales Price**—Assign the default price you charge for the item or leave it as 0.00. You can always change the price later on your forms.
- **Taxable**—Check this box if the item is taxable.
- **Income Account**—Select the income account to which you want to attribute the sale of this item.

5. When you have finished entering item information, click **Next** to continue adding items or **OK** to create the item and exit.

Editing item information

1. On the **Customers Navigator**, click on the **Items and Services** icon or from the **Lists** pull-down menu, select **Item List**.
2. In the Item List window, select the item to be edited by clicking on it once to highlight it.
3. From the **Item** menu button list, select **Edit** to open the **Edit Item** window.
4. Make changes to the desired fields.
5. When you have finished making changes, click **OK**.

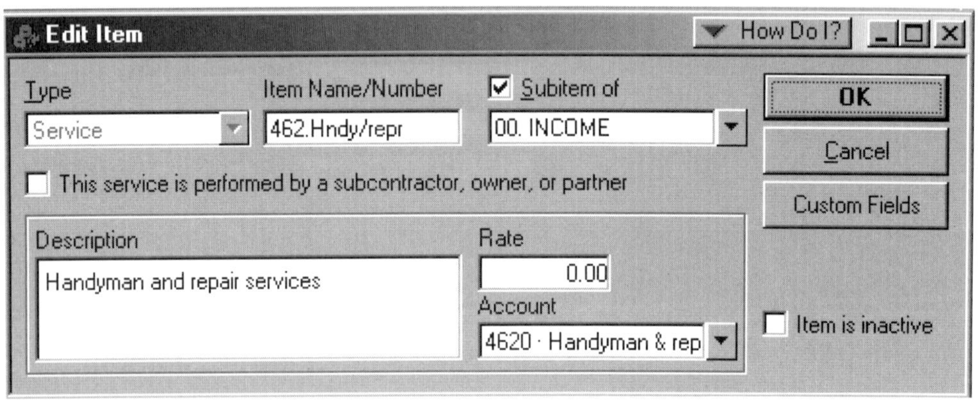

EXHIBIT 10.4 Edit Item window.

CHAPTER 11

Building a Customer:Job List

The Customer:Job feature in QuickBooks Pro 2000 is designed to help you track all financial transactions that apply to a specific customer or job. With the Customer:Job List you can link the following to specific customers or jobs:

- Estimated costs
- Purchase orders
- Bills, checks, and credit card purchases
- Employee time and payroll costs, including payroll taxes and, with special allocation procedures, various labor burden costs
- Invoices and credit memos or refunds
- Payments received
- Statements showing history of invoices, payments, and finance charges

Formulating a naming structure

A good starting point is to establish a consistent naming and summary structure for your Customer:Job List. Think of the Customer:Job List as an outline, with the customer level being the highest level of the outline, and the jobs being subcategories under the customer level. You also can add more layers of detail by creating jobs within jobs (subjobs, or even sub-subjobs) if you wish.

Taking the time to create a solid design and methodology for customer or job names at the beginning is important.

- You and your staff will use the Customer:Job List thousands of time. You want the list to be logical, easy to use, and consistent so entry errors are kept to a minimum.
- The Customer:Job List determines the structure of your reports. When you run profit and loss by job reports in QuickBooks Pro, the financial results for each name will be reported in its own column. Those totals also will roll up into the total for the next higher level. In other words, each level within the structure summarizes the financial results of the level(s) located immediately below it in the hierarchy.

Alternatively, when you create project reports, such as job estimates vs. actual detail, the results are arranged in item order as shown in Exhibit 11.1. In these reports you will see only the results of postings to the individual customer or job that you select.

Because the design of your Customer:Job List determines the structure of your reports, you can organize your Customer:Job List in a way that will provide you with the specific information you most often want to see. If you handle both fixed-price and cost-plus contracts, and want to consistently measure totals and results from all of the projects within each type of contract, you can creatively organize your Customer:Job List in a way that will let you see those results. Or if

Dreamscape Remodeling, Inc.
Job Estimates vs. Actuals Detail for Fields, Bruce, & Elaine
All Transactions

	Est. Cost	Act. Cost	($) Diff.	(%) Diff.	Est. Revenue	Act. Revenue	($) Diff.	(%) Diff.
Service								
00. INCOME								
450.Design					2,200.00	2,000.00	(200.00)	(9.09%)
460.Rem-res'l					44,800.00	38,000.00	(6,800.00)	(15.18%)
Total 00. INCOME					47,000.00	40,000.00	(7,000.00)	(14.89%)
01. DIR COSTS								
530.Labor	10,000.00	10,348.00	348.00	3.48%				
532.Lbr brdn	2,500.00	2,587.00	87.00	3.48%				
550.Subs	4,300.00	5,874.00	1,574.00	36.61%				
560.Materials	7,500.00	8,283.00	783.00	10.44%				
570.Misc cost	3,000.00	2,208.00	(792.00)	(26.4%)				
Total 01. DIR COSTS	27,300.00	29,300.00	2,000.00	7.33%				
Total Service	27,300.00	29,300.00	2,000.00	7.33%	47,000.00	40,000.00	(7,000.00)	(14.89%)
TOTAL	**27,300.00**	**29,300.00**	**2,000.00**	**7.33%**	**47,000.00**	**40,000.00**	**(7,000.00)**	**(14.89%)**

EXHIBIT 11.1 Job estimates vs. actual detail report.

you want reports that show how each of your supervisors are doing on individual jobs and in total, you can organize your Customer:Job List structure to provide you with those results.

Organizing your list

The most basic way to design a Customer:Job List for a remodeler is shown in the following example.

- Adams, P&S—basement finish (customer)
- Farmington, M—kitchen remod (customer)
- Franklin, O—garage (customer)
- Miller, T&B—new deck (customer)
- Peterson, J—bath remod (customer)
- Peterson, J—kitchen remod (customer)

This approach is simple and shows each project at the customer level. When the list is sorted it will organize itself, and subsequent reports, in alphabetical order. When you create a profit and loss by job report, you will see a total for each project, as well as a company total. If you want to see the totals for Peterson, J, you can filter your report to show just those two jobs and the calculated total. If both of these jobs are active, however, when you or your staff enter related data you'll need to be careful that you select the correct Peterson, J project. If one of these projects is no longer active, inactivate it to reduce the chance for erroneous entries. If you need help inactivating the project, click on the **How Do I?** Help button from within the Customer:Job List.

A variation on this approach might be to number each of these projects as they are started:

- 001-Peterson, J—kitchen remod (customer)
- 002-Farmington, M—kitchen remod (customer)

- 003-Miller, T&B—new deck (customer)
- 004-Franklin, O—garage (customer)
- 005-Peterson, J—bath remod (customer)
- 006-Adams, P&S—basement finish (customer)

One of the advantages to this approach is that when your list is sorted the reports will show projects in the order they were started. Some of my clients have developed numbering variations that include the year in the number. For example, jobs started in 2001 might start with "1" and the following 2 or 3 digits would indicate the job start date or the job's order in the current year's projects. The primary disadvantage to this system is that everyone in your organization needs to learn and refer to each job by the number you assign. This usually isn't a problem when you have a small number of projects, but if you handle lots of small projects you might want to use another ordering format.

Another alternative is to add a layer of project information by adding job and subjob levels as follows:

- Smith, Mark (customer)
  - Bathroom remodel (job)
    - Initial remodel (subjob)
    - Change orders (subjob)
  - Deck (job)
  - Kitchen remodel (job)
- Jones, F&M (customer)
  - Bathroom remodel (job)
    - Initial remodel (subjob)
    - Change orders (subjob)
  - Deck (job)
  - Kitchen remodel (job)

If you handle different types of construction work and want to see the results rolled up for each category, you might consider structuring your Customer:Job List something like the following:

- Fixed-price residential jobs (customer)
  - Smith, S&J (job)
    - Addition (subjob)
    - Handyman jobs (subjob)
  - Jones T&C (job)
    - Kitchen (subjob)
    - Deck (subjob)
- Cost-plus residential jobs (customer)
  - Porter, B (job)
    - Addition (subjob)
    - Bath (subjob)
  - Miles, K (job)
    - Kitchen (subjob)
    - Deck (subjob)

If you are a trade contractor or if you do a number of jobs for various builders you might consider the following type of layout:

- New Horizons Builders (customer)
  - NH West subdivision (job)
    - NHW.062 (subjob)
    - NHW.064 (subjob)
    - NHW.066 (subjob)
  - NH South subdivision (job)
    - NHS.492 (subjob)
    - NHS.495 (subjob)

- Valleyview Building Corp (customer)
  - 127 Edgewood (job)
  - 3921 Clifford (job)

If you are a developer and/or a spec home builder and purchase predeveloped lots or land to develop and build your own homes to sell you might want to create your list in a format like this:

- Perryville Landing subdivision (customer)
  - Phase 1(job)
    - PV101 (subjob)
    - PV102 (subjob)
    - PV103 (subjob)
  - Phase II (job)
    - PV201 (subjob)
    - PV202 (subjob)
    - PV203 (subjob)
- Martindale subdivision (customer)
  - MV014 (job)
    - MV014-lot (subjob)
    - MV014-home (subjob)
  - MV015 (job)
    - MV015-lot (subjob)
    - MV015-home (subjob)

Creating subjobs for lots and homes can be helpful when you are developing land or improving lots and building homes for sale. You will find that it's easier to track costs and income for lots and homes separately when you create and use subjobs for each.

If, for accountability purposes, you want to code jobs to various supervisors or lead carpenters, and sort projects and reports in that manner, you could:

- Use a letter, such as the first letter of each person's last name, as a precedent to the project. Tom Kelly's projects would be listed as k.Smith—kitchen, k.Jones—deck, and Frank Price's jobs would be listed as p.Porter—kitchen, p.Miller-addition, and so on.
- Make each supervisor a customer and enter the actual customers and projects as jobs and subjobs. Then you would see all of one supervisor's jobs listed side-by-side, or you could create a report showing only that supervisor's jobs and totals by selecting only the jobs showing that supervisor's code. If you choose, you then could share the financial results with each supervisor so they could track their progress and better control their own jobs. You also could use the financial results to reward outstanding performers and to help underachievers see exactly where they can improve.

TIP If you have an existing QuickBooks Pro file but have decided to create a new file to take advantage of some of the features outlined in this book, you can export your Customer:Job List, Employee List, and Vendor List into a special ".iif" (Intuit interchange format) file. Then you can import your lists into your new QuickBooks Pro company file by following the instructions in the QuickBooks Pro User's Guide. Before you begin to use the lists in your new file, be sure to review them and delete any names that will not be used by your new company.

Setting up your list

Lay out your potential Customer:Job List format on a yellow pad or spreadsheet and then think about the format for a couple of days. If you have a numbering system that works well, continue with it. If you think your system could be improved, or if you don't have a naming or numbering system, this is a good time to consider your alternatives.

After you feel somewhat comfortable with one or two formats it's probably time to discuss the organization, numbering-lettering system, and reporting implications with others in your company who may need to submit or enter project information, use the list, or use the management reports. I have found that different people bring some important and often specific feelings to the table that may influence your decisions. It never hurts to tap into all available brainpower!

It's wise to keep customer and job names as short as possible, and to include key identifiers very close to the beginning of the name. This makes data entry easier and faster. You also will find short or unique names useful because certain job reports are arranged with the job names in columns across the top of the report. To keep your reports to a manageable length, you will want your columns to remain fairly narrow. When you narrow the columns, you will be able to see only the first 8 to 10 letters of the project name, so design the first letters or digits to be unique and meaningful to you.

Sorting sequences

QuickBooks Pro sorts lists based on the punctuation marks, numbers, and letters you include in your customer or job names. Keep these alpha-numeric sequences in mind when you design your list. The sort order within each set of information is:

1. Punctuation marks, followed by
2. Numbers, followed by
3. Letters in alphabetical order.

The punctuation marks sort in the following order:

1. !
2. $
3. &
4. *
5. @
6. ^
7. –
8. +
9. <
10. >

To indicate that certain jobs are finished or that specific properties are sold or closed, you might want to add a punctuation mark at the beginning of the project name as a visual signal to the report reader. When sorted, this also will cause the finished jobs to sort in a specific sequence. You might want to use codes like these:

- ! = Work completed
- $ = Sales agreement signed
- & = Closing has taken place
- * = All known income and costs entered—job is complete

When you re-sort your list those projects will appear first within their categories. Keep in mind that the individual entering the data will need to remember the modified project codes, so you might want to use just the final "*" code when the job is done—no more data would need to be entered at that point anyway.

If any of your numbering sequences will reach more than 100 or 1,000, use the full three or four digits for all numbers within the sequence, such as 001, 002, 003, or 0101, 0102, 0103 to keep your numbers in order. If you don't, you will find that all numbers beginning with a 1 will group together, all numbers beginning with a 2 will group together, and so on, as shown.

- 10
- 19
- 100
- 1000
- 2
- 20
- 27
- 200
- 2000
- 3
- 4

Autofill feature

As you enter data, the Autofill feature in QuickBooks Pro will try to match what you are entering with the first matching entry it can find in the list you are accessing. This feature can be helpful, but it also can lead you astray. For example, assume that you have the following names within your Customer:Job List, with some being customers, and others being jobs or subjobs.

- Sallyforth subdivision
- Smith, J
- Smith, T
- Smith, Thomas
- Sun porch (under Adams, M)
- Sun porch (under Lincoln, A)

With the Autofill feature:

When you type	Your cursor will go to:
S	Sallyforth subdivision (first "S" in list)
Sm	Smith, J (first "Sm" in the list)
Smith, T	Smith, T (first "Smith, T" in the list)
Smith, Th	Smith Thomas (first "Smith, Th" in the list)
Su	Adams, M:Sun porch (first "sun porch" in the list)

If you want to make an entry to the Lincoln, A:sun porch, you would need to go to "Lincoln, A" and then manually select "sun porch."

Keep in mind these unique QuickBooks Pro features and make each name unique. Avoid duplicating names at any level. In one of the preceding examples, several of the jobs have an identical name—kitchen remodel or deck. If you used the sample listing and typed in "deck," you would find yourself at the first "deck" in the list—with no where else to go. If you add a unique prefix—perhaps the customer's initials, or a job number—you will avoid dead-end situations.

Adding special-purpose customers and jobs

In addition to the customer or job entries that you create for actual projects, I recommend you add a least some of the following special entities to your Customer:Job List:

- zz.Direct
  - 1.NJS-cnstr
  - 2.Int/loan fees
  - 3.Land & dev costs
  - 4.G&A and warranty

- zz.Overhead
  - 7.Sales/mrkting
  - 8.G&A/warranty
  - 9.Owners

Note that the customer-level titles have a prefix of "zz." This guarantees that when the Customer:Job List is sorted, these customers or jobs will stay at the end of the list so that they are shown just before your company total on reports.

The jobs shown below the "zz.Direct" customer correspond to expense accounts in the cost of goods sold (COGS) section of the Chart of Accounts (see chapter 8). The jobs shown below the "zz.Overhead" customer correspond to expense accounts in the expenses section of the Chart of Accounts (see chapter 9).

After creating the zz.customers, create the following jobs in your Customer:Job List, regardless of what type of company you own:

- **1.NJS-construction** is used to collect non-job-specific (NJS) construction costs such as:
  - Vacation, holiday, emergency, or bonus pay for direct-labor employees
  - Pay for staff meetings, training, shop cleanup, or company errands for direct-labor employees
  - Operating costs for and depreciation on construction equipment
  - Small tools and miscellaneous job supplies such as sand paper, saw blades, brooms, protective gear, uniforms, and so on
  - Other costs associated with direct production that cannot logically or reasonably be assigned to another, more specific job item (Refer to the subaccounts under the title account **S.560.NJS dir cost** in the COGS section of the Chart of Accounts. There also is an NJS section in the Item List.)

> NOTE
> Costs are accumulated in this NJS job. If you choose to allocate weighted or average NJS costs back to various jobs, subtract the costs can from this job and assign them back to individual jobs using journal entries (see chapter 34).

- **7.Sales/marketing** is used to track sales expense as if it was a separate function within the company. Since all of your direct labor costs must be assigned to a job for proofing and balancing purposes, if various members of your direct labor or your administrative team also double as sales ambassadors at shows or other special events, it can be especially useful to track sales-related costs. Just assign their time to the sales job and, even though it will appear as a direct labor or administrative cost in your general ledger, you will still be able to run a job report on the sales/marketing function and see your actual sales/marketing costs.
- **9.Owners** is used to segregate owner's costs for tax purposes. Since an owner's time can be split between so many functions it can be useful to create this job in order to segregate costs incurred by or on behalf of the owner.

Including special jobs for developers

I have created the following additional jobs for specific costs incurred during the life of a development. If they are applicable to your situation, add them as jobs under the "zz.customer" categories illustrated earlier.

- **2.Int/loan fees** is used to record all interest and loan fees for land and land development loans using the **0601d.Dir Int** item listed in the **as.LAND & DEV** section of the Item List (see chapter 10). Assign all fees and interest for construction loans to this job using the **d.Int/fees** item listed in the **94.OTHR CNSTR** section of the Item List.
- **3.Land and dev costs** is used to record all costs associated with land, land development, engineering, legal fees, general permits, and paperwork related to the development,

> NOTE When you establish the development, create an entry to allocate estimated total interest and fees to each unit using items from the allocations section of the Item List with a $0 check. The allocation shows up as an allocated cost for each unit, and as an allocated cost offset in the interest and fees job, with the net affect of $0 for the company. Then as work progresses run actual-to-date vs. estimated total reports and actual-to-date vs. budget-to-date reports to monitor interest and fees costs against the planned totals for the development. For a full approach to special allocations for developers, please see chapter 34.

infrastructure costs, general operations, association fees, taxes through closing, utilities through closing, general maintenance, repair costs to common spaces, and so on. Use the items listed in the **as.LAND & DEV** section of the Item List.

- **4.G&A and warranty** is used to record both general and administrative (G&A) and warranty and repair. Include the various G&A costs incurred over the life of the development, such as office supplies, administrative salaries, general professional fees, and so on. Post G&A directly to your overhead expense accounts as these costs are incurred. It is not necessary to assign them to any job. At the end of the accounting period, create a $0 check to offset the period's G&A expense and move the G&A expense into the COGS section of the Chart of Accounts.

Creating a new customer

1. Go to the **Customers Navigator** and click on the **Customers** icon, or click on the **Customers** pull down menu and select **Customer:Job List.** Then select **New** to open the **New Customer** window.

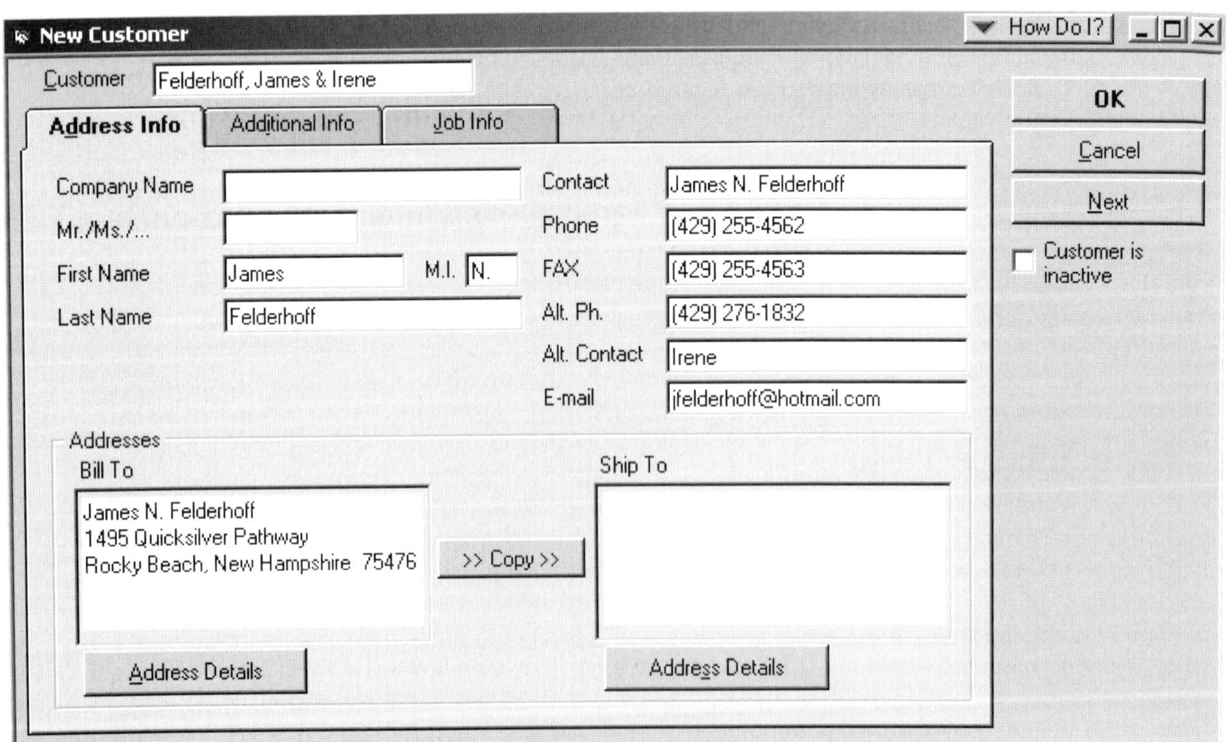

EXHIBIT 11.2a New Customer window.

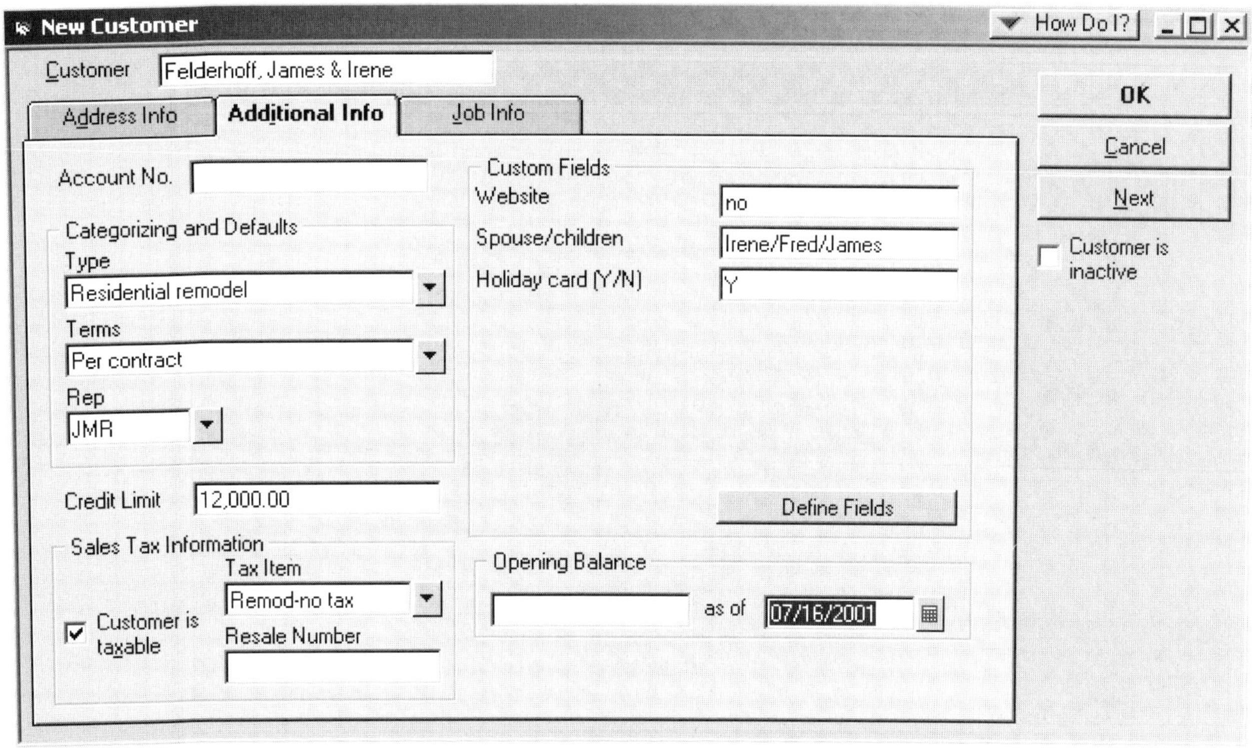

EXHIBIT 11.2b New Customer window. *Continued*

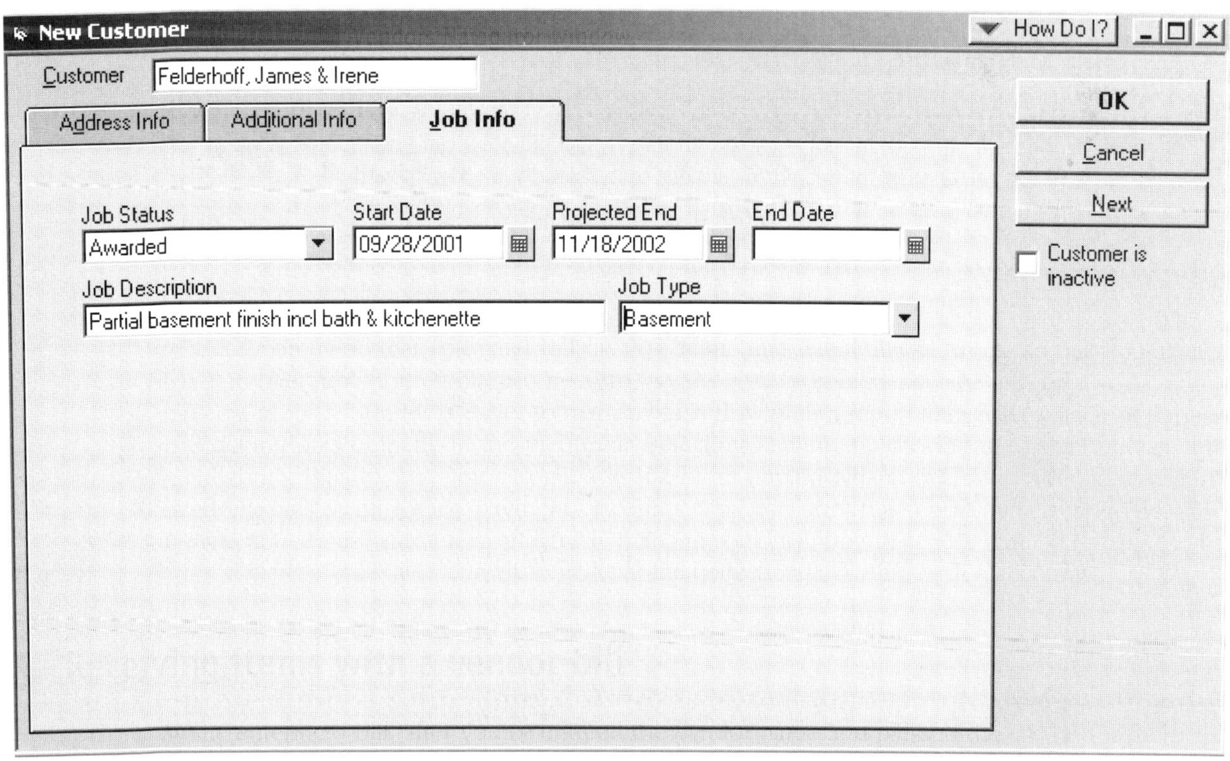

EXHIBIT 11.2c New Customer window. *Continued*

2. Three tabs will appear across the top of the window—**Address Info** (see Exhibit 11.2a), **Additional Info** (see Exhibit 11.2b), and **Job Info** (see Exhibit 11.2c). Under the **Address Info** tab, enter the requested information in the appropriate field. Most of the information in these fields is self-explanatory. If you need help, click the **How Do I?** Help button, and select **Add a New Customer**.
3. Click on the **Additional Info** tab and enter information in the following fields, as applicable:
   - **Account**—If you assign account numbers to customers, enter the number that you want to associate with this customer (seldom applicable to construction).
   - **Type**—Select a customer type from the list or add a new type.
   - **Terms**—Select or set up payment terms.
   - **Rep**—If you employ salespeople, you can use this field to track customers by salesperson.
   - **Credit Limit**—If you enter a credit limit, QuickBooks Pro will warn you when customers' invoices exceed their credit limit.
   - **Tax Item**—Select the tax item or tax group applicable to the customer. If the customer is non-taxable, select the appropriate non-tax type item, for example re-seller or institutional.
   - **Customer Taxable**—Place a checkmark in this field, so that the correct type of tax or non-tax item is tracked for the items sold.
   - **Resale Number**—Type in the customer's resale number, if applicable.
4. When you have finished entering customer information, click **OK.**

You can continue to add new customers to the Customer:Job List or return to the Customers Navigator.

Editing customer or job information

1. Go to the **Customers Navigator** and click on the **Customers** icon, or click on the **Customers** pull-down menu and select **Customer:Job List**.
2. Select a customer or job name from the list. Double-click on the name to go to the Edit Customer window.
3. Make changes to the desired fields.
4. When you have finished making changes, click **OK**.

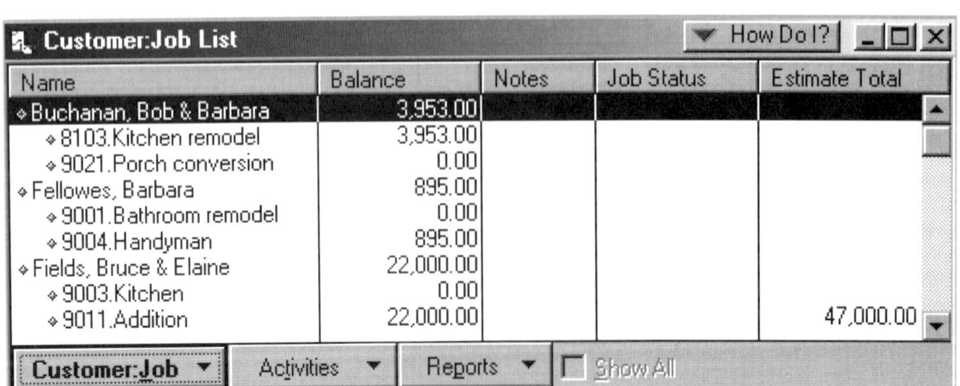

EXHIBIT 11.3a Customer:Job List window.

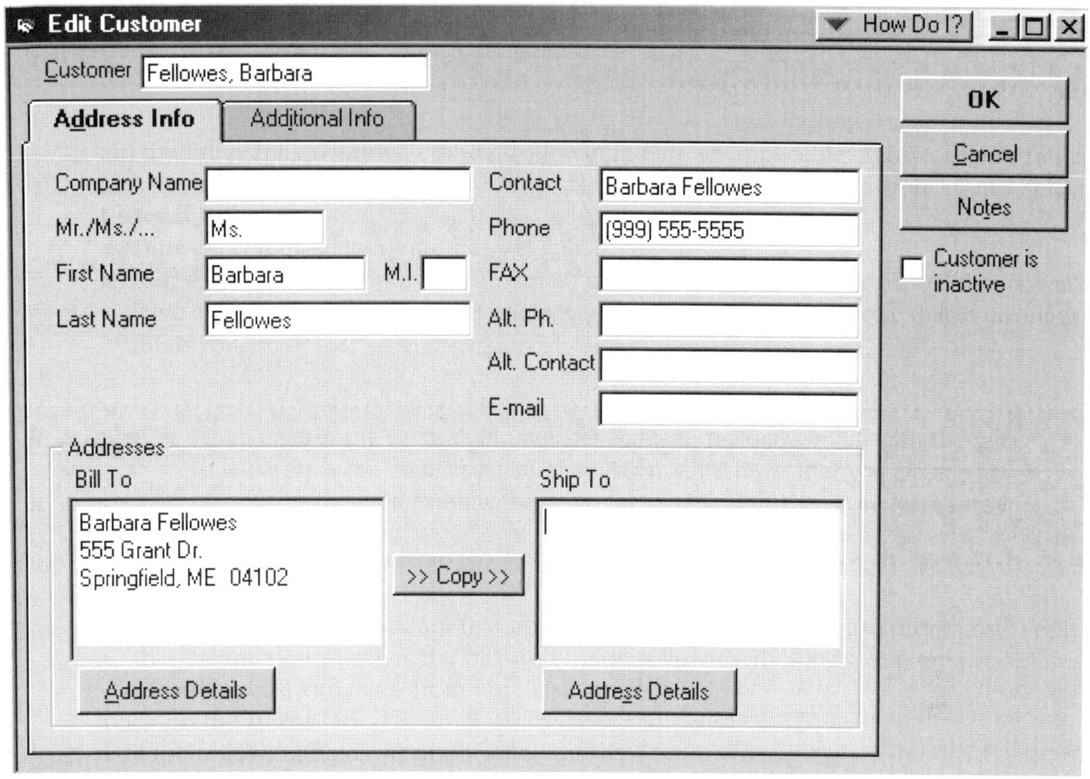

EXHIBIT 11.4 Edit Customer window.

Creating custom fields

You can establish additional fields for tracking and sorting by clicking on the Define Fields button on the New Customer, New Job, Edit Customer, or Edit Job window.

1. Click on **Define Fields**.
2. Type a label for the field name(s) you want to create.
3. Select the list you want this field to appear in.
4. Click **OK** to create the new field(s).

Adding jobs for a customer

After you have entered a customer, you can create jobs under that customer. If your customer and job are one and the same, you can enter job information on the third tab of the New Customer or Edit Customer window. If you choose to create jobs that are under the customer, the third tab disappears from the customer window and will be available only for the individual jobs. Also, please note that QuickBooks Pro only allows *one estimate per job*.

Follow these steps to enter one or more jobs for a customer.

1. Go to the **Customers Navigator** and click on the **Customers** icon, or from the **Lists** menu, select **Customer:Job List**.
2. Select a customer name in the **Customer:Job List** window by highlighting the name.
3. From the **Customer:Job** pull-down menu, select **Add Job**.
4. In the New Job window, enter the job name for the customer.

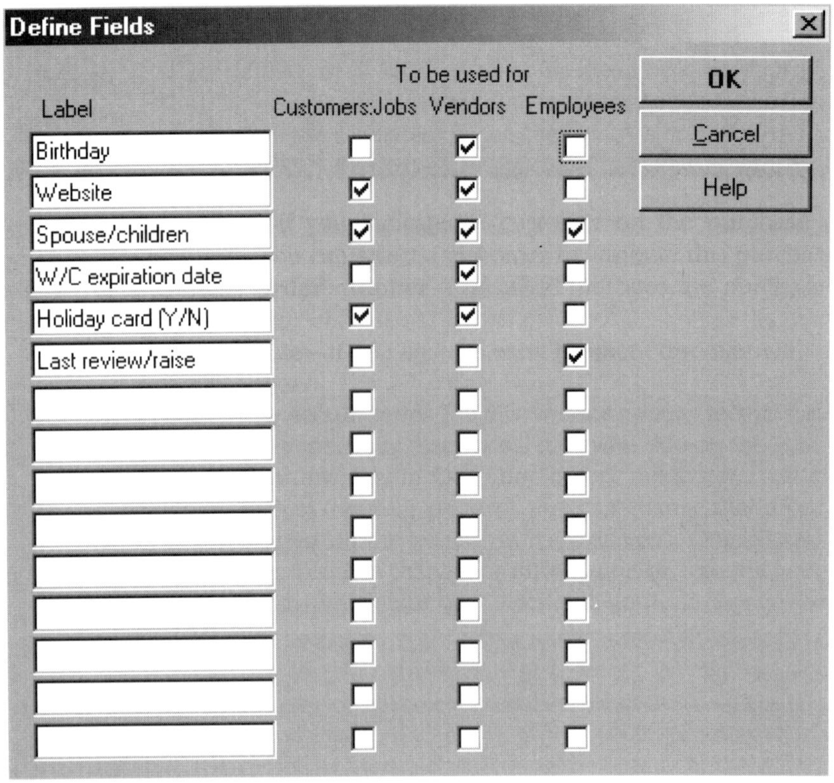

EXHIBIT 11.5 Define Fields window.

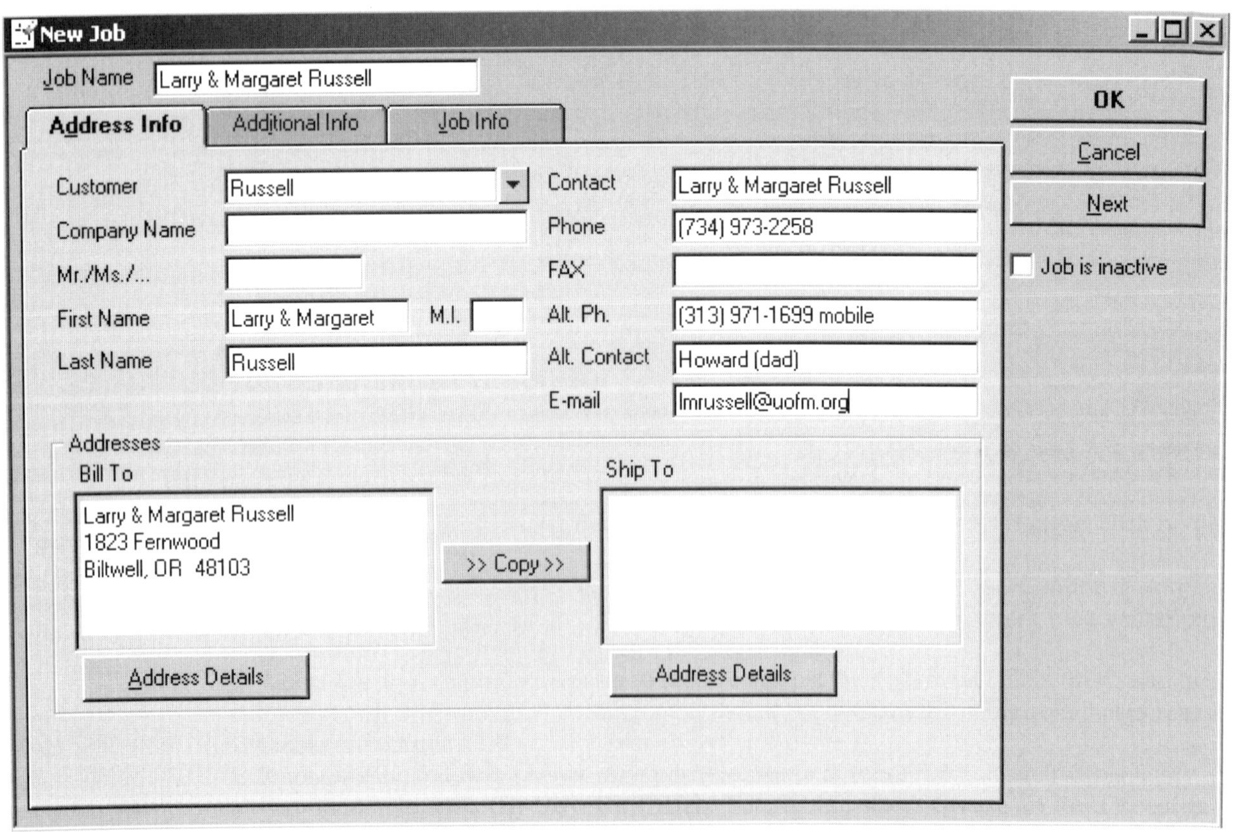

EXHIBIT 11.6 New Job window.

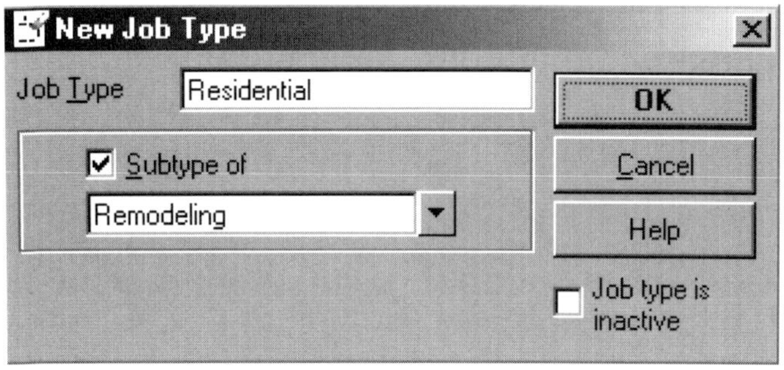

EXHIBIT 11.7 New Job Type window.

5. Enter the following information, as desired.
   - **Start Date**—The date the job started.
   - **Projected End**—The date the job is scheduled to be completed.
   - **End Date**—The actual date the job ended.
   - **Job Description**—Enter a description for the job.
   - **Job Type**—You may enter a job type, such as bathroom or kitchen, or custom home or model name. This field also has a pull-down list of job types that you can set up and then select from to keep your job types consistent.
     - Click on the **Job Type** button and select **Add New**.
     - Enter the job type(s), including subtype(s), under a job type.
     - Click **OK** to enter the job type in the **Job Type** pull-down list.
6. Click **OK** to enter the job for the customer.
7. At this point you also may edit the customer's information under the **Address Info** and **Additional Info** tabs.
8. Click **OK** to exit the window and save the customer:job information.

Printing the Customer:Job List

Under the Reports pull-down menu are several options for ways of viewing customer data by creating reports using your Customer:Job List. You can also select **Print List** from the **Customer:Job** pull-down menu. Experiment to find the reports that work best for you.

CHAPTER 12

Creating additional lists

In addition to the main lists used by QuickBooks Pro 2000—Chart of Accounts, Item List, and Customer:Job List—there are several smaller lists and sublists you can create that will make data entry faster and more consistent. With the exception of the Vendor List and Employee List, these are pull-down lists that you select from to enter information into a field on a QuickBooks Pro window. In this chapter I'll review the set up procedures for the Vendor List (See chapter 25 for information on setting up the Employee List.) and describe the general way in which you create the pull-down lists, as well as examples of the lists you may want to populate first.

Building a Vendor List

A vendor in QuickBooks Pro is any organization or individual to which you pay money including:

- Product suppliers or service companies
- Financial institutions
- Tax agencies
- Trade contractors
- Consultants
- Utility companies

QuickBooks Pro has several features that make identifying and paying vendors simpler using the Vendor List. As with all lists, think about how you want to enter your vendors. Use a consistent method that is easy to understand and perpetuate.

Using descriptive names

QuickBooks Pro enables you to put enter vendors using a descriptive name for reference purposes, as well as the vendor's formal name for payment. This makes it easier to identify the correct vendor, even when you pay one vendor for multiple items.

For example, you may owe Bank One for a Ford truck loan and a Chevy car loan. In the vendor name you can enter a descriptive name, such as Bank—Ford truck loan or Bank—Chevy car loan. In the Print on Check As field in the Address Info tab, however, you would enter Bank One.

After a vendor is in the system and a transaction is entered, QuickBooks Pro recalls the last transaction each time you access the vendor. If you select Bank—Ford truck loan, QuickBooks Pro automatically populates the next transaction using the information from the last transaction for Bank—Ford truck loan. For example, if your last transaction was a payment on your Ford truck loan, the next time you go to make a payment on the loan using the name Bank-Ford truck loan,

QuickBooks Pro's Autofill feature will repeat everything from the previous transaction, including account, name, and payment amount. You may either use the same information as is if it is a re-occurring monthly payment, or edit the information for the new transaction. This feature speeds up data entry and minimizes entry errors by eliminating the need to re-key information for repetitive transactions.

QuickBooks Pro also automatically enters the vendor's account number on your check if you put a checkmark next to the preference: Autofill payee account number in check memo in the Checking preference (see chapter 13) and enter the information under the Additional Info tab when you create the vendor.

Identifying 1099 vendors

When you enter a vendor you must decide if you will need to send the vendor a 1099 at the end of the year. If the answer is yes, you must fill in either the vendor's Employer's Identification Number (EIN) or Social Security number, and check the Vendor eligible for 1099 box. QuickBooks Pro then will select this vendor when it is time to print 1099s. Once you check the box you also must enter the vendor's full address, including city, state, and zip code. The system will not allow you to exit the vendor entry until all the required information is entered.

Creating a new vendor

1. From the **Lists** menu, select **Vendor** and then choose **Vendor List**, or on the **Vendors Navigator**, click the **Vendors** icon.
2. From the **Vendors** menu button list, select **New** to open the New Vendor window. Click on the **Address Info** tab and edit the fields, as applicable. Most of the information in these fields is self-explanatory. If you need help, click the **How Do I?** Help button, and select **Add Vendors, Notes, and Types**.
3. Click on the **Additional Info** tab and enter the following information, as applicable:

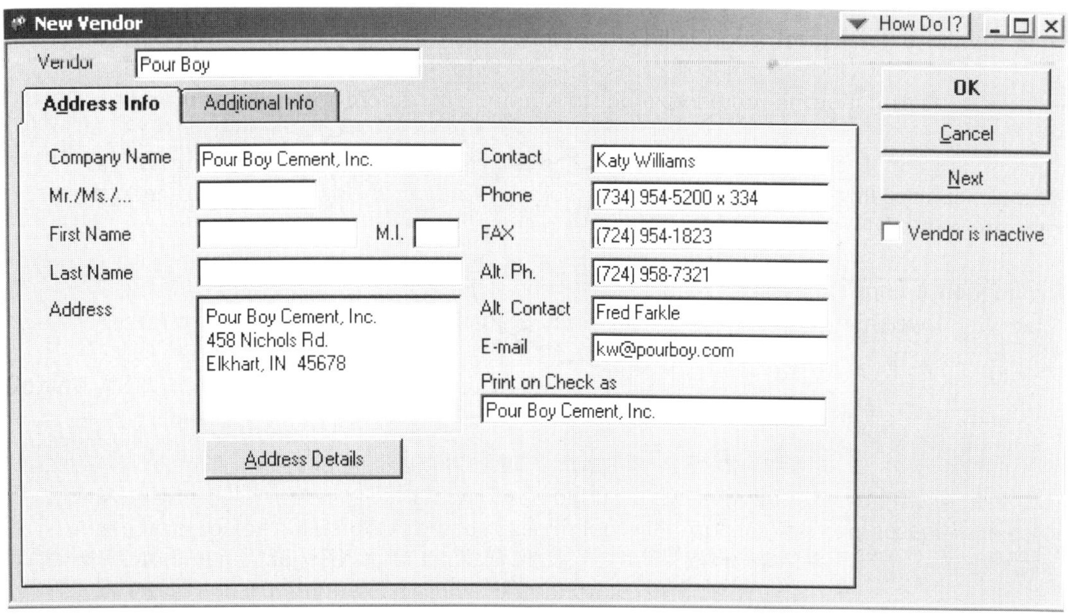

EXHIBIT 12.1 New Vendor Address Info window.

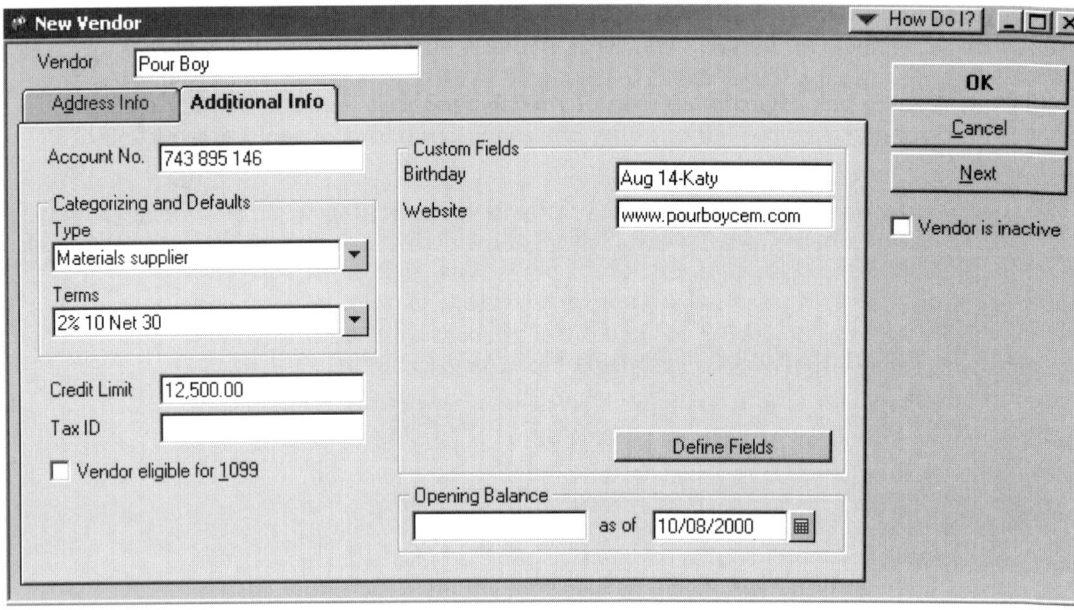

EXHIBIT 12.2 New Vendor Additional Info window.

- **Account**—Enter the account name or number that you want to associate with this vendor. This could be a number that the vendor has assigned to your company or a unique name-number assigned by your company.
- **Type**—Select a type from the list or add a new type. Setting up pull-down lists is described below.
- **Terms**—Select from the pull-down list or set up a new term based on the vendor's payment terms.
- **Credit Limit**—If you enter a credit limit, QuickBooks Pro will warn you when you issue a bill that exceeds your limit with the vendor.
- **Tax ID**—If you may potentially need to file an IRS form 1099-MISC for this vendor, enter the company's employer identification number (EIN) or a social security number if the vendor is an individual.

TIP The best way to obtain this information from vendors is to require it before you pay them the first time.

- **Custom Fields**—You can establish fields for additional information related to vendors in this section.

TIP Some of my clients indicate that they create a field for the date they need to check their trade contractors' workers' compensation insurance renewals.

- **Opening Balance**—Do not enter an opening balance for a new vendor.

4. When you have finished entering vendor information, click **OK**.
5. Continue to add new vendors by clicking on **Save & New** or return to the previous window by clicking on **Save & Close**.

Editing vendor information

To edit vendor information, follow these steps.

1. From the **Lists** menu, select **Vendors** and then choose **Vendor List**, or on the **Vendors Navigator**, click on the **Vendors** icon.
2. Select a vendor name from the list by double-clicking on the name. Make changes to the desired fields.
3. When you have finished making changes, click **OK**.

Creating customer and vendor profile selection lists

Using the Vendor List as an example, you used two fields that accessed selection-type lists that you create in QuickBooks Pro—a vendor-type list and a list of terms that may be applicable to your vendors.

Vendor Profile List

By categorizing your vendors you can take advantage of QuickBooks Pro filtering features that let you extract financial information about different types of vendors, or that lets you select certain vendor types for purposes such as creating mailing labels. In this pull-down list you can enter various categories that might apply to your particular vendors, such as materials supplier, trade contractor, employee reimbursement, trade organization, and so forth. Or you may categorize your vendors by product line, or type of service. Create vendor types that make sense for the way you do business.

Terms List

The Terms field lets you enter payment terms, such as when bills or invoices are due, using:

- A date-driven series of options, such as net due before a certain date, discount percent if paid by a certain date, or a specific date
- A standard series of options, such as amount due "within" a certain time period of 20 days, for example, and a discount percent if paid within that time period

You name each term according to the detailed options you have chosen. Examples include:

- 2/10, net 30 (2 percent discount if paid within 10 days, otherwise the balance is due within 30 days)
- Net 20 (the balance is due within 20 days)
- Due upon receipt
- 5th (due by the 5th of the month, due by the 5th of next month if a new bill is entered within 2 days of the bill due date)

When you set up a vendor or customer you may select a term from the pull-down list you have created. Your selection then becomes the default term associated with that vendor or customer. You also can edit the terms when you prepare the transaction. If you enter a credit limit for a vendor or customer, QuickBooks Pro will show a warning window if you exceed the credit limit when you enter a bill for the vendor or invoice for a customer.

To create new terms in the Terms List follow these steps.

1. From the **Lists** menu, select **Customer or Vendor Profile Lists** and then select **Terms List**.
2. From the **Terms List** window, click on the **Terms** button and select **New**.
3. In the **Terms** field, enter the name you want to assign to this specific term.
4. From the **New Terms** window, enter the information relating to this specific term

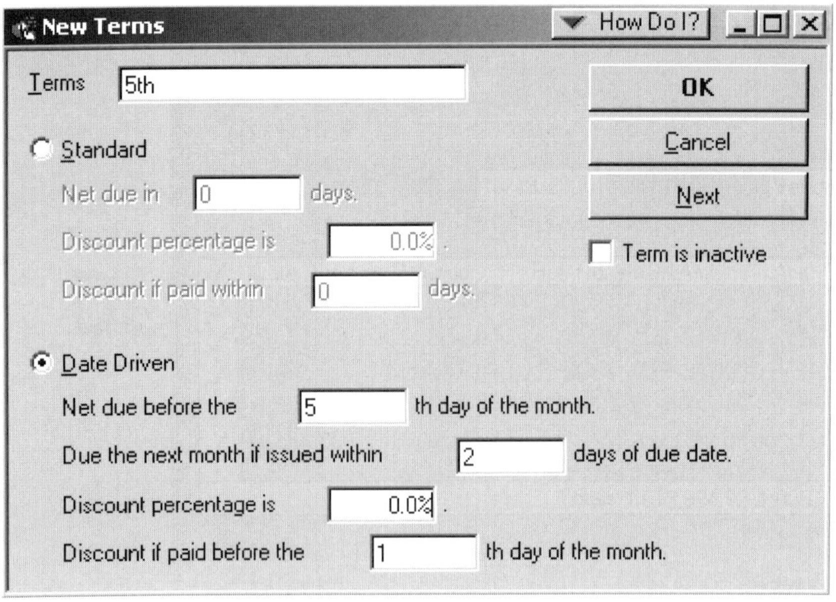

EXHIBIT 12.3 New Terms window.

5. When have finished entering the information for the new term, click **Next** to set up another term, or **OK** to return to the previous window.

Creating custom fields

If there is additional information you would like to keep or track, from vendors' birthdays or names of your customers' children to an employee's most recent pay raise date, you can create special customized text fields to capture the information. QuickBooks Pro lets you create up to a total of 15 custom fields in the Vendors, Customer:Job, and Employee lists, and indicate for each field whether it applies to vendors, customers, and/or employees. The custom fields appear on the Additional Info tab in the Create and Edit windows of the Customer:Job, Vendor, and Employee lists, and are created by clicking on the Define Fields button.

To create custom fields, follow these steps.

1. From a Create or Edit window in the Vendors, Customer:Job, or Employee lists click on the **Define Fields** button.
2. Enter the name you want to use for the information you will enter into the field. Then, place a checkmark in the column of the list(s) in which you want to use this field.
3. Click **OK** when you are finished.

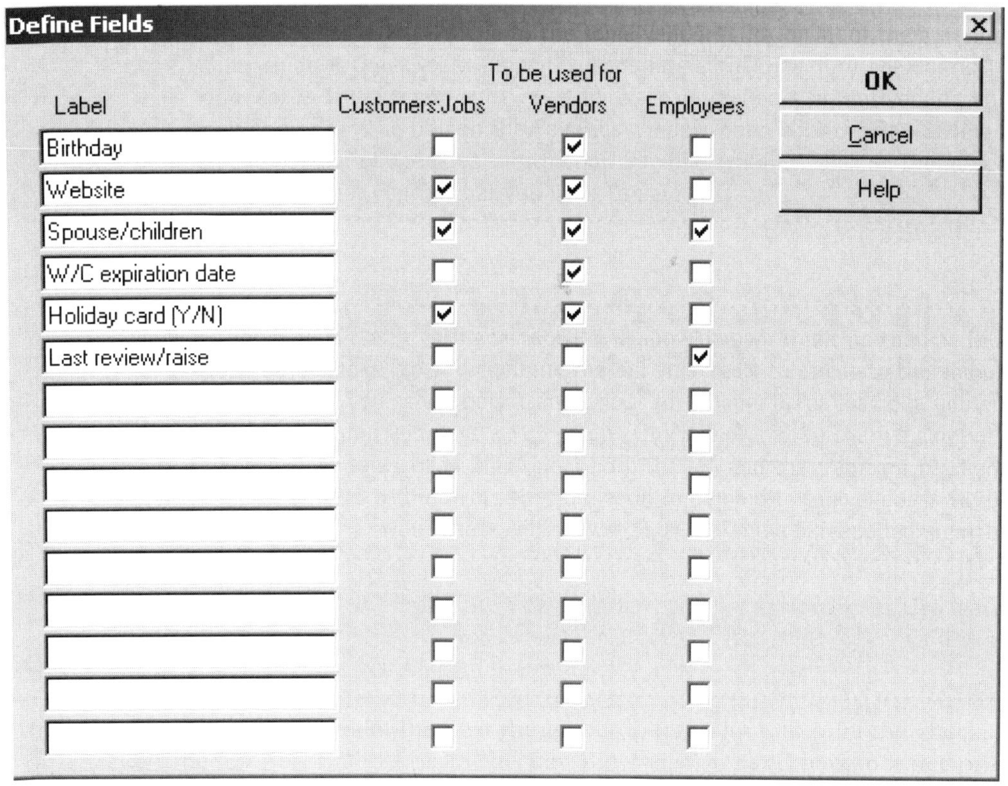

EXHIBIT 12.4 Define Fields window with example of custom field names.

CHAPTER 13

Setting preferences

Setting your company's file preferences properly is critical to the effective use of your entire accounting system. This chapter describes how to configure QuickBooks Pro 2000's company preferences to match your accounting system to the way your company does business.

There are two types of preferences that you can modify in QuickBooks Pro.

- The Company Preferences tab choices relate to the structure of data, reports, and features you do or do not want to use. These choices apply to all users. Only the QuickBooks Pro File Administrator can access or change Company Preferences.
- The My Preferences tab choices relate to how an individual user wants to use QuickBooks Pro's keystroke, desktop, or reporting functions or information.

Accessing QuickBooks Pro preference windows

Each of the preference windows is accessed using the following steps.

1. In the pull-down **Edit** menu, click on **Preferences**.
2. The system will default to the General Preferences window. The preference categories are arranged alphabetically in the scroll bar (see Exhibit 13.1) at the left side of the window. Select a preference category by clicking with your left mouse button on one of the preference icons.
3. When you have modified the preference window click on the next preference you wish to change, and respond with **Yes** to the Save Changes screen that appears, or click **OK** to exit the Preferences window and save your changes.

Modifying the preferences

QuickBooks Pro separates the preferences into several categories, each of which is represented by an icon on the left of the General Preferences window. Clicking on an icon will change the window. The individual preference windows will appear with the system's default settings. The preferences I recommend for each category are shown on the preference window from QuickBooks Pro in the accompanying exhibits.

General Preferences

Step #1 changes the general preferences under the My Preferences tab and step #2 changes the general preferences under the Company Preferences tab. Exhibits 13.2 and 13.3, respectively, illustrate the changes.

Step #1

1. Click on the **General** icon.
2. Click on the **My Preferences** tab.
3. Place a checkmark before the following preferences. I recommend you do not place a checkmark before the other options.
   - **Pressing Enter moves between fields**
   - **Beep when recording a transaction** (change if desired.)
   - **Warn when editing a transaction**
   - **Warn when deleting a transaction or unused list item**
   - **Automatically recall last transaction for this name**
4. Under Desktop options, place a checkmark before **Save when closing company**.
5. Click on the next preference you wish to change, and respond with **Yes** to the Save Changes screen that appears, or click **OK** to exit the Preferences .window.

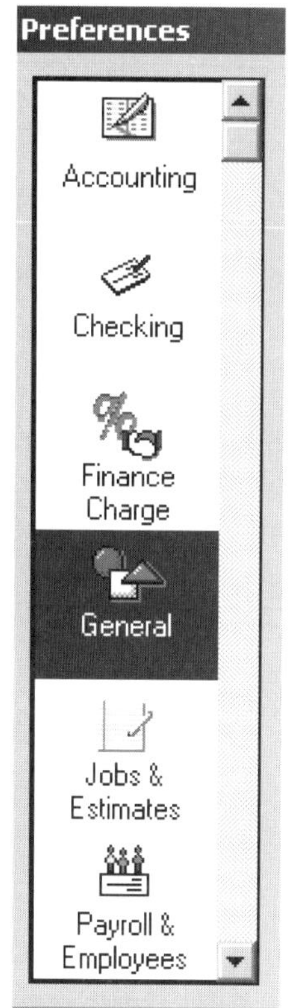

EXHIBIT 13.1 Scroll bar from Preferences window.

Step #2

1. Click on the **Company Preferences** tab.
2. Place a checkmark before the following preferences:
   - **Time format**—Show portions of an hour as decimal (change if desired)
   - **Always show years as 4 digits**
3. I recommend you do not change the other preferences.
4. Click on the next preference you wish to change, and respond with **Yes** to the Save Changes screen that appears, or click **OK** to exit the Preferences .window.

Accounting Preferences

1. Click on the **Accounting** icon (see Exhibit 13.4).

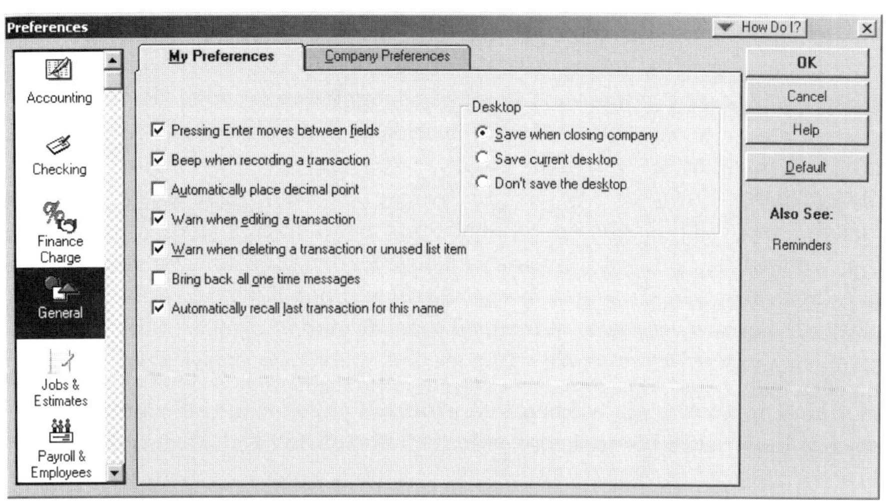

EXHIBIT 13.2 General Preferences window with My Preferences tab.

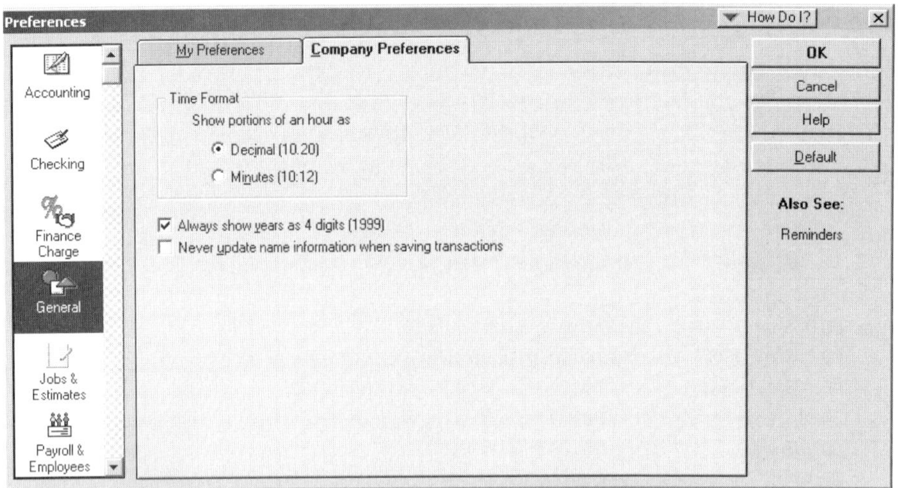

EXHIBIT 13.3 General Preferences window with Company Preferences tab.

2. Click on the **Company Preferences** tab. (There are no options under My Preferences in the Accounting window.)
3. Please place a checkmark before the following preferences:
   - **Use account numbers**
   - **Show lowest subaccount only**
   - **Require accounts**
   - Check **Class tracking** only if you have a specific plan to use classes. My construction clients rarely need the class tracking features, so I recommend leaving this option unchecked for now.
   - **Use audit trail**
4. Do not change the other preferences.
5. Click on the next preference you wish to change, and respond with **Yes** to the Save Changes screen that appears, or click **OK** to exit the Preferences .window.

Checking Preferences

1. Click on the **Checking** icon (see Exhibit 13.5).

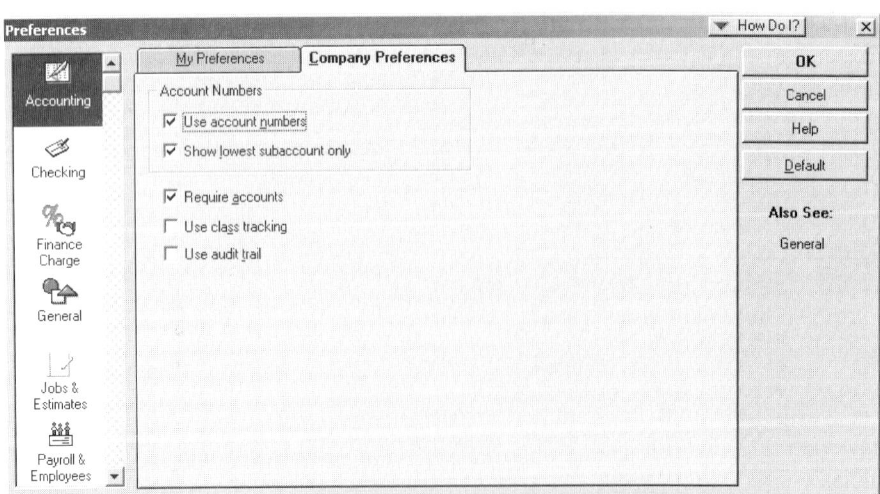

EXHIBIT 13.4 Accounting Preferences window.

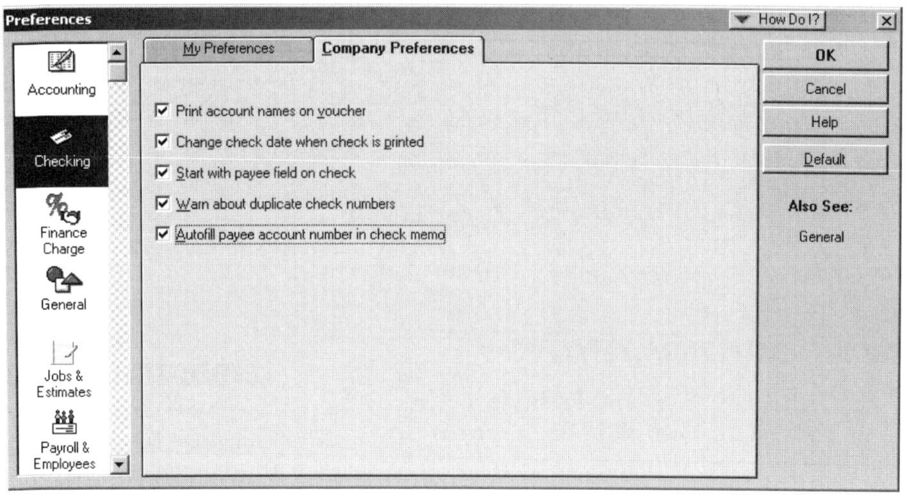

EXHIBIT 13.5 Checking Preferences window.

2. Click on the **Company Preferences** tab. (There are no options under My Preferences in the Checking window.)
3. Place a checkmark before all five preferences:
   - **Print account names on voucher**
   - **Change check date when check is printed**
   - **Start with payee field on check**
   - **Warn about duplicate check numbers.**
   - **Autofill payee account number in check memo.**
4. Click on **OK**.

Finance Charge

1. Click on the **Finance Charge** icon.
2. Click on the **Company Preferences** tab. (There are no options under My Preferences in the Finance Charge window.)
3. Enter a finance charge percentage in the **Annual Interest Rate (%)** field if you apply finance charges.

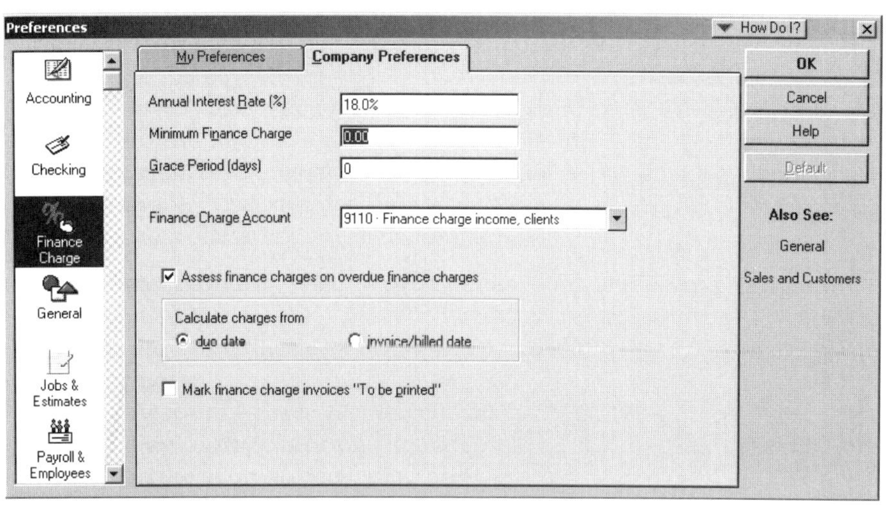

EXHIBIT 13.6 Finance Charge Preferences window.

4. Enter a minimum finance charge as a dollar amount in the **Minimum Finance Charge** field if you apply a minimum finance charge.
5. Enter a grace period as a number of days in the **Grace Period** field, if applicable.
6. Select a **Finance Charge Income Account** from the pull-down list.
7. Place a checkmark before the preference: **Assess finance charges on overdue finance charges**, if applicable.
8. Click on **Calculate changes from due date**.
9. Click on the next preference you wish to change, and respond with **Yes** to the Save Changes screen that appears, or click **OK** to exit the Preferences window.

Jobs and estimates preferences

1. Click on the **Jobs and Estimates** icon.
2. Click on the **Company Preferences** tab. (There are no options under My Preferences in the Jobs and Estimates window.)
3. QuickBooks Pro displays its job status fields and default descriptions, which you can change if you believe you will be using this feature. An example of how you are able to modify the wording relating to a project's status is shown in Exhibit 13.7.
4. Click **Yes** for the preference: **Do You Create Estimates?**
5. Place a checkmark before the preference: **Warn about duplicate estimate numbers**.
6. Depending on your accounting situation, click **Yes** or **No** for the preference: **Do You Do Progress Invoicing?**
7. Check **Yes** for the **Don't print items that have zero amounts** preference. This preference is available only if you chose yes for the progress invoicing preference. Place a checkmark before the preference: **Warn about duplicate estimate numbers**.
8. Click on the next preference you wish to change, and respond with **Yes** to the Save Changes screen that appears, or click **OK** to exit the Preferences window.

Payroll and employees preferences

If you have employees and you pay them with a payroll check prepared by you or by an outside payroll service, complete this screen as indicated.

1. Click on the Payroll and Employees icon (see Exhibit 13.8).
2. Click on the **Company Preferences** tab. (There are no options under My Preferences in the Payroll and Employees window.)

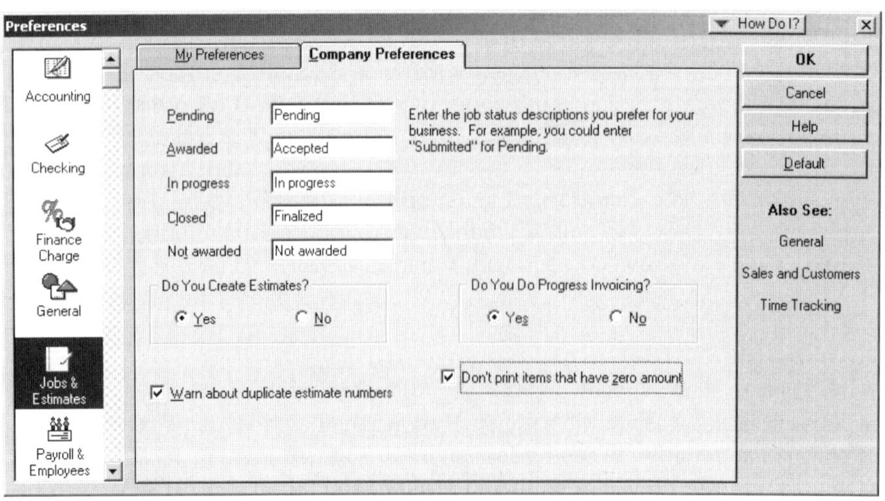

EXHIBIT 13.7 Jobs and Estimates Preferences window.

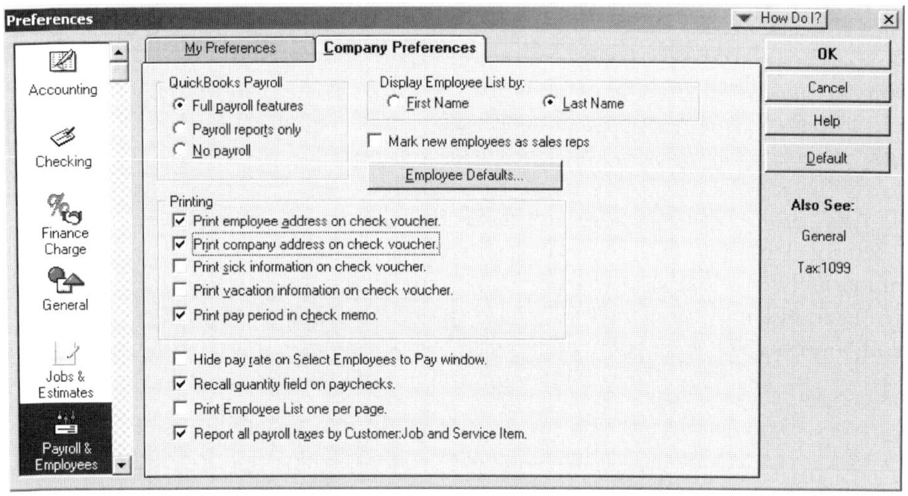

EXHIBIT 13.8 Payroll and Employees Preferences window.

3. Place a checkmark before the following preferences:
   - **QuickBooks Pro Payroll—Full payroll features** (Click **No payroll** only if you do not have anyone on payroll.)
   - **Display Employee List by Last Name**—This usually is the easiest way to access employee information, but choose the option you prefer.
   - **Print employee address on check voucher**
   - **Print company address on check voucher**
   - **Print pay period in check memo**
   - **Recall quantity field on paychecks**
   - **Report all payroll taxes by Customer:Job and Service Item**—This is the very powerful feature that assigns payroll taxes to your job costs, and that also allows you to create other special compensation-related allocations.
4. At this time you will want only to complete the screen you see. The **Employee Defaults** button is addressed in chapter 25, after you have created your Payroll Item List.
5. Click on the next preference you wish to change, and respond with **Yes** to the Save Changes screen that appears, or click **OK** to exit the Preferences window.

Purchases and vendors preferences

1. Click on the **Purchases and Vendors** icon (see Exhibit 13.9).
2. Click on the **Company Preferences** tab. (There are no options under My Preferences in the Purchases and Vendors window.)
3. Place a checkmark before the following preferences:
   - **Inventory and purchase orders are active** (I highly recommend that you use purchase orders for accounting and control purposes even if you do not give them to vendors. See chapter 16 for further information.)
   - **Warn if not enough inventory to sell**
   - **Warn about duplicate purchase order numbers**
   - **Bills are due __ days after receipt**—Enter the number of days, as default, you want QuickBooks Pro to add to the Date field when you enter a new bill. For example if you enter a bill dated January 7th, if you indicate 10 days in this preference, QuickBooks Pro will, by default, make the Bill Due date January 17th. You then can edit the Bill Due date if you wish.
   - **Warn about duplicate bill numbers from same vendor**—This is a valuable control to help prevent duplicate payments

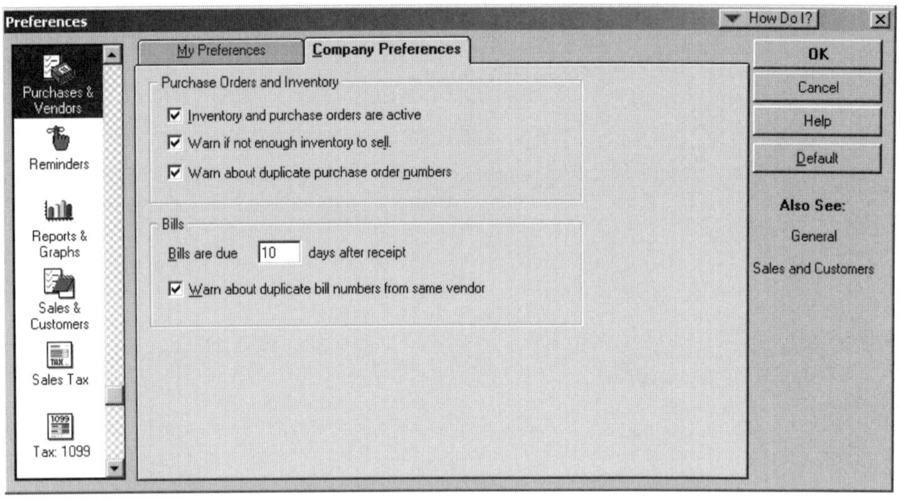

EXHIBIT 13.9 Purchases and Vendors window.

4. Click on the next preference you wish to change, and respond with **Yes** to the Save Changes screen that appears, or click **OK** to exit the Preferences window.

Reminders Preferences

1. Click on the **Reminders** icon.
2. Click on the **My Preferences** tab.
3. If you don't want to be bothered with seeing the Reminders List every time you open your file, do not place a checkmark in front of the preference: **Show Reminders List When Opening a Company File**.
4. Click on the **Company Preferences** tab.
5. Indicate your preference of Show Summary, Show List, or Don't Remind Me for each option.
6. Enter the number of days ahead of time that you want the system to remind you of the task in the Remind Me column.
7. Click on the next preference you wish to change, and respond with **Yes** to the Save Changes screen that appears, or click **OK** to exit the Preferences window.

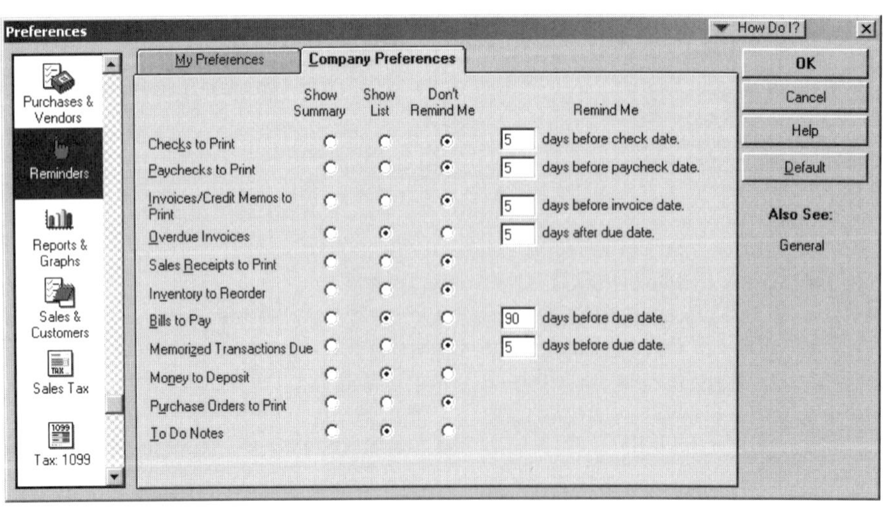

EXHIBIT 13.10 Reminders Preferences window.

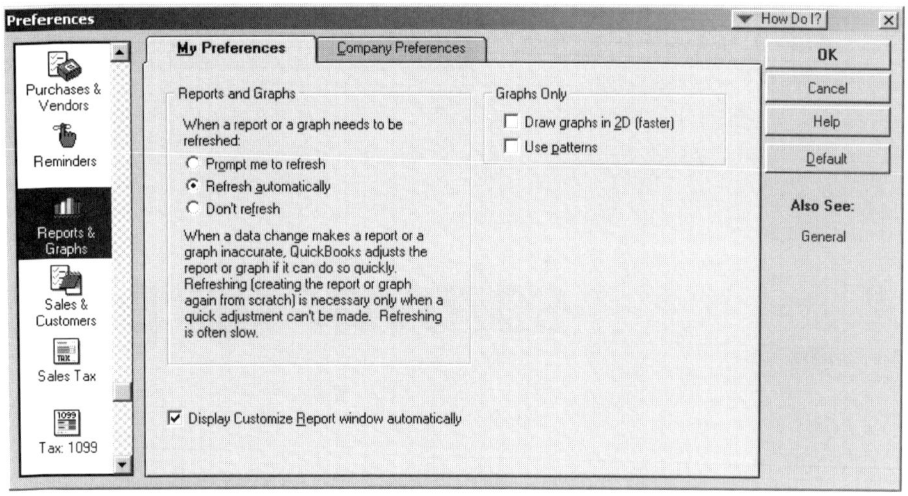

EXHIBIT 13.11 Reports and Graphs Preferences window with My Preferences tab.

Reports and graphs preferences

1. Click on the **Reports and Graphs** icon.
2. Click on the **My Preferences** tab.
3. Click in front of the preference: **When a report or a graph needs to refreshed—Refresh automatically**.
4. Place a checkmark before the preference: **Display Customize Report window automatically**—This lets you choose dates and other options before you prepare a report. If this is not checked, QuickBooks Pro runs the report based on default preferences. You then will need to re-run the report based your needs.
5. Click on the **Company Preferences** tab.
6. Click before the preference: **Summary Reports Basis—Accrual** (Do not change to Cash even if you use cash-basis accounting. I advise this because some accounts contain entries that are not able to convert properly to provide you with a true cash-basis report. If you select cash-basis reports, you may receive misleading results.

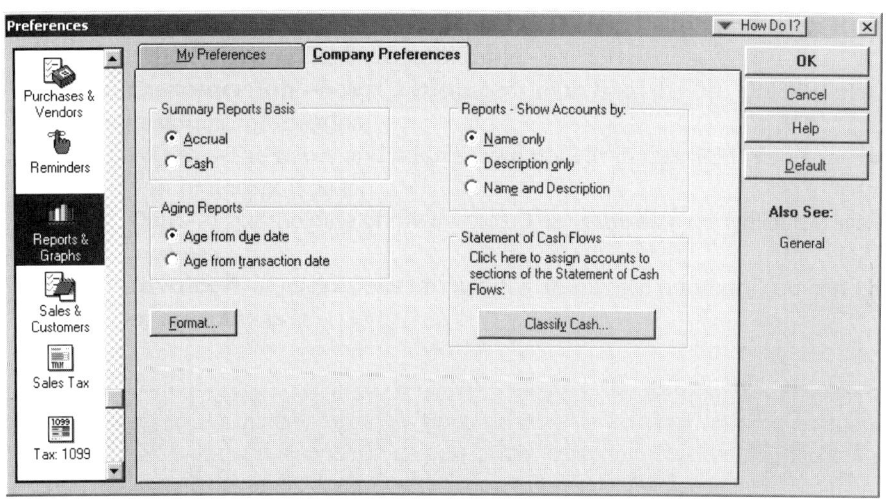

EXHIBIT 13.12 Reports and Graphs Preferences window with Company Preferences tab.

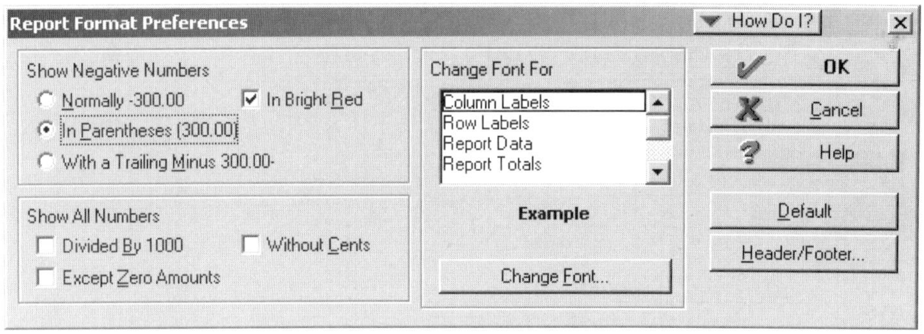

EXHIBIT 13.13 Report Format Preferences.

7. Click before the preference: **Aging Reports—Age from due date**.
8. Click before the preference: **Reports—Show Accounts by—Name only**.
9. Click on the **Format** button.
10. To make your reports easier to read, click before these preferences:
    - **Show Negative Numbers**
    - **Parentheses (300.00)**
    - **Bright Red**.
11. Click the **Change Font** button to modify the various fonts used in your reports. I do not recommend changing your fonts at this point.
12. Click on the next preference you wish to change, and respond with **Yes** to the Save Changes screen that appears, or click **OK** to exit the Preferences window.

Sales and customers preferences

1. Click on the **Sales and Customers** icon.
2. Click on the **Company Preferences** tab. (There are no option under My Preferences in the Sales and Customers preferences.)
3. Enter information in the **Usual Shipping Method, Default Markup Percentage**, and **Usual FOB** fields as appropriate.

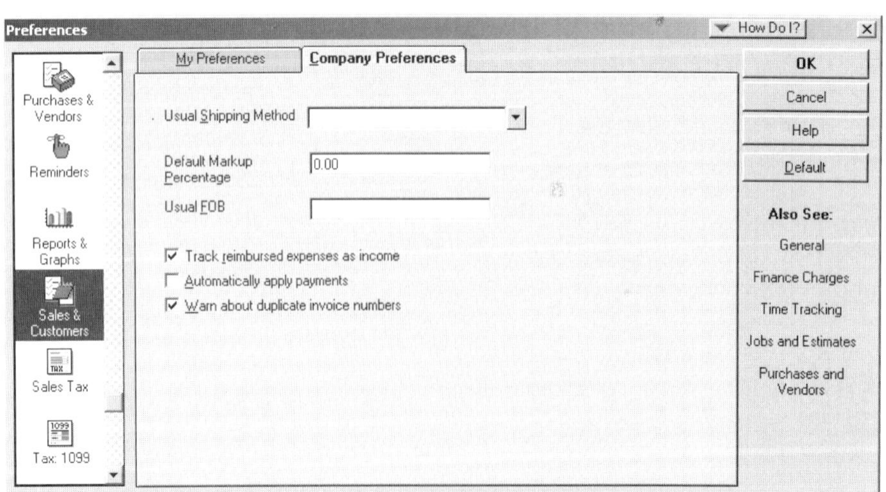

EXHIBIT 13.14 Sales and Customers Preferences window.

4. Place a checkmark before the following preferences:
   - **Track reimbursed expenses as income**
   - **Warn about duplicate invoice numbers**
5. Because of the way QuickBooks Pro handles credit memos, I recommend that you not use the **Automatically apply payments** preference.
6. Click on the next preference you wish to change, and respond with **Yes** to the Save Changes screen that appears, or click **OK** to exit the Preferences window.

Sales tax preferences

1. Click on the **Sales Tax** icon (see Exhibit 13.16).
2. Click on the **Company Preferences** tab. (There are no options under My Preferences in the Sales Tax preferences.)
3. Answer the preference questions appropriate for your business. Consult with your tax advisor if you are unsure of the appropriate response and to set up the sales tax items appropriate for your business.
4. Place a checkmark before the preference: **Mark taxable amounts with "T" when printing**.
5. Click on the next preference you wish to change, and respond with **Yes** to the Save Changes screen that appears, or click **OK** to exit the Preferences window.

Tax: 1099 preferences

1. Click on the **Tax: 1099** icon (see Exhibit 13.16).
2. Click on the **Company Preferences** tab. (There are no options under My Preferences in the Tax: 1099 preferences.)
3. If you say **Yes** to the preference: **Do you file 1099 forms?**, use the account drop-down lists to select the account numbers to which you normally assign trade contractor costs for Box 7. For example, select your Trade contractor account, and if you pay an individual to clean your office, also select your Janitorial/cleaning service account. In addition, select your Legal fees account for Box 13A. QuickBooks Pro provides default threshold amounts or you can contact your accountant for additional information.
4. Click on the next preference you wish to change, and respond with **Yes** to the Save Changes screen that appears, or click **OK** to exit the Preferences window.

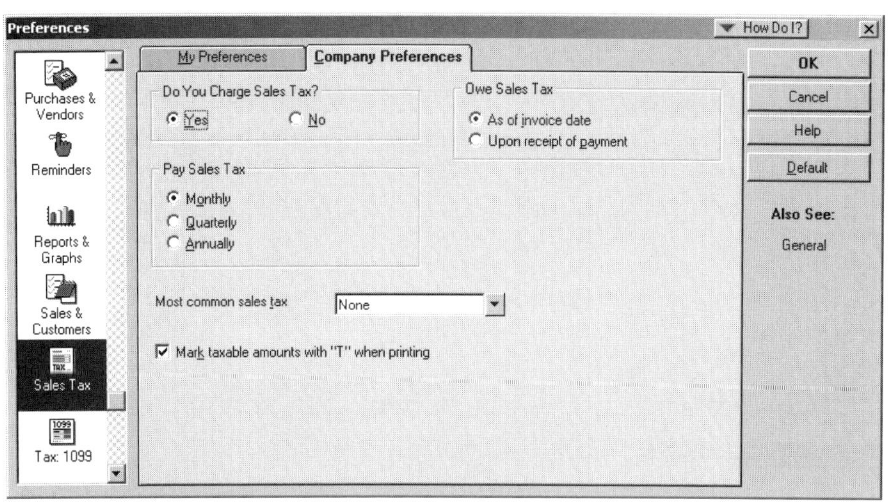

EXHIBIT 13.15 Sales Tax Preferences window.

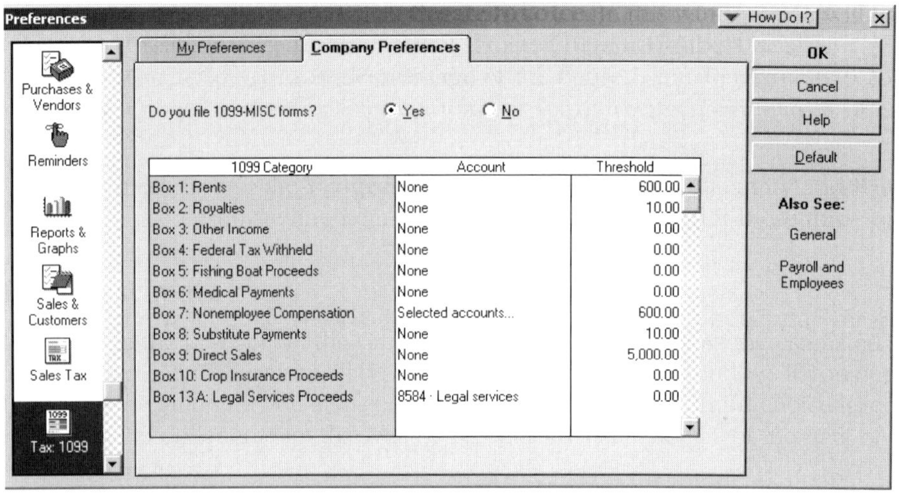

EXHIBIT 13.16 Tax: 1099 Preferences window.

Time tracking preferences

1. Click on the **Time Tracking** icon.
2. Click on the **Company Preferences** tab. (There are no options under My Preferences in the Time Tracking preferences.)
3. If you say **Yes** to the preference: **Do You Track Time?**, it means that you can enter time spent on jobs in a timesheet format. Then you can automatically access that information for payroll, time reports, or other cost-assignment allocations. Indicate the first day of your company's work week.
4. Click on the next preference you wish to change, and respond with **Yes** to the Save Changes screen that appears, or click **OK** to exit the Preferences window.

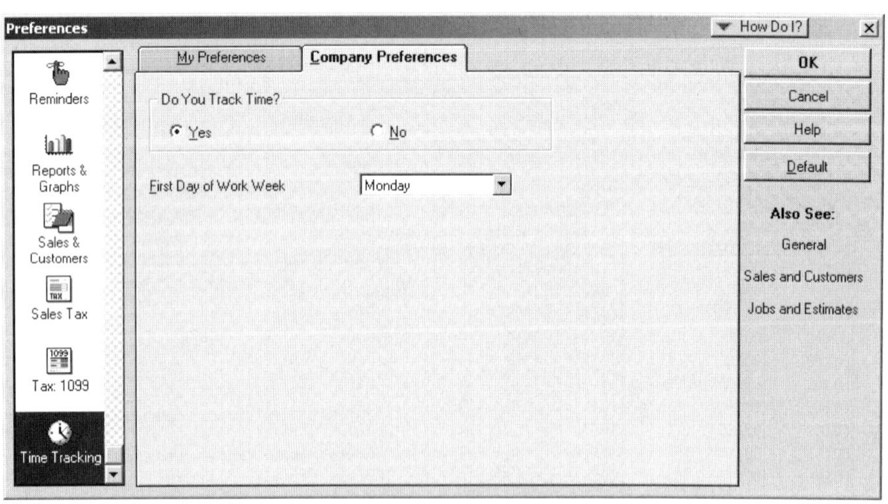

EXHIBIT 13.17 Time Tracking Preferences window.

CHAPTER 14

Entering beginning balances

It is not necessary to begin your new accounting system on the first day of your fiscal year, though it is advisable that you begin on the first day of a new period, either a month or quarter. Regardless of when you transition to QuickBooks Pro 2000, you must have a good set of current beginning balances and year-to-date detail. Decide on a starting date and push to get the detail you need as close to that date as possible.

If you have never established a solid balance sheet and have *any* assets or liabilities, you will need the services of a knowledgeable accountant to help you establish accurate beginning figures.

Transitioning to QuickBooks Pro

Getting opening balances into QuickBooks Pro can be difficult. I *do not* recommend that you follow the QuickBooks Pro start-up routine. Instead I have developed specific procedures for my clients that I have found simplifies a complex task and minimizes the possibility of errors with checks and balances.

To make the transition as simple and accurate as possible you will need the following information when you are ready to begin using your QuickBooks Pro financial management system:

- Accurate balance sheet account balances, including assets, liabilities, and equity
- List of outstanding checks and deposits
- List of outstanding accounts receivable
- List of outstanding accounts payable
- List of 1099 vendors, including their Employer Identification Number (EIN) or Social Security number, complete mailing address, and year-to-date earnings
- List of employees and their year-to-date earnings and withholdings
- Your state unemployment and disability rates, if applicable
- Chart of Accounts
- Item List

TIP This is not a task to turn over to anyone who is not properly trained!

Entering summary balance sheet balances

Entering opening balances for your balance sheet accounts is not difficult, except for those accounts that have functioning subledgers within QuickBooks Pro.

In the Chart of Accounts, create the following temporary accounts:

- Accts receivable(temp)—Account type: Other current asset
- Accts payable (temp)—Account type: Other current liabilities

Begin by creating a journal entry dated the day prior to your new file startup date. For example, if your startup date is April 1, date your journal entry March 31. You will enter the correct debit and credit balances for each of your asset, liability, and equity accounts. Enter the accounts receivable and accounts payable balances to the temporary accounts you just created. (You will be coming back to enter the detail in the following steps so that you will have detailed aging reports and accurate outstanding items in your bank account.) The amount left over should equal your year-to-date net income. If it doesn't, your postings are not yet correct. Find the problem, and when you have the correct leftover amount, post it to the Opening balance equity account.

Entering income statement balances

Next, create a similar journal entry for the balances of the income and expense accounts. The leftover balance also should equal your year-to-date net income. When you have the correct amount left over, post it to the Opening balance equity account. This will zero out the first entry. Once these entries are made (and the opening balance equity is zero), the opening balance sheet accounts are entered successfully in the system.

Establishing an accurate beginning bank balance

There are several steps you will need to take to enter an accurate, detailed beginning bank balance.

Outstanding checks

The first step is to enter all checks that were outstanding as of the start date. We'll create a temporary vendor to accomplish this.

1. In the **Vendors Navigator**, select **Vendors.** Click the **Vendor** button and select **New**. Create a vendor named **zz.outstanding check**.
2. Click **OK**.
3. In the **Banking Navigator**, select **Checks**.
4. At the top of the window select the bank account (see Exhibit 14.1).
5. Change the number of the check to be the number of the first outstanding check. Then enter the date the outstanding check was written, and the amount of the outstanding check.
6. On the **Expenses** tab, select the checking account the outstanding check was written against (the bank account selected in #4 above). QuickBooks Pro will display a warning screen. Bypass the screen by clicking OK. The amount that shows should be the same amount that was entered in the top portion of the window.
7. Click **Save & New**.
8. Repeat steps 5–7 for all of your outstanding checks.

Outstanding deposits

QuickBooks Pro isn't very cooperative about entering outstanding deposits, so I have found a way to fool the system. What you are going to do is create a new temporary account that is a subaccount of your checking account called **O/Sdeposit(temporary)**. You also will need to create a new vendor called **zz.Outstanding deposit**.

Step #1 enters the outstanding deposits. Step #2 fools the system by merging the subaccount with the main bank account.

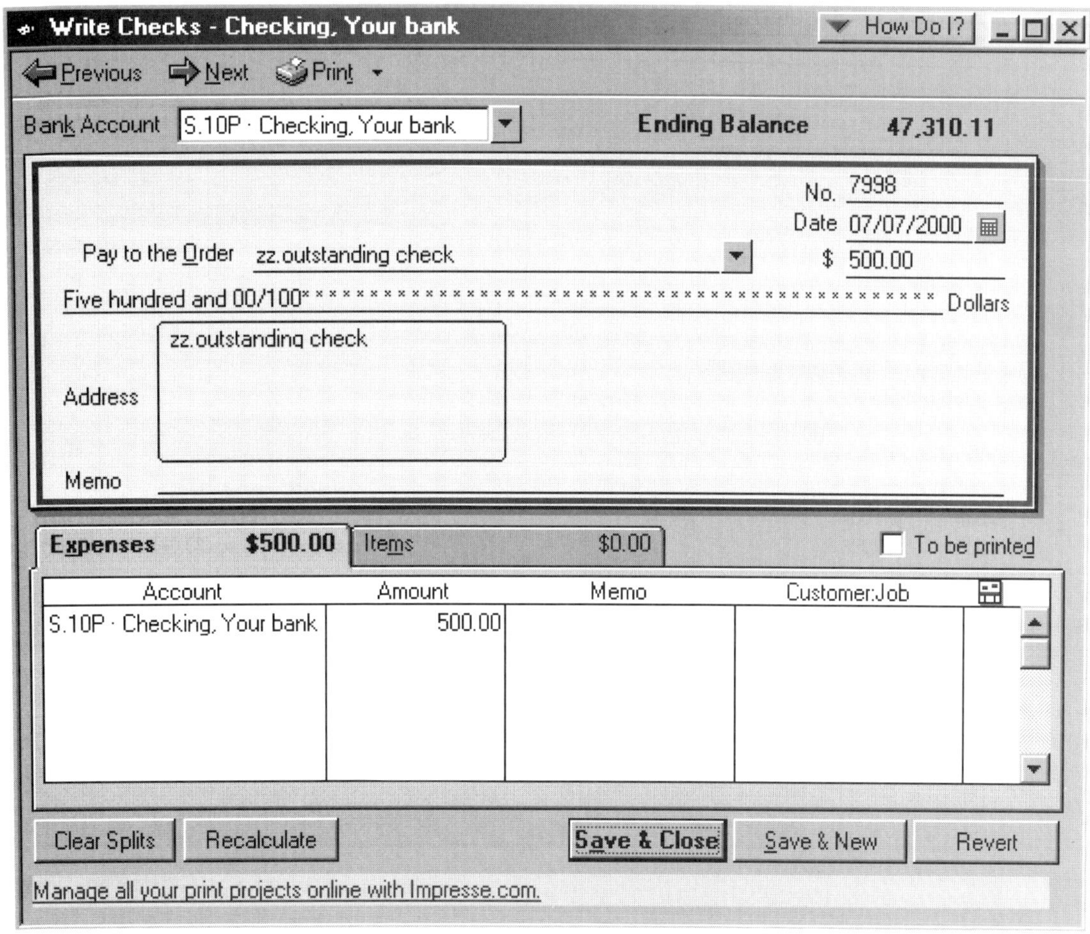

EXHIBIT 14.1 Write Checks window.

Step #1

1. In the **Vendors Navigator**, select **Vendors**. Click the **Vendor** button and select **New**. Create a vendor named **zz.Outstanding deposit**.
2. Click **OK**.
3. Go to **Lists**, click on **Chart of Accounts**, select **New Account** and create a subaccount of your checking account called **O/Sdeposit(temp)**.
4. Click **OK**.
5. In the **Banking Navigator**, select **Deposits** (see Exhibit 14.2).
6. Pick out the bank account and enter the correct date of the outstanding deposit.
7. Indicate that money was received from the **zz.outstanding deposit** vendor in the column titled **Received from**.
8. Tab to the **From Account** column and select the **O/Sdeposit(temp)**.
9. Put in the amount.
10. Click on **Save and New.** Enter the rest of your outstanding deposits.

Step #2

1. In **Lists**, select **Chart of Accounts**.
2. Edit the temporary subaccount **O/Sdeposit(temp)** by changing the number of the account so that it is the same as the main bank account's account number.

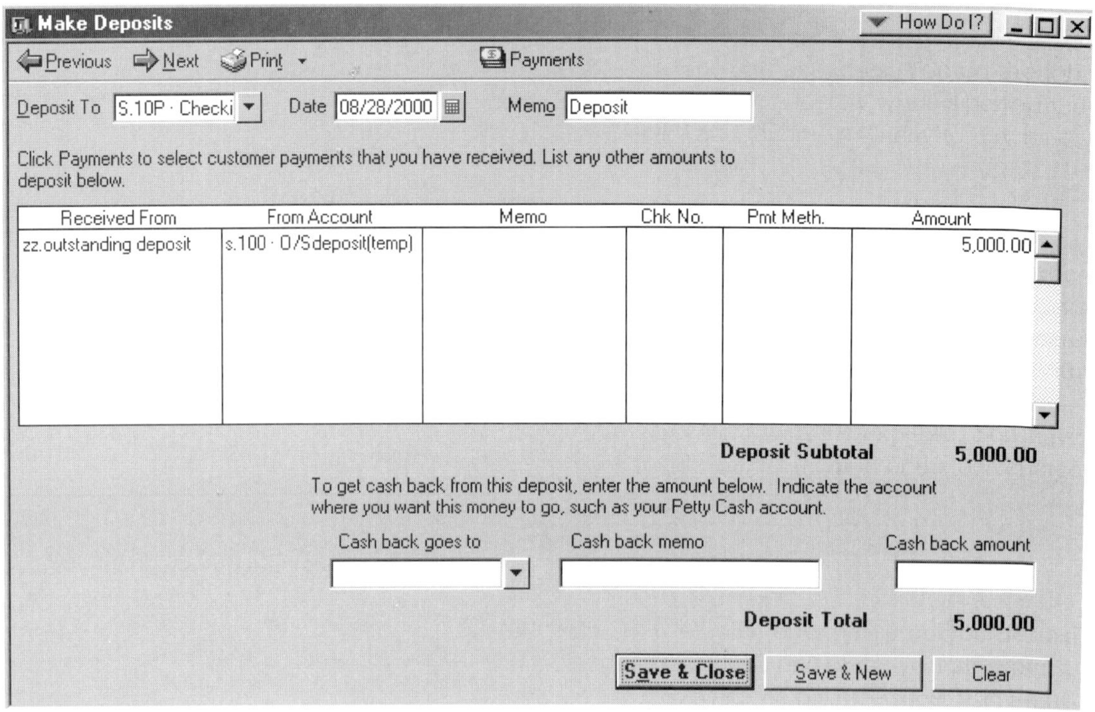

EXHIBIT 14.2 Make Deposit window.

3. Say **OK**.
4. The system will ask if you want to merge the two accounts. Say **OK**. (That's how you fool the system.)

Finalize your cleared bank balance

Follow these steps to ensure that the outstanding checks and deposits are entered correctly.

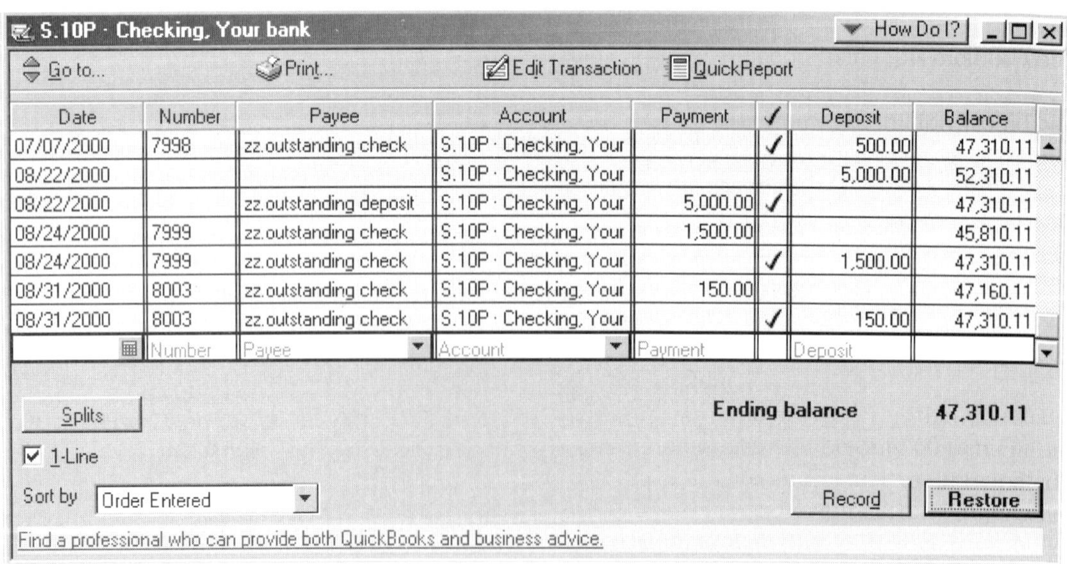

EXHIBIT 14.3 Check Register window.

1. In the **Banking Navigator**, select **Check Register** (see Exhibit 14.3).
2. Manually clear the deposit side of the zz.outstanding check transactions and the payment side of the zz.outstanding deposit transactions. Also, manually clear the total balance you entered in the original journal entry. Your opening cleared bank balance, as shown in the first field of your QuickBooks Pro bank reconciliation, now should be accurate. The list of outstanding items is in the system.

NOTE To manually clear a check, follow these steps.

1. In the **Banking Navigator**, select **Check Register.**
2. Select the bank account. There will be a column between payment and deposit.
3. "Hover" over that column. A checkmark will appear. Click to enter the checkmark.
4. Click on **Record** to save the checkmark.
5. The system will ask if you want to change the transaction. Say **Yes**.
6. In the **Banking Navigator** select **Reconcile**.
7. Pick out the bank account to see the opening balance and compare it to the cleared balance on your bank statement. All the outstanding checks and deposits should be recorded accurately and displayed on the reconciliation screen. Verify that the amounts match. If they do, click the **Leave** button.
8. You may now enter new transactions to your bank account.

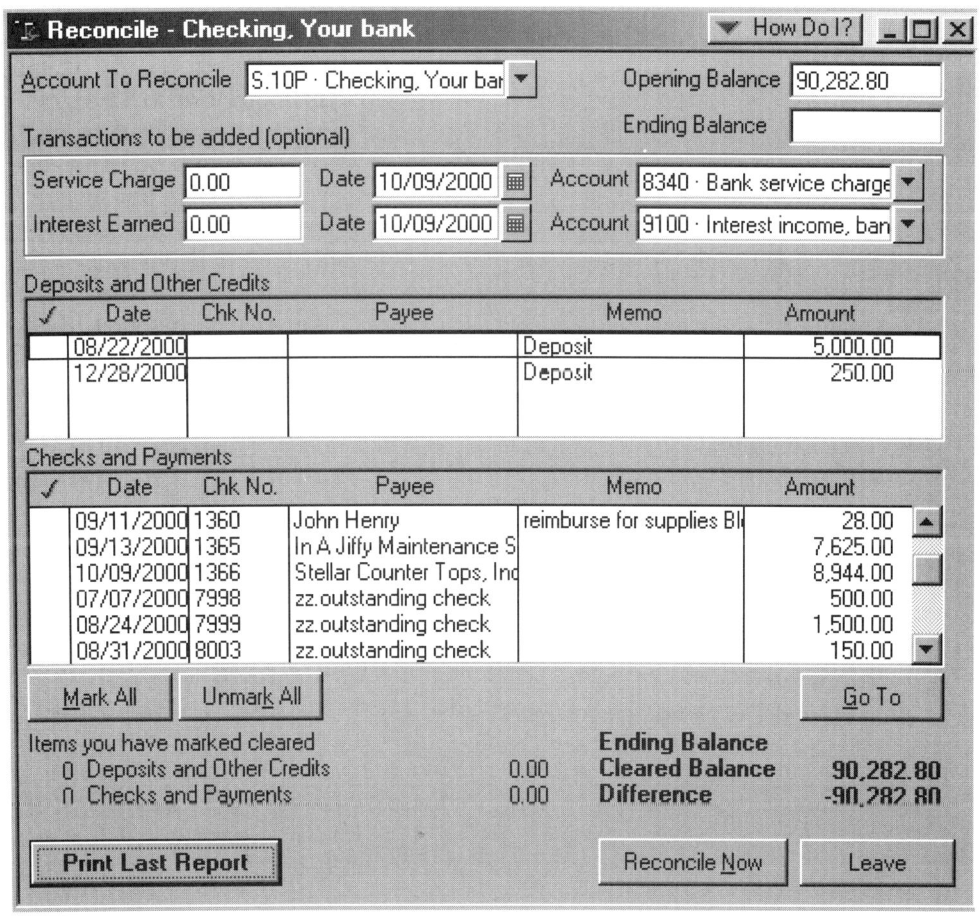

EXHIBIT 14.4 Reconcile window.

Entering accounts receivable

To get your outstanding accounts receivable entered and aged properly follow these steps. Step #1 creates a special temporary item to use to move the opening balance for each of your outstanding invoices from the temporary account into the QuickBooks Pro accounts receivable subsystem. Step #2 enters the outstanding invoices. Step #3 checks that the invoices are entered correctly.

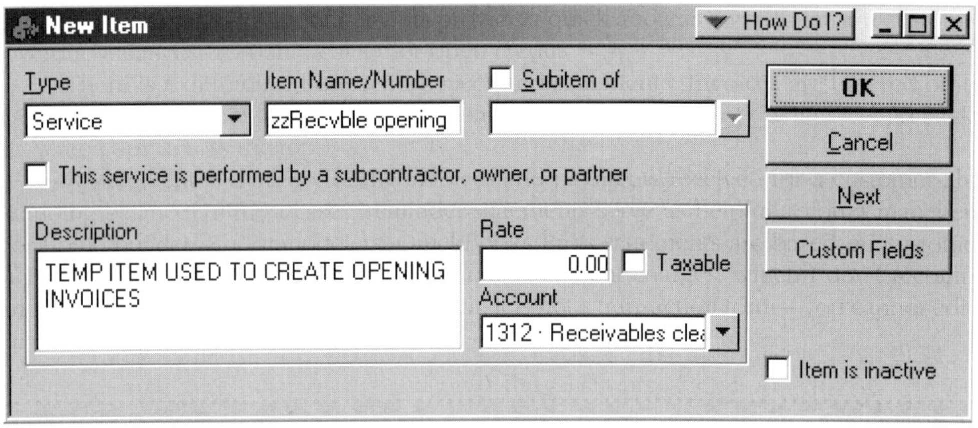

EXHIBIT 14.5 New Item window.

Step #1

1. Under **Lists**, select **Item List**. Click the **Item** button and select **New**.
2. Select **Service** for the item type.
3. Create an item called **zz.Recvble opening**.
4. In the description say, "**Temp item used to create opening invoices**."
5. Do not check the box that says "**This service is provided by a subcontractor, owner, or partner**."
6. Do not make this item a subitem.
7. Leave the Rates box empty.
8. Use account number **1312.Receivables clearing (temp)**.
9. Click **OK**.

Step #2

1. In the **Customers Navigator**, click **Create Invoices** to enter the outstanding accounts receivable (see Exhibit 14.6).
2. Enter the following information:
   - Correct (actual) customer:job
   - Actual date of original invoice
   - Original invoice number
   - Terms for that customer
3. Pick out item **zz.Recvble opening**.
4. Fill in the correct amount of the receivable.
5. Select **Save & New** until you have entered all outstanding invoices.

Step #3

1. Under **Lists** select **Chart of Accounts.** Scroll down to account **#1312 Receivables clearing (temp)**. The account should have a zero balance. A balance indicates that the

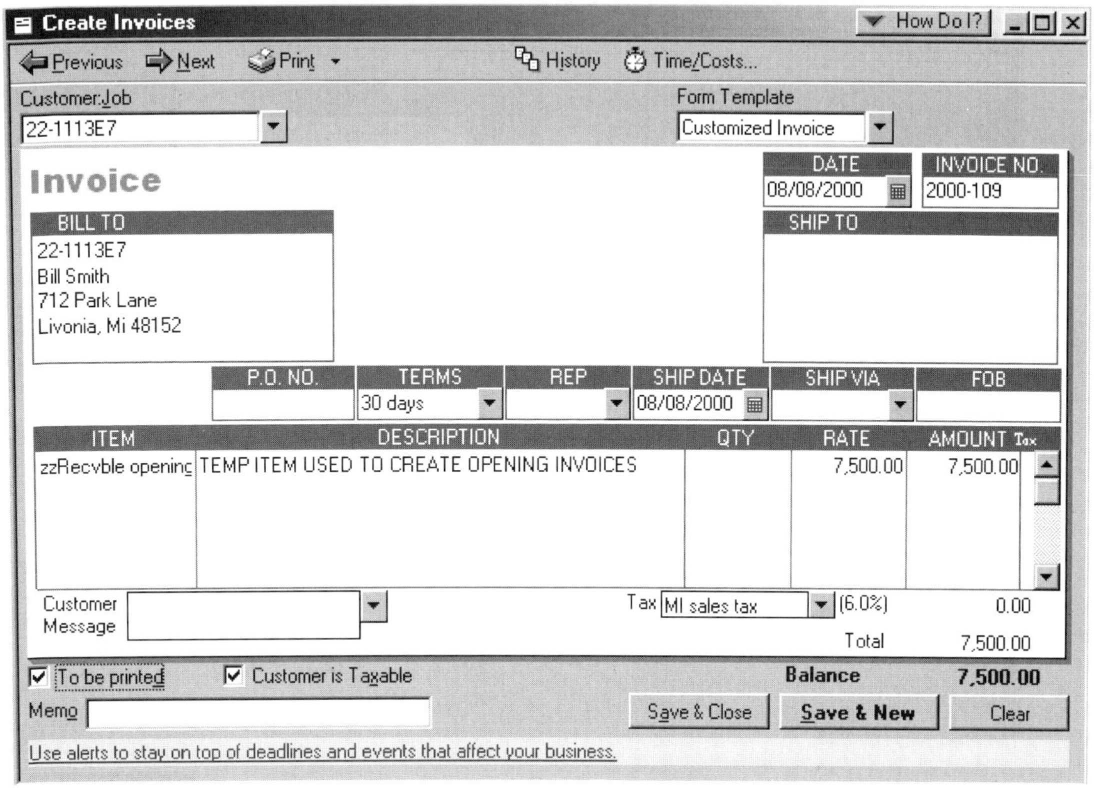

EXHIBIT 14.6 Create Invoices window.

detailed information did not agree with the amount you used in your opening journal entry.

2. Now run an accounts receivable aging report. Set the date to be the start date of your QuickBooks Pro system. The receivables should show in a columnar format by age with rows for each customer:job (see Exhibit 14.7).
3. After you have confirmed that opening balances are correct, inactivate your temporary account and the temporary item.

> **NOTE** All outstanding invoices must be dated on or before the QuickBooks Pro start date for the opening balances to work as intended. For example, if your start date is April 1, the invoices must be dated before March 31.

Entering accounts payable

To correctly enter accounts payable, follow these steps. Step #1 enters the outstanding bills. Step #2 verifies that they were entered correctly.

Step #1

1. In the **Vendors Navigator**, select **Enter Bills** (see Exhibit 14.8).
4. Enter the following information:
   - Actual vendor
   - Date of the bill

Dreamscape Homes & Remodeling, Inc.

A/R Aging Summary

As of March 31, 2002

	Current	1- 30	31- 60	61 - 90	>90	TOTAL
b630 Borgsdorf						
b817 Bailey	12,189.77		17,000.00			29,189.77
b819 Goldman			6,236.00			6,236.00
J707 Tripper		(36.00)			36.00	
J801 Hickory				500.00		500.00
J821 Kuberton			2,500.00			2,500.00
s701 Neidminster						
s802 Schwartz			5,000.00			5,000.00
s803 McMannon-Pierce				(33,000.00)		(33,000.00)
z724 Bailey			5,733.32			5,733.32
z807 Branch		(8,500.00)				(8,500.00)
z823 Anderson			6,570.17			6,570.17
zz.Small jobs						
z850-small jobs		100.00	550.00			650.00
Total zz.Small jobs		100.00	550.00			650.00
TOTAL	12,189.77	(8,436.00)	43,589.49	(32,500.00)	36.00	14,879.26

EXHIBIT 14.7 Accounts Receivable Aging Summary.

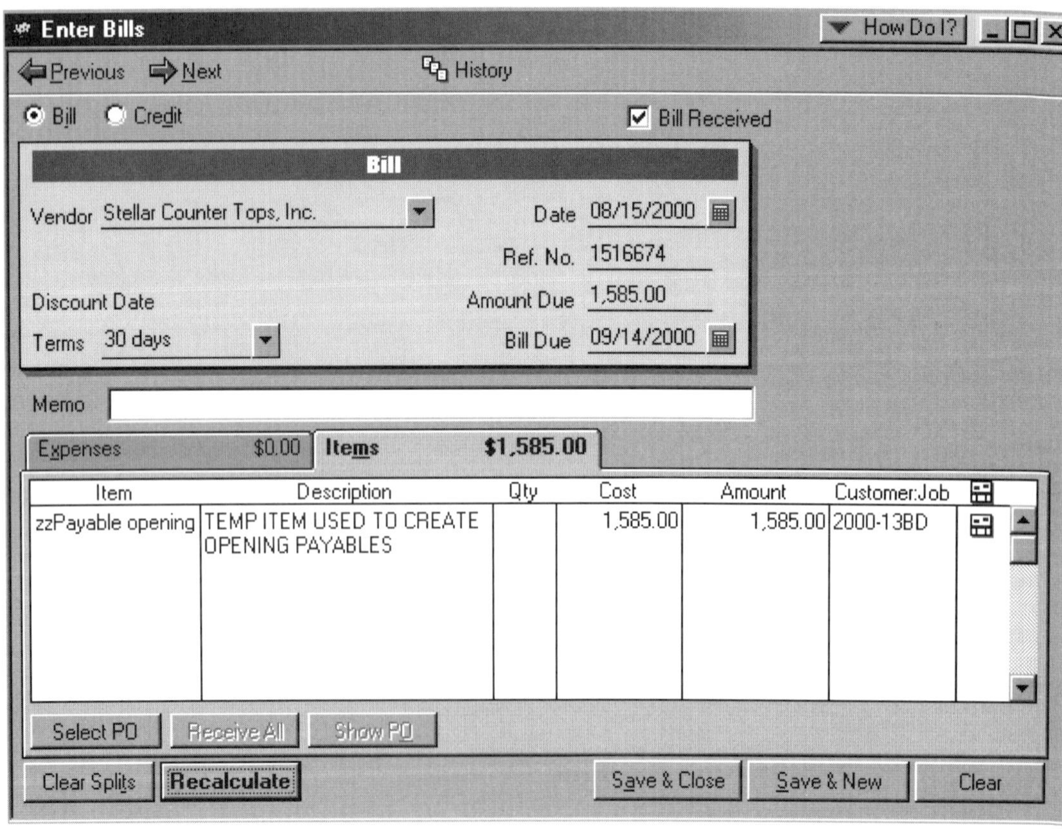

EXHIBIT 14.8 Enter Bills window.

- Vendor's bill number or other reference number
- Amount due
- Terms
- Bill due date

5. Under the **Expenses** tab enter the account **#2400 Payables clearing (temp)**. The amount that is owed will show at the bottom.
6. Don't enter a customer:job.
7. Select **Save & New** until all outstanding bills are entered.

Step #2

1. Under **Lists,** select **Chart of Accounts.** Scroll down to account **#2400 Payables clearing (temp)**. The account should have a zero balance. A balance indicates that the detail entered did not agree with the amount you used in your opening journal entry.
2. You also can run an accounts payable aging summary to see the payables that are owed and the age of the bills.
3. After you have confirmed that opening balances are correct, inactivate your temporary account.

Dreamscape Homes & Remodeling, Inc.
A/P Aging Summary
As of March 31, 2002

	Current	1 - 30	31 - 60	61 - 90	> 90	TOTAL
AT&T Universal MC			4,000.00			4,000.00
Clearwood Cabinet Design		2,466.80				2,466.80
Detroit Edison		63.08				63.08
Downhome Tool & Equipment, Inc.		61.83				61.83
Hometown HBA			24.00			24.00
Hometown Lumber		6,196.12	11,642.16			17,838.28
LKT Wholesale		10.89				10.89
MBNA-Mastercard		1,000.00				1,000.00
Precision Electric Inc.		111.90				111.90
Seamus MacDee, Architect			322.00			322.00
Smiley Electric		809.55	273.45			1,083.00
Wolfstone Glass, Inc.		14.84	1,050.12			1,064.96
TOTAL		10,735.01	17,311.73			28,046.74

EXHIBIT 14.9 Accounts Payable Aging Summary.

Adding 1099 vendors

Any time you deal with employees and 1099 vendors, the cleanest time to transition to a new accounting system is the first of the year. Since this isn't always possible, I have created procedures for entering 1099 vendors at the end of a period, either a month or quarter.

Before you can enter 1099 vendors in the system you must have their Employer Identification Number (EIN) or Social Security number, full address, and an up-to-date list of actual payments you have made to them this calendar year. To enter their year-to-date balances for this calendar year, follow these steps.

1. In the **Banking Navigator**, select **Make Journal Entry** (see Exhibit 14.10).
2. Enter the same date as the opening balance journal entries.
3. Click on the **Account** pull-down list and select the **S.55P.Trade Contractors** cost of goods sold (COGS) account.

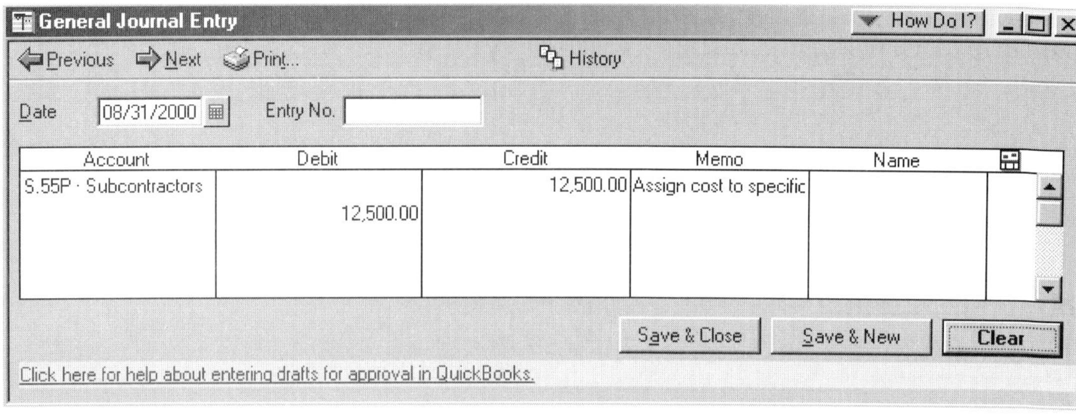

EXHIBIT 14.10 General Journal Entry window.

4. Enter the total amount paid to all 1099 vendors as of the transition date. For this example, you have five 1099 vendors to whom you actually have paid a total of $12,500 this calendar year-to-date.
5. Enter the $12,500 as a credit to the **S.55P.Trade Contractors** account.
6. Under memo say, "Assign cost to specific 1099 vendors."
7. Enter nothing under name.
8. In the **Make Journal Entry** window, tab to the next line and pick out the same account.
9. Enter as a debit the amount paid to the first 1099 vendor on your list, for example $5,000.
10. Tab over and enter the vendor's name in the name field. Continue this process for each vendor. When all vendors are entered, the grand total of all debits should equal the initial credit amount.
11. Click **OK** when all entries are made.

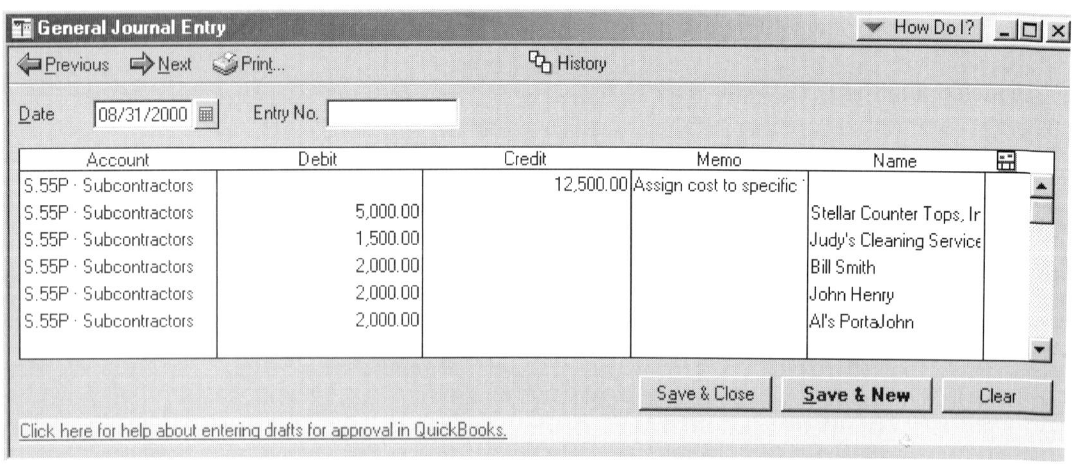

EXHIBIT 14.11 General Journal Entry window with debits matching credits.

SECTION THREE

THE BUILD PHASE

Now that you've created your Chart of Accounts, Item List, Customer:Job List, and the many other lists that frame your accounting system, it's time to fill these "buckets" with information. The following chapters walk you through the processes of:

- Creating estimates
- Generating purchase orders
- Entering and paying bills
- Sending invoices
- Receiving payments
- Reconciling your accounts

Each of these processes takes place within one of QuickBooks Pro 2000's Navigators—Customers, Vendors, or Banking. The functions that can be performed within each Navigator are listed below. An icon has been placed next to the how-to instructions to indicate the Navigator.

Customers Navigator:

- Create invoices and estimates
- Enter cash sales
- Create credit memos or refunds
- Generate statements
- Receive payments
- Accept credit card payments
- Apply finance charges
- Make deposits
- Write letters
- Access and/or modify Item List

Vendors Navigator:

- Enter and pay bills
- Pay sales tax
- Create purchase orders
- Receive items
- Adjust inventory
- Print 1099s
- View the Purchase Order List
- Access and/or modify the Item List

Banking Navigator:

- Write checks and view the check register
- Make deposits
- Transfer funds
- Enter credit card charges
- Reconcile accounts and make journal entries
- Set up and access online financial and banking services
- Access and/or modify the Chart of Accounts.

The processes within these Navigators represent tools that you can use to your advantage if you set up and use your system properly. All these tools will work together to create the final product—your management reports. Just as you subdivide your construction projects into manageable phases, it's equally useful to subdivide the major financial aspects of your business into the primary components of buying, selling, and banking. These subdivisions, especially when shown on the Navigator flowcharts, make it easier to conceptualize the flow of information and learn the accounting processes.

CHAPTER 15

Creating estimates

Estimates serve two purposes. From a business perspective, estimates allow you to plan for the costs and time it will take to complete a job. Estimates help you price the job so that you are compensated for your time, knowledge, and effort at a rate that keeps your business operating. From an accounting standpoint in QuickBooks Pro 2000, estimates are the first step to an accurate and detailed job-cost system. Information you will learn from reports that are estimate vs. actual based may enable you to fine-tune your estimates in the future to become more profitable and competitive.

In QuickBooks Pro estimates are created using items from the Item List you created in chapter 10. Your computerized estimate then potentially can be converted into invoices against which you receive and deposit payments.

If you set up your Item List to mirror the processes and broad categories of expenses associated with your typical development, remodeling, or home building jobs you will be able to create estimates in QuickBooks Pro that mirror your basic hand-written or spreadsheet-type estimates. For example, under the Site preparation item you could create a subitem for direct labor at 10 hours with a quantity of 3 people and a rate of $7 and another subitem for direct labor at 8 hours with a quantity of 2 people and a rate of $10.

The bottom line, however, is that if you have an effective estimating program, either a spreadsheet or a commercial estimating program, that program probably is the best place to work out the details of your estimate. To simplify the process of entering your estimates in QuickBooks Pro set up the items in the Item List so that they roll up under the same title names as your spreadsheet or estimating program. For example, if your estimating program has a title called Site preparation, create a title item in the Item List named Site preparation. If you have subheads under Site preparation in your estimating program create the same title item subheads in the Item List. For example, you may have subtotals for tree removal, grading, and fill dirt in both programs. This makes it easier to transfer your estimated dollars from one program to the other.

I agree that this appears to entail some duplicate entry work. The benefits of having an estimate from a quality estimating program *and* entering the estimate information in QuickBooks Pro, however, is worth the few minutes it will take to re-key the data. Unfortunately I have not found an easy way to import estimates from either an Excel spreadsheet or a commercial estimating program into QuickBooks Pro without considerable effort. This is one area where data entry is currently required.

Tracking job-cost information

So, do I hear you asking why enter estimates into QuickBooks Pro, if QuickBooks Pro isn't the best place to create estimates? The most compelling reason to do so is job costing. If you enter your estimates using items in the Item List you will be able to create management reports that will:

- Tell you how the job is progressing
- Let you track actual costs against your estimated costs
- Show you where the differences are
- Automate your invoicing procedures and assist in purchasing procedures

Even if you do not give quotes or estimates to your customers, I strongly recommend that you enter your estimate in QuickBooks Pro so that you can get management information from the job-cost reports. Go back to the control loop in chapter 1. Estimates set the standard—the first step in the process. They form the basis for the strategic planning that affects your company's bottom line.

TIP If you aren't doing detailed estimates now, start by selecting the items from the Item List that represent each stage of your job, and *then* build an estimate in QuickBooks Pro. This will represent a major step you can immediately take towards managing your job costs.

Creating an estimate

To create an estimate, follow these steps.

1. In the **Customers Navigator**, click on the **Estimates** icon or under the **Customers** menu select **Create Estimates** (see Exhibit 15.1).
2. Select the appropriate form from the Template Form list and enter information as appropriate. Many of these fields are self-explanatory. If you need help, click on the **How Do I?** Help button and select **Create Estimates**.
   - **Customer:Job**—Select a customer or job from the Customer:Job List.
   - **Name/Address**—After you select a customer or job, the default name and address assigned to the customer appears in the Name/Address field. You can make changes to the information or accept it as is.
   - **Date**—Type the date of the estimate.
   - **Estimate #**—QuickBooks Pro assigns the next available estimate number. You can make changes or accept it as is.
   - **Item**—Select item(s) from the Item List.
   - **Description**—After you select an item, the description assigned to the item appears in this field. You can make changes to the information or accept it as is. Changes do not change the default description you have assigned to the item in the Item List.
   - **QTY**—Type a quantity, if applicable.
   - **Price**—After you select an item, the price assigned to the item appears in this field. You can make changes to the information or accept it as is.
   - **Amount**—The amount is calculated automatically based on the information in the QTY and Price fields. This represents the total cost.
   - **Markup**—After you select an item you can enter a markup in dollars or percentage, if you are using a cost-plus estimate.
   - **T**—A **T** appears automatically if the item is taxable. In order to calculate and add tax, however, the customer also must be taxable. A retail or handyman business potentially might encounter taxable situations. Check with your accountant for your state's sales tax requirements.
   - **Customer Message**—Use or create a message from the pull-down list to convey marketing information, send a thank you, or alert the customer to late fees for late payment.
   - **Tax**—Select the correct sales tax item to attribute the correct tax or non-tax status to this job.

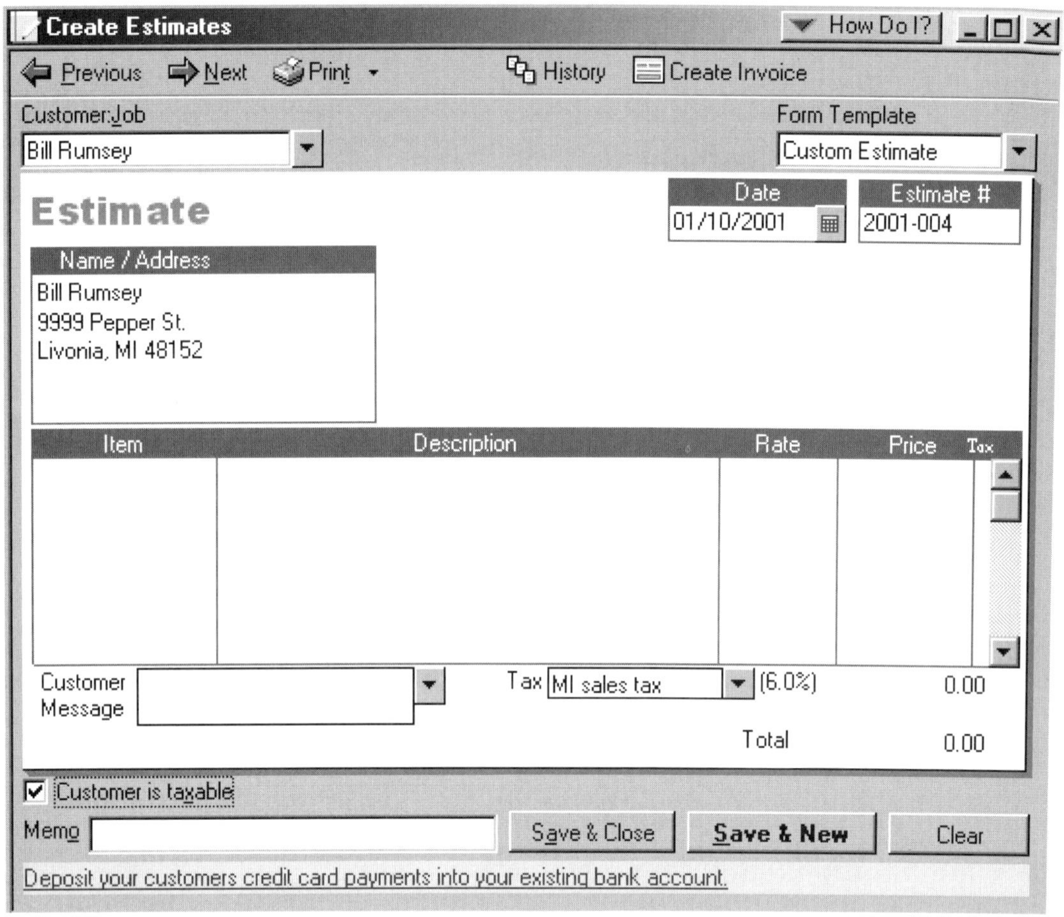

EXHIBIT 15.1 Create Estimate window with top information filled in.

- **Customer is Taxable**—After you select a customer or job, the default tax information from the Customer:Job List appears. You can make changes to the information or accept it as is.

> **NOTE** When you set up your Item List and Vendor List you can indicate whether or not items or vendors are taxable. If you never encounter situations where you charge customers sales tax, say **No** to the preference: **Do you charge sales tax?** in the Sales Tax preferences. In unique situations you can edit the invoice to indicate that an item is taxable. See your tax advisor about the implications of this change.

2. Click **Save & Close** to create the estimate.

Distinguishing between estimate types

You want your estimate to reflect the way you are going to invoice. Let's look at two types of jobs—fixed-price and cost-plus (time and materials).

Fixed-price jobs

With a fixed-price job you don't invoice for individual items. Instead you charge a single price for the job as a whole. As an example, let's use a fixed-price home building contract for $225,000. Under the terms of the contract $75,000 is due before work starts, $75,000 is due half way through the job, and $75,000 is due at completion. During the life of the job you will be generating three invoices to the customer.

TIP Use the tab key to quickly move through the estimate form.

To create a fixed-price estimate, follow these steps.

1. Click on the **Items** arrow on the estimate form and select the correct items from the Item List. For our example, select the group of income items called Custom homes under the Income title item. Within this group, select items Draw D, Draw E, and Draw F. The items you select will appear on the Create Estimates window and populate your estimated income.
2. In Draw D do not enter a quantity, cost, amount, or markup. Tab to the far right column called Total and enter $75,000 for the amount of the first invoice.
3. Repeat the step above for items Draw E and Draw F.
4. Lay in your estimated expenses by clicking on the **Items** button and selecting the cost items that match the single-line items or subtotals on your estimate. For example, select Exterior site preparation and enter $3500 in the cost column. Select Exterior site preparation materials and enter $700. Continue until all cost items are entered. When you enter the cost items the Amount and Total columns will show the amount of the item.

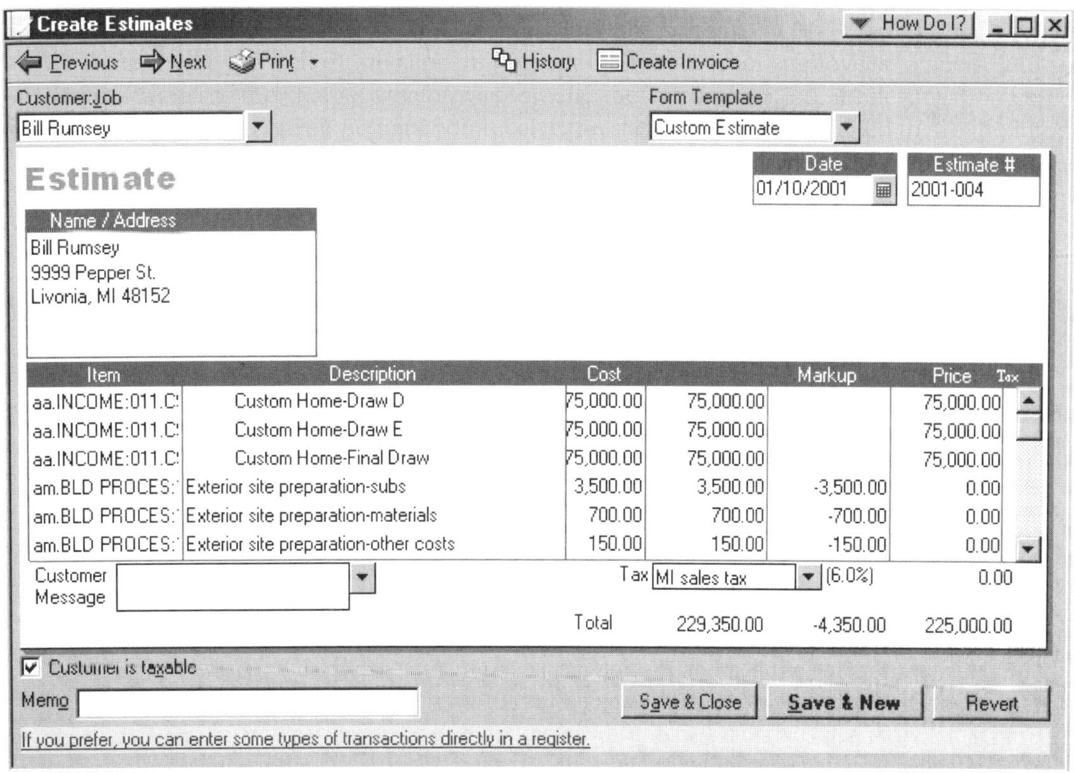

Item	Description	Cost		Markup	Price	Tax
aa.INCOME:011.C:	Custom Home-Draw D	75,000.00	75,000.00		75,000.00	
aa.INCOME:011.C:	Custom Home-Draw E	75,000.00	75,000.00		75,000.00	
aa.INCOME:011.C:	Custom Home-Final Draw	75,000.00	75,000.00		75,000.00	
am.BLD PROCES:	Exterior site preparation-subs	3,500.00	3,500.00	-3,500.00	0.00	
am.BLD PROCES:	Exterior site preparation-materials	700.00	700.00	-700.00	0.00	
am.BLD PROCES:	Exterior site preparation-other costs	150.00	150.00	-150.00	0.00	

EXHIBIT 15.2 Create Estimate with fixed-price display.

5. Because this is a fixed-price contract that should not show the costs of individual items, change the Total column amount to $0 by doing one of two things.
   - Subtract $3500 from the Markup column, or
   - Enter $0 in the Total (price) column. (QuickBooks Pro will enter the negative amount in the Markup column.)

TIP The latter option is the quickest if you are entering several items. Enter the expense for each item in the Cost column and then move down the Total column and enter $0 for each item.

6. Click **OK** when you are finished.

If you create your fixed-price estimates using these steps, only the total estimated income, or price, will print on the estimate. The total cost will show only on the window.

Cost-plus contracts

Follow the same steps as those given for creating a fixed-price contract, except skip steps 1-3. Select the specific items you will use for the job, and enter an amount in the Cost column.

To calculate the final price you are going to charge the customer, you may either:

1. Enter the amount you are going to charge the customer in the Total (price) column, or
2. Indicate a markup in the Markup column.

If you want to use the Markup column to have QuickBooks Pro compute the price you will charge the customer, you may enter either actual dollars or a percentage in the Markup column and QuickBooks Pro will calculate the Total (price).

TIP If you always mark up direct labor 40 percent, for example, select each of the direct labor items one after another. Then you can move down the Markup column and quickly enter 40% for all the direct labor items.

Editing an estimate

I normally don't recommend that you edit any line items in an estimate. If you change an estimate after you give it to a client, how do you know which version is correct? Instead of changing a line item in your estimate, I recommend that you create add-ons or deletions at the bottom of the estimate as follows.

- Use a Subtotal item to create a subtotal at the end of your original estimate.
- Arrow down several lines to use a blank line and the description field to create a descriptive title for the detailed changes to follow. For example, you may note something like CHANGE ORDER #1 DATED JUNE 24 (see Exhibit 15.3).
- Arrow down and enter the items that make up the change order.

This way you have the original and any changes noted and explained separately.

1. Under **Customers** select **Customer:Job List** or on the **Customers Navigator**, click the **Customers** icon.
2. Highlight the customer or job for which you want to edit the estimate by clicking once on the name. Select **View Estimate** from the **Customer:Job** button.

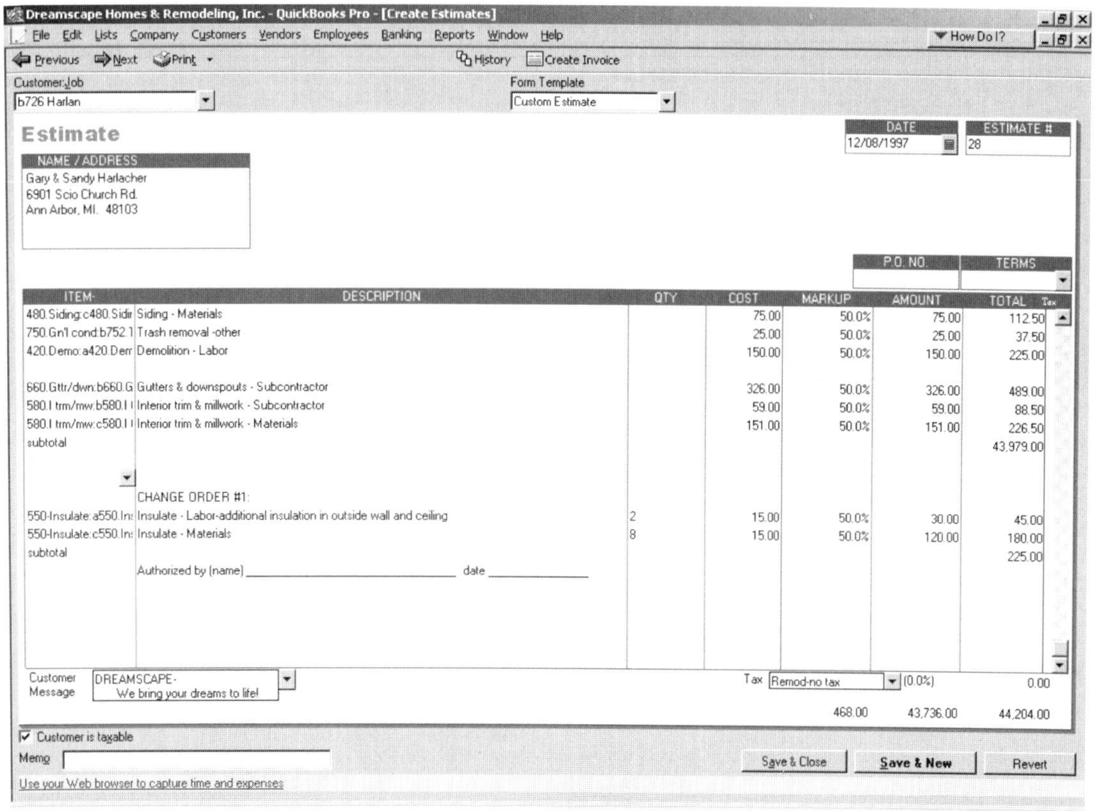

EXHIBIT 15.3 Edit Estimate window.

3. When you have finished making changes click **Save & New** to save the changes and then create a new estimate, or click **Save & Close** to save the changes and return to the Customers Navigator window.

Using job-cost reports

Once you've begun using estimates, you can run job estimate vs. actual reports that will show you where you are in the invoicing process if you are using fixed-price invoicing. More importantly, the report will show you where your areas of exposure are on the job. If you are working on a cost-plus contract you will see the actual costs in the left-hand column of the report. The two right-hand columns will show the estimated income and the actual amount you have invoiced, respectively. So, if you set up items properly and create estimates that are set up in the same way you will be invoicing, you can see how you actually are doing against your estimate at any point in the process.

Customizing estimate templates

The estimate templates in QuickBooks Pro are easily modified to accommodate your logo (it must be a bitmap file—.bmp—format). You also can change fonts, lines, titles, size, and so on. You can change the name of your estimate to a proposal by editing the form. With a little bit of effort you also can line things up so the estimate will print on your preprinted proposal form.

Your templates determine what information from the estimate will be printed on the form and what will show only on the screen. Or, one template may show only cost columns from the estimate, while another shows only the income columns. You might want multiple internal templates, and one or more external templates in which you choose to present different information depending on the recipient.

> **NOTE** You can customize any QuickBooks Pro form from the form screen by selecting **Customize** from the Form Template field, or by using the Templates icon in each Navigator window. **Refer to the How Do I?** Help button on the Templates list.

To create a new estimate template, follow these steps.

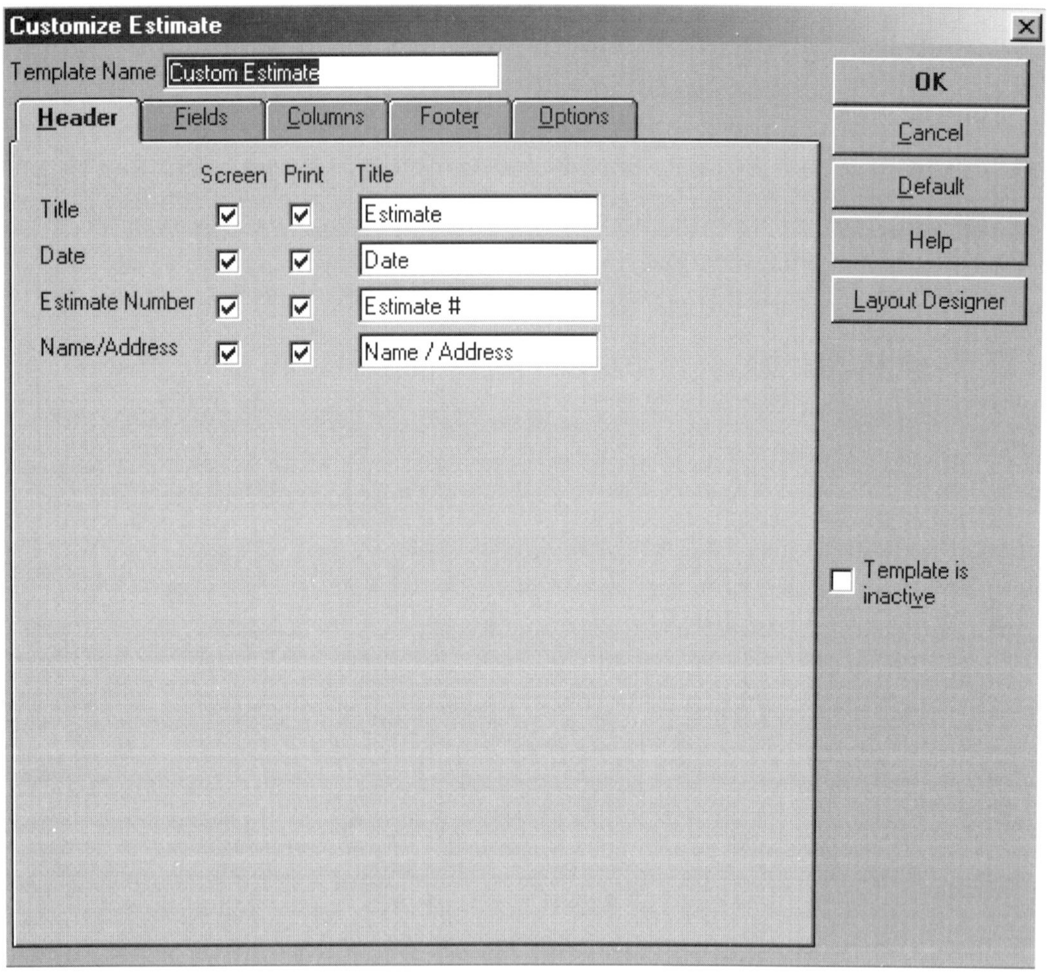

EXHIBIT 15.4 Customize Estimate window.

1. Under **Lists** select **Templates** or in the **Customers Navigator**, click on the **Templates** icon.
2. Under the **Template** button select **New** and then **Estimate**.
3. Enter a name for your template in the **Template Name** field.
4. Move through the five tabs on the Customize Estimate window to customize your estimate.
5. Click **OK** when you are finished.

TIP QuickBooks Pro has a special feature that can be especially useful for spec home builders and developers who build the same models multiple times. Let's say you create an estimate for the Brookstone model. Name the estimate Brookstone and memorize the estimate. Once a template is memorized you can pull it up and assign it to a new job without having to re-enter all the information. For example, if the new job includes a deck, you can edit the estimate to include items for the deck, without having to recreate the rest of the estimate.

NOTE You can only have one estimate per job. So if you want to have an alternative scenario for 1234 Church Street-Kitchen, for example, create another job and name it 1234 Church Street-Kitchen 2. Then create the alternative estimate. When a decision is made, delete the estimate, and then the job, that won't be used and continue on.

Printing an estimate

1. Select the estimate you want to print.
2. Click **Print**.
3. Click **Save & New** to save the current estimate and create another, or click **Save & Close** to save the current estimate and return to the previous window.

CHAPTER 16

Generating purchase orders

Purchase orders can be an important management tool. In addition to clarifying and documenting exactly what it is you want to purchase, purchase orders tell your vendor what it is you agree to purchase, at what price, for what project, due dates materials or service are needed, quality specifications, and delivery information. Any misunderstandings can be resolved at the paperwork stage—before too much of one item or not enough of another is delivered to the jobsite.

Purchase orders also are legal documents. Effectively laying out your purchase orders can help define your relationship with a vendor. Language on your purchase orders should specify:

- Delivery date(s)
- Quality of work or materials
- Standards that must be met
- Who is to perform the work or manufacture a product
- Price the vendor will charge
- Confirmation requirements and time periods that a vendor must follow
- Other legal terminology required by your attorney

Preparing purchase orders

In addition to these important management reasons for using purchase orders, there are three accounting reasons to prepare purchase orders in QuickBooks Pro 2000, even if you don't send them to your vendors.

> **TIP** Immediately after you are awarded the job, while you are still closely in touch with the cost details from your estimating process, create purchase orders for every vendor that will be involved, even if you are not going to actually issue the purchase order to the vendor. For example, create a purchase order for a permit from the city, though it is unlikely you actually will mail them the purchase order. You will soon see why this is worthwhile.

Bill entry and job-cost details

Say, for example, you get a bill that says lumber was dropped off at 1234 Question Lane. The bill doesn't indicate the total quantity of lumber dropped off or how much of the lumber relates to flooring, decking, or framing. How is your accounting assistant supposed to break down the bill for job-cost purposes? At a minimum the phone calls and time spent researching the bill will cost you money. Even if information is received it may not be accurate and might be posted using the wrong job or item numbers by the bookkeeper, who might not understand the building process.

If you have a purchase order on the system, however, when your accounting assistant begins to enter a bill for a vendor who has open purchase orders, the system will cue him or her to enter the bill against the purchase order. This enables the transfer of all purchase order information to the bill form with the click of a button. The job name, item, job stage, and other pertinent information are now all in your job-cost system, exactly as you intended, minimizing the possibility of errors and dramatically reducing research time.

Price documentation

The second reason to create a purchase order every time you agree to buy materials or services is that the purchase order documents the price for your accountant. He or she will be able to see if you are being charged an amount that is different from the anticipated or agreed upon price. If the amount is different your accountant can make the appropriate person aware of the difference.

I recommend that you implement an internal policy regarding variances using either a dollar amount or percentage. If the total on the bill exceeds the allowed variance, then management must approve payment. This creates another control in the control loop and provides immediate feedback to the owner.

TIP By creating an item in the Item List called Variance from Purchase Order under each item group, you can track and measure all differences between the purchase order and the bill. Enter all amount differences under these variance items, and when you run a management report all the variances will stand out. Industry experts estimate that companies that check the dollar amount on the actual bill against the purchase order can increase their bottom line from 5 to 7 percent.

Outstanding orders

The third way your purchase order system can save you time and stress is to help you monitor outstanding purchase orders. You regularly can review an outstanding purchase order report, organized by job if you wish, with due dates included, so that you can place reminder, monitoring, or confirmation calls to vendors before a materials shortage brngs your job to a halt.

Creating a purchase order

To create a purchase order, follow these steps.

1. In the **Vendors** menu select **Create Purchase Orders** or on the **Vendors Navigator** click on the **Purchase Orders** icon (see Exhibit 16.1).
2. Enter information in the following fields, as applicable. Many of these fields are self-explanatory. If you need help, click the **How Do I?** Help button, and select **Create a Purchase Order.**

TIP Use the tab key to quickly navigate through the purchase order.

- **Vendor**—Select a vendor from the pull-down list or select **Add New** to create a new vendor.

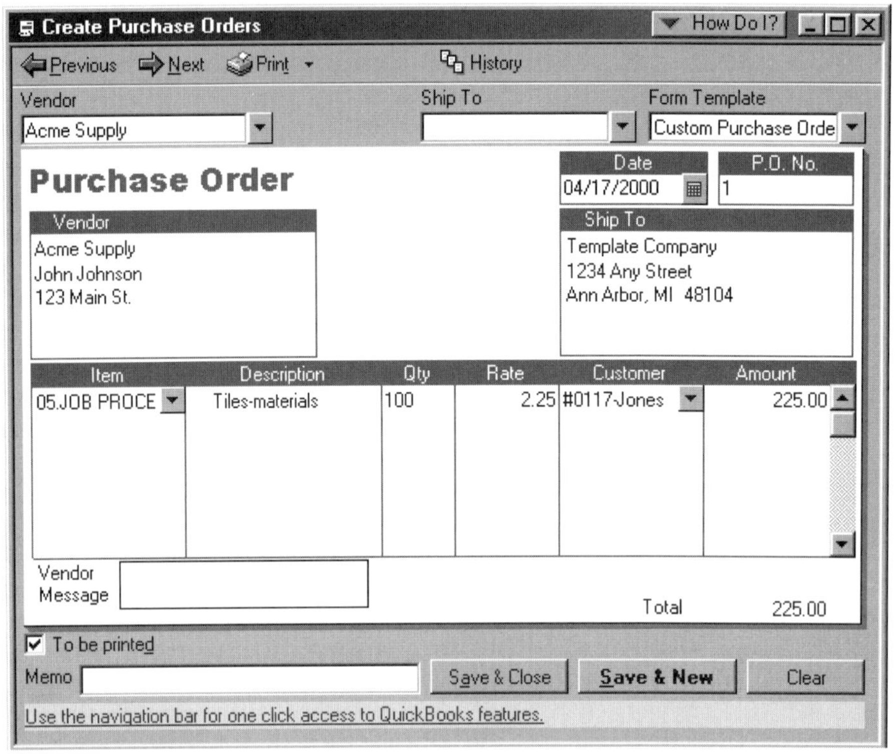

EXHIBIT 16.1 Create Purchase Orders window.

- **Vendor box**—The default name and address information for this vendor appears here. You can make changes to the information or accept it as is.
- **Date**—Type the date of the purchase order.
- **Purchase Order #**—QuickBooks Pro assigns the next available number. You can change the information or accept it as is.
- **Ship To**—Your business address appears as the default in this field. If you wish you can modify this information. You also can select the jobsite or another location by selecting a name form the pull-down list.
- **Item**—Select an item from the Item List or type in an item name.
- **Description**—After you select an item, the default description appears in this field. You can change the information or accept it as is.
- **Quantity**—Type a quantity, if applicable.
- **Rate**—After you select an item, the default amount that you pay for the item appears in this field. You can change the information or accept it as is.
- **Customer**—Select a customer or job to attribute this purchase to. You can choose for this information to not appear on the printed purchase order when you customize your template.
- **Amount**—The amount is calculated automatically based on the information in the Quantity and Rate fields.
- **Vendor Message**—Type a message in this field such as, "Please note the new ship-to address," or "Please confirm the availability and pricing of each item within 24 hours by faxing to (fax number)."
- **Memo**—This memo box at the bottom of the purchase order is the line that appears in the Purchase Order selection window when you are entering the associated bill. The memo also can be displayed on the open purchase orders by job report. Therefore, this is a good place to include additional identifying information, especially if the vendor's bill won't include your purchase order number.

3. To print a copy of the purchase order, click **Print**.
4. When you are satisfied with your purchase order, click **Save & New** to save the current purchase order and create another, or click **Save & Close** to save the current purchase order and return to the previous window.

Editing a purchase order

With QuickBooks Pro, you can easily edit the purchase order following these steps.

1. In the **Vendors** menu, select **Purchase Orders List**, or on the **Vendors Navigator**, click on the **PO List** icon.
2. Find the purchase order number you want to edit, and click on it twice to open the purchase order form.
3. Make the required changes.
4. When you are satisfied, click **Save & New** to save the changes and create a new purchase order, or click **Save & Close** to save the changes and return to the previous window.

Designing purchase order templates

QuickBooks Pro allows you to customize your purchase orders by creating different templates. You can add your logo, change the font, size, and layout, and even align the template so it prints on your preprinted purchase order forms.

A common error for remodelers and custom home builders is the removal of the client's name from the purchase order. Think about the information you want your vendors to have, as well as the information you need back from the vendor on his or her invoice, and design your purchase order template(s) accordingly. Remember that you can select the information you would like to print, and that which you only want to see on your screen.

TIP

You can customize any QuickBooks Pro form using the Templates function in each Navigator window. Refer to the How Do I? Help button on the Templates list or consult the QuickBooks Pro User's Manual for instructions on how to customize QuickBooks Pro forms.

Benefiting from purchase order tricks

After working with remodelers and custom and spec home builders for a number of years I have come up with a few QuickBooks Pro tricks that relate to remodeling and building activities.

Multiple purchase orders to the same vendor

If you are purchasing different items from the same vendor over a period of time for one job you can prepare all the purchase orders at one time. Use your estimate to guide you in the creation of the individual purchase orders with the specific items and delivery dates. Then print out all the purchase orders and mail or fax them on the appropriate dates. Or you may leave the purchase orders on the system and print them just before the mail date by clicking on the To Be Printed box. Either way set up a reminder system so you don't forget to send the purchase order in time for the order to be assembled and delivered.

Repetitive purchase orders

Developers and spec home builders can memorize purchase orders for a model home to re-use later. Once you finalize all the purchase orders for the Winchester model, for example, memorize the Winchester purchase orders under the Winchester memorized transaction group. The next

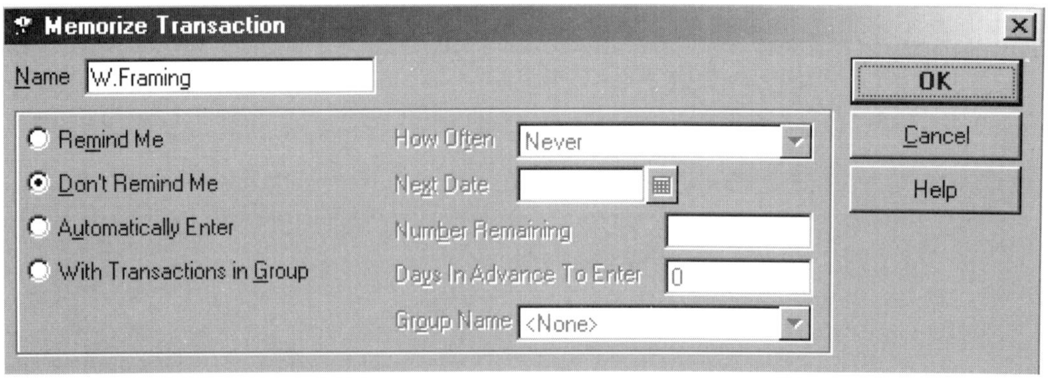

EXHIBIT 16.2 Memorize Transaction window.

time you need purchase orders for a Winchester model all you need to do is post the transaction group of memorized purchase orders and modify them as necessary.

> **TIP** To memorize a purchase order, while you have the purchase order up on the screen, go to Edit, then Memorize Purchase Order. Give the memorized purchase order a name, such as W.Framing (Winchester framing materials), not a specific job name. Click on Don't Remind Me as shown in Exhibit 16.2. Then click OK. Now go to your memorized transaction list and double click on W.Framing. The W.Framing purchase order will appear in the window.

Special job number

When you access memorized purchase orders, such as those in the Winchester group, the job number assigned to the purchase order when you memorized it will appear. Therefore, when you memorize your initial set of purchase orders for the Winchester model home, give all the line items in the purchase orders the job number zz.NewPOs. The zz prefix will force the name to the bottom of your Customer:Job List.

QuickBooks Pro then can post the entire Winchester group of memorized purchase orders. Then, run an open purchase orders report and filter down to the job name zz.NewPOs. You will be able to see all the purchase orders you just created. Double-click on each one and enter the correct job name. Then click Save & Close. The transaction is now assigned to the actual job. You still have to touch the purchase orders, but you don't have to create them from scratch, as shown in Exhibit 16.3.

Printing vendor mailing labels

QuickBooks Pro allows you to produce mailing labels to streamline sending statements or groups of invoices. Consult the QuickBooks Pro User's Guide or click on the How Do I? Help button for mailing label specifications.

1. In **File** select **Print Forms** and then **Mailing Labels**.
2. In the Select Mailing Labels to Print dialog box (see Exhibit 16.4), complete the following fields, as applicable. Many of these fields are self-explanatory. If you need help, click the **How Do I?** Help button.
   - **Name**—Select a name or a group of names from the list.
   - **Customer Type**—Select all or specify one type from the list.
   - **Vendor Type**—Select all or specify one type from the list.
   - **With Zip Codes**—You can further sort the list by specifying names or types with a certain zip code.
   - **Sort mailing labels by**—Select a method from the list.

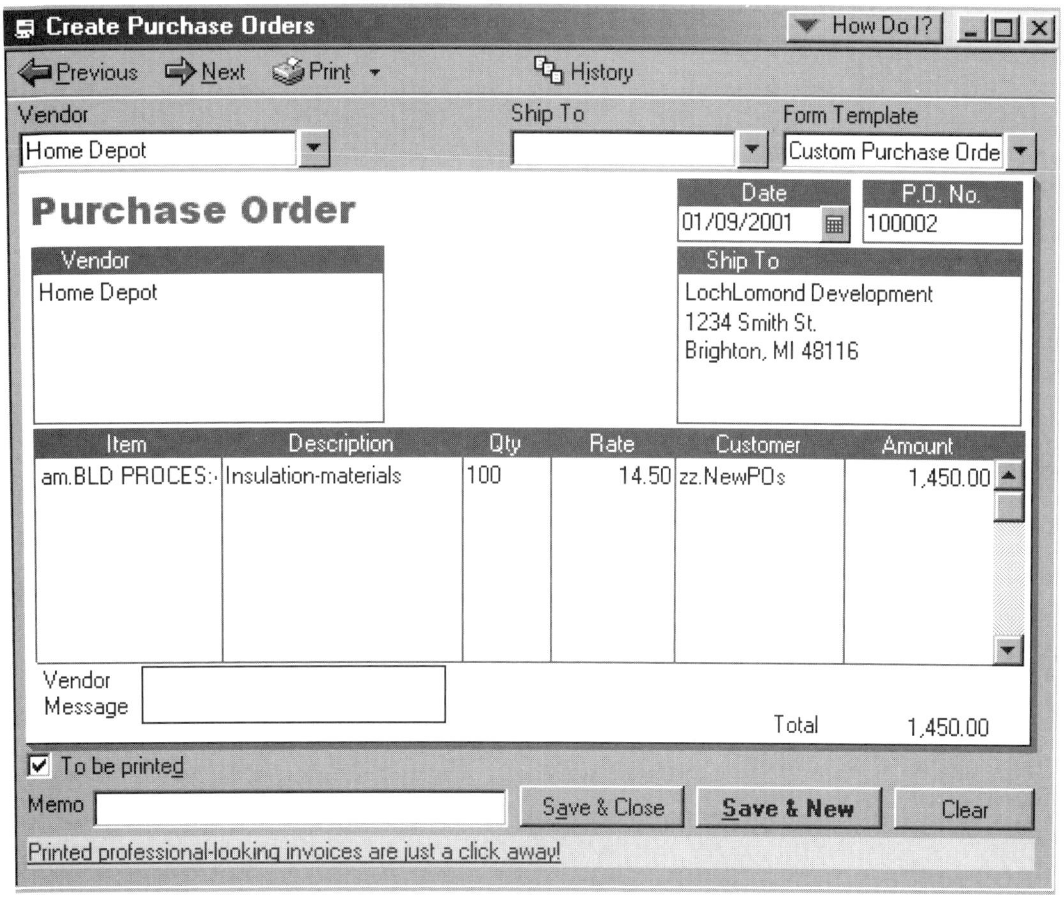

EXHIBIT 16.3 zz.NewPOs created in Create Purchase Orders window.

- **Print Ship to Addresses**—If you are sending invoices leave this box unchecked.
- **Check No.**—Type the number if paid by check.

3. When you have finished making changes, click **OK.**
4. In the Print Labels dialog box, select the correct label type and quantity.
5. When you have finished making your selections, click **Print.**

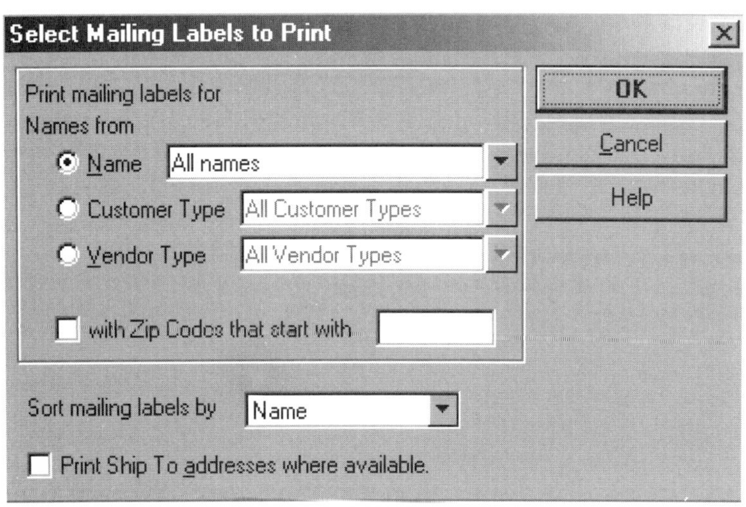

EXHIBIT 16.4 Select Mailing Labels to Print window.

CHAPTER 17

Entering bills

The process of entering bills in QuickBooks Pro 2000 creates accounts payable entries and also enters an expense or direct cost that you have incurred but have not yet paid. The accounts payable account increases and the corresponding expense account(s) increases as of the date on the bill. In other words, the date the lumber company puts on their bill is the date you normally enter in the system as the bill date. This means that your direct cost account and accounts payable increase as of that date, regardless of when you actually pay the bill.

NOTE Treat a credit memo from a vendor the same way you would a bill. Click the button for Credit in the Enter Bills window and enter the requested information, exactly as you entered it on the original bill. For example, if you used the item roofing materials for the Jones' custom home, use the same item and customer. Enter the amount of the credit as a positive number. This will put the credit into the payables system, so that it can be applied against a bill.

Creating bills from purchase orders

Remember the first reason in chapter 16 for using purchase orders? Using purchase orders puts information on the system that then can be immediately and easily converted to bills. When you enter a bill in QuickBooks Pro, one of the steps in the process is to match the bill to an existing purchase order previously issued to the vendor.

Sometimes you will receive a bill with the delivery of materials. It is just as likely, however, that you will receive materials or services first and then receive the bill at a later date. QuickBooks Pro has processes for entering both types of receipts. On the Vendors Navigator you will see a flowchart (see Exhibit 17.1). At the bottom you will see the process you follow for purchases. After a purchase order is prepared, the road of choices splits, giving you two options—Receive Items with Bill and Receive Items, then Receive Bill. The second option lets you enter bills in two steps. If you have received items and haven't yet seen the bill you can enter what has actually been received and later convert the information (an item receipt) to a bill when you get the vendor's bill. This lets you track items as they are received, and keeps the open purchase orders report accurate.

There is a third option when entering bills—entering bills without a purchase order. When the telephone, utility, or other overhead (or non-direct cost) bills come in for payment, go straight to Enter Bills, also shown in Exhibit 17.1. In this chapter we will look at all three processes.

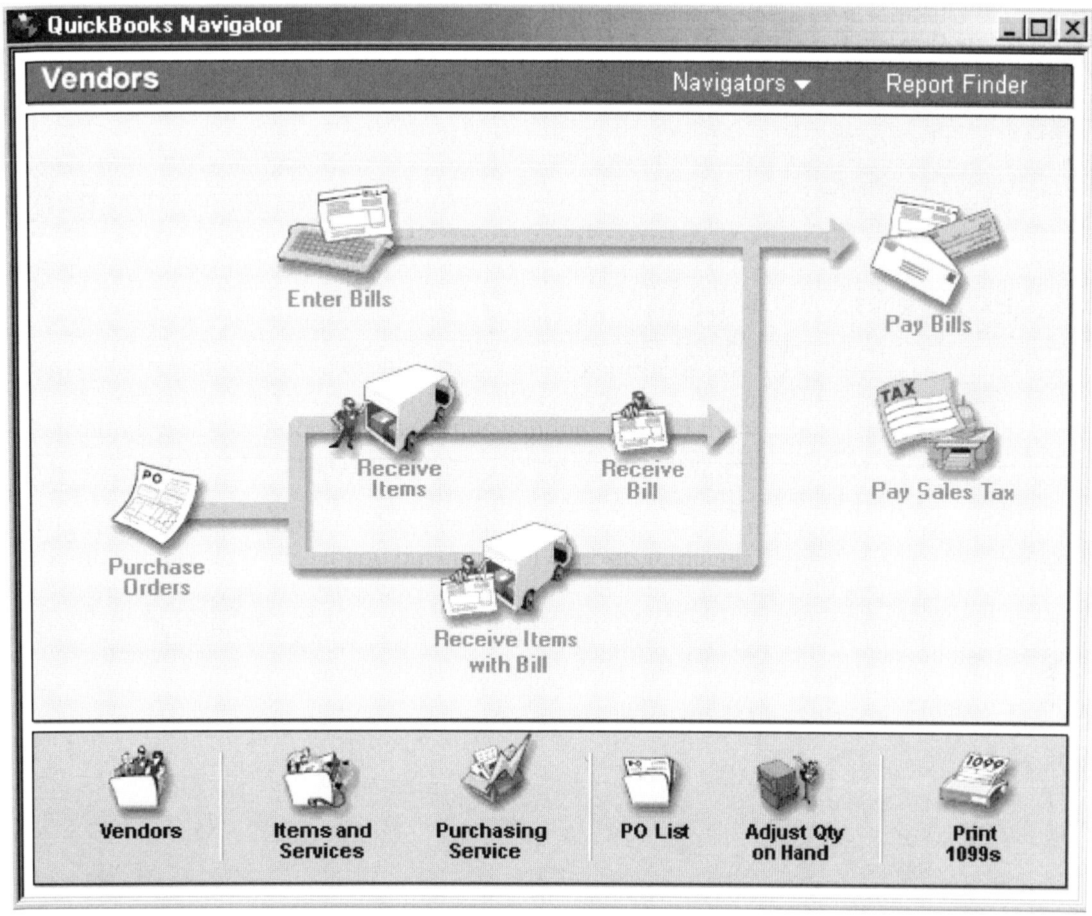

EXHIBIT 17.1 Vendors Navigator window.

TIP If you receive a bill that is dated in the previous accounting period just after you have closed your books, date the bill in the current period. Use the first day following the closed period as the bill date and enter the original date of the bill in the memo field to have a complete record.

I recommend that you enter bills as they come in, several times a week, so that you always have an overview of the bills you need to pay, as well as immediate access to up-to-the-minute job-cost reports. I find it is easier to manage cash flow with a report than with a stack of papers. By clicking on Company in the menu bar, selecting Reminders, clicking the Detail button, and highlighting Bills to Pay, you can get a listing of all unpaid bills in due-date order so you don't overlook bills or miss early-payment discounts. You can print this report by selecting File, then Print List. Use this list to more effectively manage your cash flow. The Reminders List can also be a good way to motivate yourself to make some collection calls on overdue accounts!

Receiving items with a vendor bill

While entering a bill, after you've entered the vendor name and pressed the tab key, if there are existing purchase orders on the system for the vendor, the system will respond: *Open purchase orders exist for this vendor. Do you want to receive against one or more of these orders?* If you say yes, a list of the existing purchase orders for that vendor will appear in the window. Use the purchase order number or other identifying information on the vendor's bill to match the bill to an existing purchase order.

> **TIP** If the vendor ordinarily doesn't include your purchase order number on its bill, or if you didn't actually send the purchase order to the vendor, it is a good idea to include a description in the memo field of the purchase order to simplify this matching process.

Place a checkmark in front of the correct purchase order and click OK. The system picks up the information from the purchase order and drops it into the bill, including the item(s), description, quantity, total amount, customer:job, and so on. You do not need to re-key the information to enter the bill.

To enter a vendor bill when the items are received with a bill and there is an existing purchase order, follow these steps.

1. On the **Vendors Navigator**, click the **Receive Items with Bill** icon.
2. Select a **Vendor** from the Vendor List.
3. If existing purchase orders exist for the vendor, the system will ask: **Do you want to receive against one or more?** Click **Yes**. In the Open Purchase Orders window place a checkmark next to the purchase order you want to receive the bill against.
4. If there are no existing purchase orders, you will see the Enter a Bill window (see Exhibit 17.2).

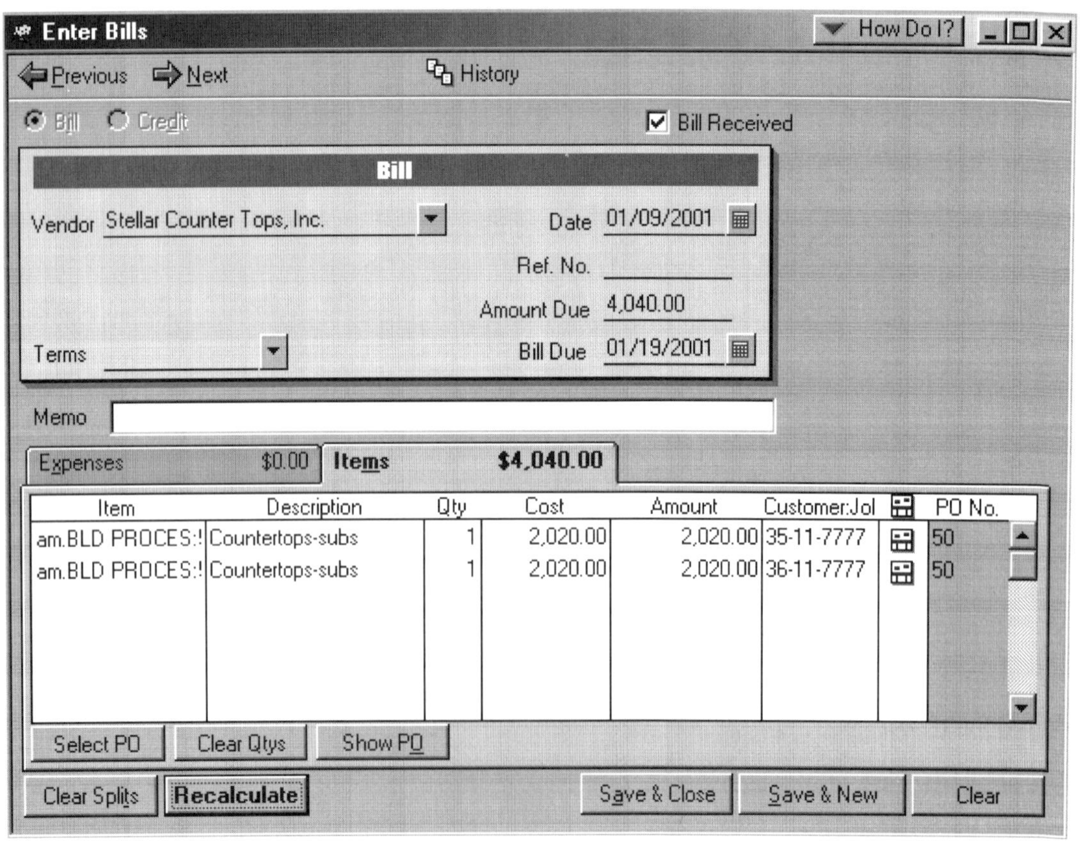

EXHIBIT 17.2 Enter Bills window.

TIP To see the purchase order click on the **Show Purchase Order** button at the bottom of the Enter Bills window.

5. On the Enter Bills window, edit the following fields, as applicable. Many of these fields are self-explanatory. If you need help, click the **How Do I?** Help button and select **Enter a Bill.**
   - **Date**—Type the date of the bill.
   - **Ref. #**—Enter the vendor's bill (invoice) number. If you are entering a bill that doesn't have an actual bill number, you might say July bill—or something else that is meaningful to you.

NOTE If you put in the vendor's bill number and later get another bill with the same number from the same company the system will tell you that you already have received a bill with that number from the vendor. This keeps you from entering duplicate bills.

   - **Amount Due**—The amount that appears is based on the purchase order cost. Verify that the amount matches the bill, and if not, follow established company procedures regarding cost variances from purchase orders.
6. Click on the **Items** tab.
7. Edit the following fields, as applicable:
   - **Item**—The items from the purchase order appear in this column. If the bill is for a partial delivery, delete the lines for items that weren't received or change the quantity. The purchase order will remain open with the items and/or quantities that weren't delivered, and the bill you entered in the system will match the vendor's bill for items delivered.

TIP Highlight the item(s) that weren't received, then hold down the Control key and press the Delete key to remove the entire line of items from the bill. The items will stay open on the purchase order. Alternatively, if you ordered six of one item and only four were received, you can change the quantity field to four, which will leave the purchase order open for the remaining two items. To see what items haven't yet arrived, run an open purchase orders by job report.

   - **Description**—The description from the purchase order for the item appears in this field. You can change the information or accept it as is.
   - **Quantity**—Type the quantity that you are receiving with this bill if different than that from the purchase order.
   - **Rate**—The amount that you agreed to pay for the item as indicated on the purchase order appears in this field. You can change the information or accept it as is.
   - **Amount**—This amount is calculated automatically based on the information in the Quantity and Rate fields.

NOTE If the amount that the system fills in is different than what is on the vendor's bill, you have two options. You either can change the amount in the Amount column, or you can leave the amount the same and add a new item from the Item List called Variance from Purchase Order to track the difference. If the purchase order is for $2,020 and the bill is for $2,040, for example, insert the variance item and put in

NOTE Continued

the difference of $20. You also can enter a negative amount if the bill is for less than the purchase order. This makes it easy on job-cost reports to see the variance between the estimated cost on the purchase order and the actual cost on the bill.

- **Customer**—If you indicated a customer on the purchase order, it appears here. Otherwise, you can select a customer to attribute this purchase to, if applicable.
- **Purchase Order number**—This field indicates the purchase order number the system converted.
- **Discount Date**—If you have an early payment discount with this vendor the discount date will appear.
- **Terms**—The default terms for this vendor appear in this field. For example, if your terms for this vendor say that the bill always is due on the 10th of the month. If you set up the vendor this way in QuickBooks Pro, the system will enter the 10th of the following month in the Bill Due field. You can change the information or accept it as is.
- **Bill Due**—Based on the terms you have entered, QuickBooks Pro calculates the date the bill is due. You can change the information or accept it as is. If the system generated date is the 10th, which falls on a weekend, you can change the date to the 9th or 12th, for example.

NOTE

When you change the amount of the bill in the top portion of the window, it won't necessarily change the amount on the bottom of the window. The amount at the bottom is the total of the line items or accounts that the system copies from the purchase order. Make sure the total amount due is correct in the Amount Due field.

TIP

There is a neat little button called Recalculate at the bottom of many entry screens. If the total at the bottom of the window is different than the amount on top click on the button to have the system change the total on the top to match the details you have on the bottom, if the amount on the bottom is appropriate.

8. When you have finished making changes, click **Save & New** to save the changes and enter a new bill, or click **Save & Close** to save the changes and return to the previous window.

Receiving items without a vendor bill

QuickBooks Pro allows you to receive items without a bill. To track items as they are received, but before you receive a bill from the vendor, follow these steps.

1. In the **Vendors** menu select **Receive Items**, or on the **Vendors Navigator**, click on the **Receive Items** icon.
2. In the Create Item Receipts window select a **Vendor** from the list. Confirm if you are receiving an item against an open purchase order. Place a checkmark next to the purchase order that you want to receive the items against.
3. When you select a purchase order to receive against the system will fill in the Create Item Receipts window with the information on remaining outstanding items from that purchase order (see Exhibit 17.3).
4. Enter/edit the following fields, as applicable. Many of these fields are self-explanatory. If you need help, click the **How Do I?** Help button, and select **Create an Item Receipt.**
   - **Date**—Type the date that you received the item.
   - **Ref. #**—Type a reference number, like the packing slip number.

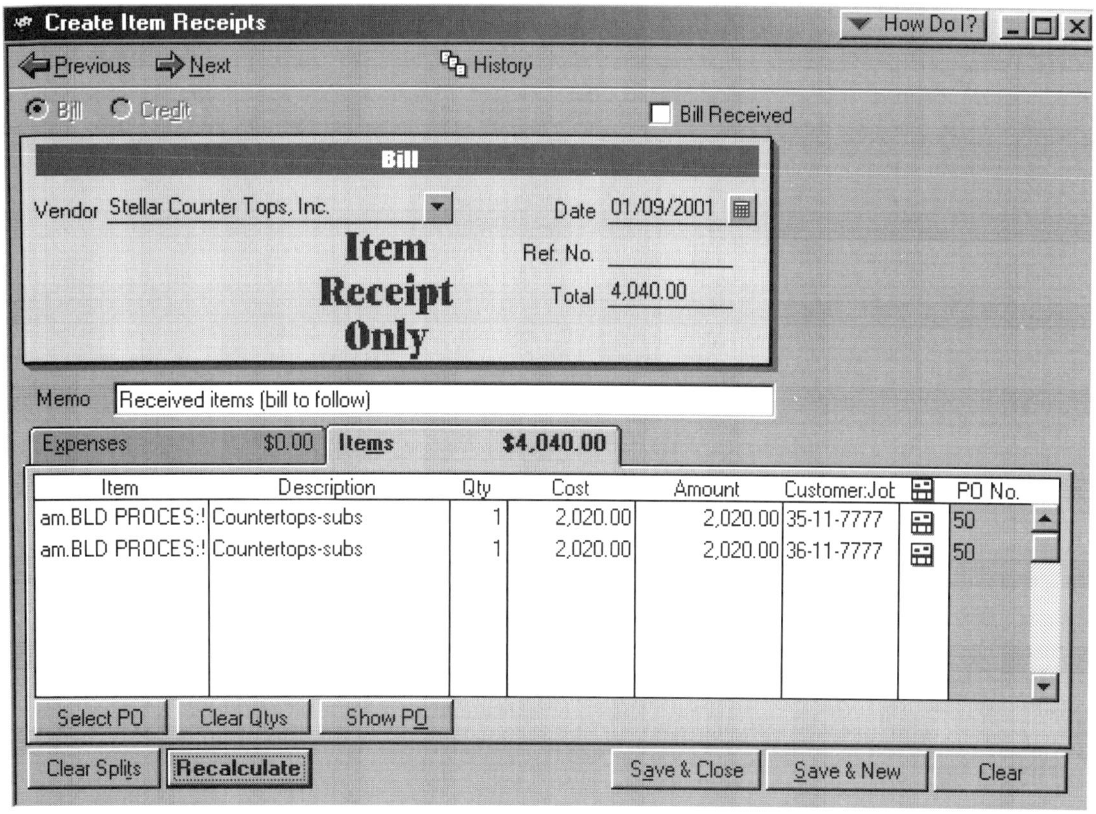

EXHIBIT 17.3 Create Item Receipts window.

- **Total**—The amount of the purchase order will appear here. You can modify the amount if necessary.

To receive items, follow these steps.

1. Click on the **Items** tab.
2. Enter or edit the following information, as applicable.
   - **Item**—The items from the purchase order appear in this column. If you need to delete an entire line item, you will find **Delete Line** under the **Edit** menu.
   - **Description**—The description from the purchase order appears in this field. You can change the information or accept it as is.
   - **Quantity**—The quantity that you ordered appears here; modify if necessary.
   - **Rate**—The amount that you agreed to pay for the item as indicated on the purchase order appears in this field. You can change the information or accept it as is.
   - **Amount**—This amount is calculated automatically based on the information in the QTY and Rate fields.
   - **Customer**—If you indicated a customer or job on the purchase order it appears here. Otherwise, you can select a customer or job to attribute the item to.
3. When you are satisfied with the item receipt, click **Save & New** to save the form and receive additional items, or click **Save & Close** to return to the previous window.

NOTE

This process puts your item receipt into payables because, since you have received materials, you owe for them, bill or not. When you run financial reports, the accounts payable amount will include the costs shown on the item receipt as of the date of the receipt. When the bill comes, the item receipt is converted into a regular bill, so the item receipt payable on the item receipt date has

NOTE Continued

"disappeared" and "reappeared" as a bill payable on the bill date. If the amounts changed between the item receipt and the bill, the expenses and accounts payable data on any financial statements you previously prepared will change in your QuickBooks Pro accounts, so be sure to discuss this situation with your CPA.

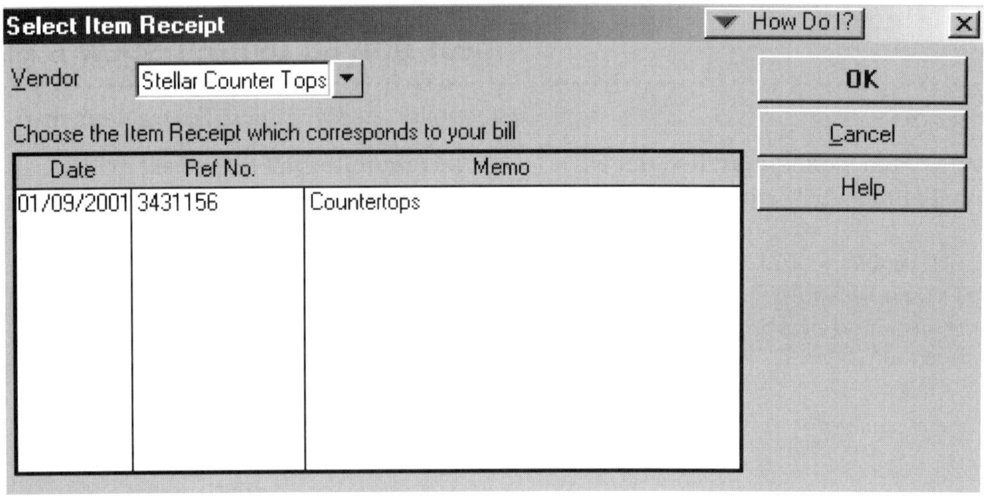

EXHIBIT 17.4 Select Item Receipt window.

Creating a bill from an item receipt

In the Create Item Receipt window there is a small box labeled Bill Received. If you click and unclick that box it turns the window from an item receipt into a bill and back to an item receipt. The item receipt is therefore just a bill-in-waiting, as described in the previous note.

To create a bill from an item receipt, follow these steps.

1. Under Vendors select **Enter Bill for Received Items** or on the **Vendors Navigator**, click the **Receive Bill** icon.
2. On the Select Item Receipt window, select a **Vendor** from the pull-down list to show all open item receipts for that vendor.
3. Select the correct receipt by highlighting it with your mouse and clicking **OK**.
4. When you select an item receipt, the Enter Bills window will open with the item receipt information filled in.
5. Edit the following fields, as applicable. Many of these fields are self-explanatory. If you need help, click the **How Do I?** Help button, and then select **Enter a Bill**.
   - **Date**—Type the date of the bill.
   - **Ref. #**—Type the bill number or the month for recurring bills.
   - **Amount Due**—The amount appears here. This number is based on the amount of the item receipt that you are converting. You can change the information or accept it as is.
   - **Discount Date**—If you have an early payment discount with this vendor the percentage will appear in this field.
   - **Terms**—The default terms for this vendor appears in this field. You can change the information or accept it as is.
   - **Bill Due**—Based on the terms QuickBooks Pro calculates the date the bill is due. You can change the information or accept it as is.

TIP

Additional costs, such as shipping charges, can be handled one of two ways. If what you ordered is a direct cost, then the shipping charge is, too, and should be

TIP
Continued

entered using direct cost (COGS) items. You can add the cost of the shipping directly into the costs of the item(s) by changing their amounts, or you can add the same item to the window again and enter the amount of the shipping. Or if you prefer, you can create an additional COGS account for shipping charges, create an item for shipping charges linked to that account, and then use that item on your forms. In all instances, the shipping costs should be assigned to the job.

Entering bills without an existing purchase order

If a purchase order was not created, the biggest decision you will need to make is whether the materials or services on the bill are direct costs, which should be assigned using COGS items, or other expenditures, which, should be assigned directly to accounts in the Chart of Accounts. The third option is that the purchase order contains both.

If the bill is job related or another direct cost, click on the Item tab and enter the information. If the bill is not a direct cost, click on the Expense tab and enter the requested information. If the bill includes both job costs and overhead costs, use both tabs.

NOTE

Under the Items tab you will see the item description, quantity, cost, amount, customer:job, and so on. There also will be a small Invoice icon to the right. If the item is a billable cost to the job, and you want to be able to select the item when you create an invoice for the job, leave the icon alone. If you do not want to directly charge the item when you invoice the job, click on the icon. A red X will appear on the icon, marking the item as not billable to the job but leaving the item still assigned to the job for job-cost reports. You don't have to cross off the icon, but if you don't the item will show on your unbilled costs by job report.

Option #1

1. Click on the **Expenses** tab.
2. Edit the following fields, as applicable.
   - **Account**—Select or modify the account for the bill. Choose from asset, liability, equity, expense, or other expense account types. Do not select income or COGS account types.
   - **Amount**—Type or modify the amount of the expenditure if the information is missing or different.
   - **Memo**—Type a comment, if any.
   - **Customer**—Do not select a customer:job, because if you do you will see this expenditure listed as "No Item" on your detailed job-cost report. If this cost relates to a customer or job, go to Option #2, below.
3. When you are satisfied, click **Save & New** to save the changes and enter a new bill, or click **Save & Close** to save the changes and return to the previous window.

Option #2

1. Click on the **Items** tab.
2. Edit the following fields, as applicable.
   - **Item**—Select the correct items for the bill from the Item List
   - **Description**—The default item description will appear. You can accept it as is, or modify it as desired.
   - **Qty**—Enter the number of units, if applicable.
   - **Cost**—Enter the cost per unit, if applicable.
   - **Amount**—Type or modify the amount for each item.

- **Customer**—Select a customer or job. If you do not select a customer or job, your job-cost detail reports will not be accurate. Every item should be assigned to a customer or job.

3. When you are satisfied, click **Save & New** to save the changes and enter a new bill, or click **Save & Close** to save the changes and return to the previous window.

NOTE

Let's say you get a bill from the lumber company that includes items for jobs, which are direct costs, and bark mulch to landscape your office, which is an expense. You can split the entry between the two tabs. If there is nothing in the Expense tab, the amount for that tab will be $0. As you enter the account for the bark mulch, the amount in the Expense tab will change to equal the cost of the mulch. The system will calculate the totals under each tab so you can see that the total due for the two tabs together equals the amount due on the lumber company's bill.

You also can split a bill that includes several items for different jobs. Split the line items between the different jobs and check the total to make sure it matches the bill. Once you enter all the items, click on the Recalculate button and the system will enter the total amount due at the top. If the total doesn't match the bill then you have to adjust the amounts at the bottom until it does. The system will not let you save the bill if the total on the top doesn't match the total on the bottom.

CHAPTER 18

Paying bills

In QuickBooks Pro 2000, you use the Pay Bills feature when you are ready to cut and mail checks to pay for bills that you have entered through the Enter Bills process. You actually pay a bill on QuickBooks Pro before you create a check—confusing, but true!

As we discussed in chapter 17, when a bill is entered QuickBooks Pro creates a payable in the accounts payable subledger. When a bill is paid, QuickBooks Pro reduces the payable created when the bill was entered, and also reduces the balance in your bank account by the amount of the check you are writing to pay the bill. This is why you pay a bill before you create a check. Think of it as recording a manual check in your check register and adjusting the balance before you write and mail the check.

Selecting bills for payment

To pay bills, follow these steps.

1. In the **Vendors Navigator**, click on the **Pay Bills** icon, or under **Vendors**, select **Pay Bills**.
2. In the Pay Bills window (see Exhibit 18.1), enter the following information, as applicable. Most of this information is self-explanatory. If you need additional help, click on the **How do I?** Help button and select **Pay Bills**.
   - **Payment date**—Enter the payment date, usually the day you perform the Pay Bills process. The payment date will change to the actual date that you print the check if you selected the preference: **Change check date when check is printed** in the Checking preferences. This preference lets you pay bills one day and print the checks the next while recording the actual date of the check in the Check Register.
   - **Select Bills to Be Paid**—Click **Due on or before the due date** and enter a date to change how you view the list of bills to be paid, or select Show All Bills. I usually like to see bills due in the next 30 to 60 days.
   - **Sort Bills By**—You can sort by due date, by discount date, by vendor, or by the amount due, with the due date or vendor my preferred sorting method.
   - **Discount**—I don't use QuickBooks Pro's discount feature here or in the Enter Bills process. I prefer to enter the discounted bill amount as the bill amount due. Then if the discount is lost because the bill wasn't paid in time, I record the lost discount in account **9290.Discounts lost and late fees** in the expense section of the Chart of Accounts.
3. The center of the window shows your unpaid bills that meet the sorting criteria and the sort order that you selected above. Click in the far left column next to the bills you want to pay.

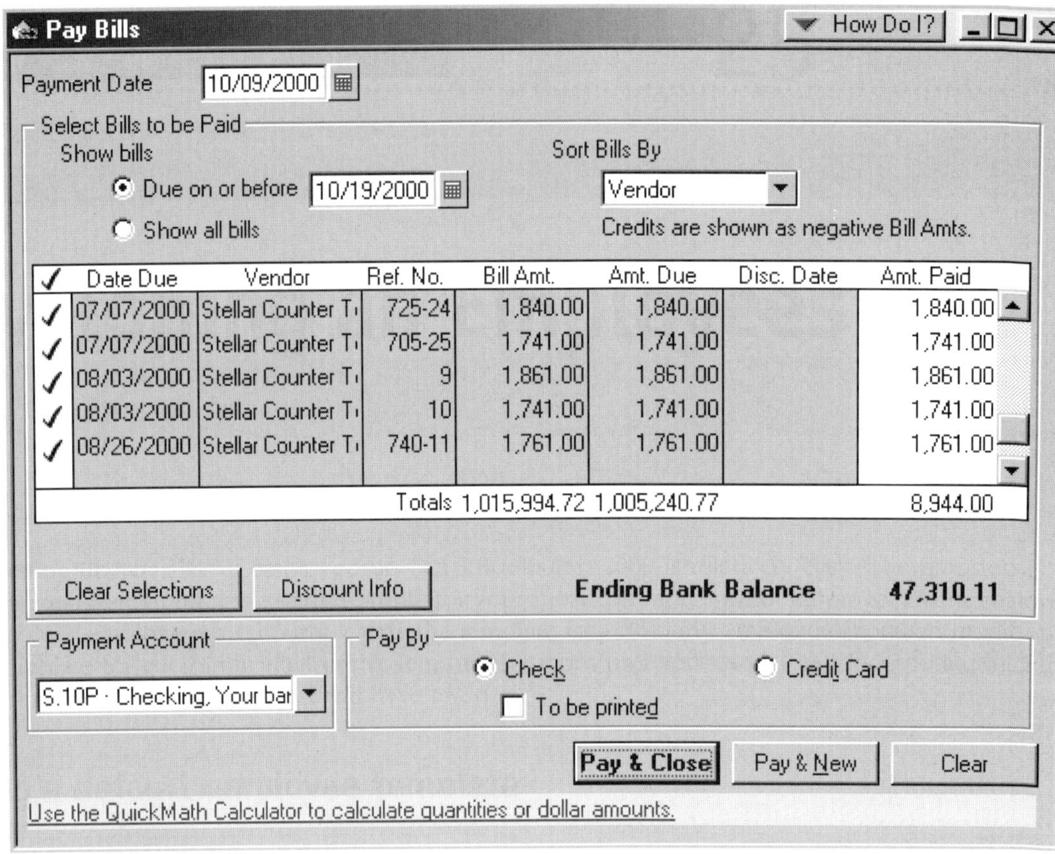

EXHIBIT 18.1 Pay Bills window.

4. Tab to the **Amount Paid** column if you want to pay less than the full amount due. You now can change the amount.
5. In the **Payment Account** field select the account you will use to pay the bills—either a checking account or a credit card account. The system will show the ending balance in the account after the selected bills are paid. For example, if you have $5,000 in the account and select bills that total $3,000, the ending bank balance in the window will be $2,000.
6. Click on the button in front of **Check** for the payment type, and then click on the button **To Be Printed**. If you will be using a credit card to pay the bills, click on the button in front of **Credit Card.**

NOTE If you don't click on **To Be Printed** when you select **Check** QuickBooks Pro will start automatically assigning check numbers to the bills you have selected. Then if you change your mind and pay other bills before these your checks will be out of sequence. Also, if you don't always enter your checks in order, QuickBooks Pro can easily assign an incorrect check number.

7. When you are finished selecting bills for payment click on **Pay & Close** at the bottom of the window.

TIP

If you have several bills due to the same vendor, click on all of them in one Pay Bills window. QuickBooks Pro will consolidate all of them into one check. If for any reason you want to cut two or more separate checks to that vendor, click on one or more of the bills in the Pay Bills window, then click the **Pay & New** button at the bottom of the page. On the new Pay Bills window, click on the next group of bills for that vendor that you want to pay.

NOTE

If you have a credit with a vendor it will show up as a credit on the Pay Bills window. QuickBooks Pro will combine the bill and credit to calculate the amount due the vendor if you select both transactions for payment. For example, you enter a bill for $500 and you have a $25 credit (which shows as a negative amount) for the same vendor. In Pay Bills select both the bill and the credit for payment. QuickBooks Pro will indicate that the total to be paid to the vendor is $475. The voucher copy o f your check will show a $500 bill and a $25 applied credit.

Printing checks

After you select bills for payment the next thing you do typically is to print the checks. Even if you checked the To Be Printed box in the Pay Bills window, and have not yet printed the checks, the checks still have been entered in your bank account and its balance has been reduced.

You can access the Check Register two ways:

1. Under **Lists**, click on **Chart of Accounts**. Select the correct checking account and double click on it.
2. In the **Banking Navigator** click on the **Check Register** icon and select the correct checking account.

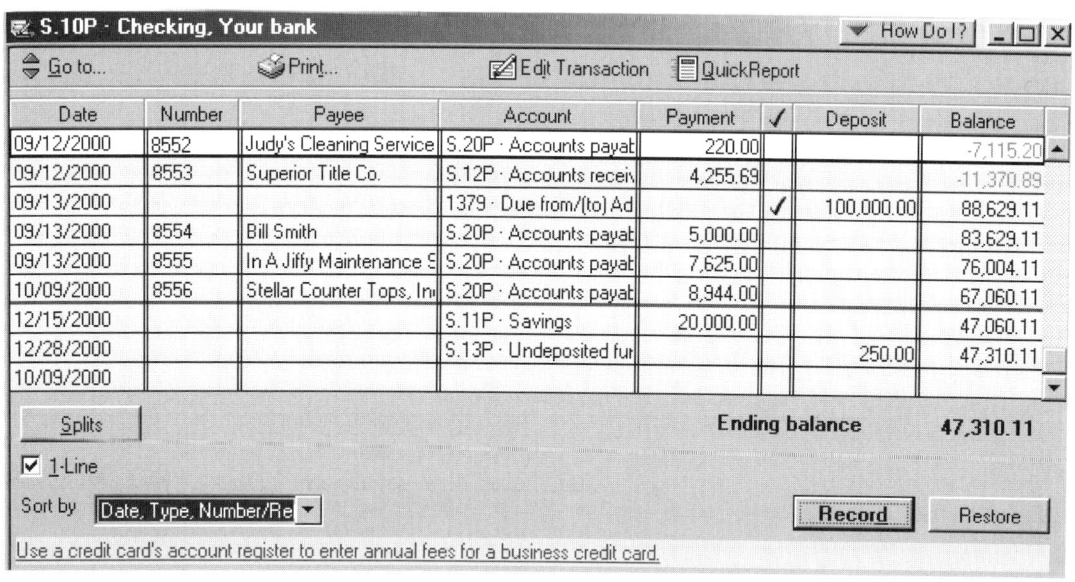

EXHIBIT 18.2 Check Register window.

The checks for the bill payments show in the Check Register window. Instead of showing a check number, however, the words "To Print" will appear in the Check Number field. You have two options for assigning check numbers.

Option A

If you already have written the checks manually and are now entering the checks to record the payment, enter the correct check number in the Check Number field. You can do this directly in the Check Register. Click **Record**.

Option B

If you are going to print the checks using QuickBooks Pro, get the correct number of blank checks you will need out of the locked drawer in your office. Check that the account number is correct and note the number of the first check. To have QuickBooks Pro print the checks, follow these steps.

1. Under **File** select **Print Forms**.
2. From the list of forms select **Checks**. A window will pop up showing the checks waiting to be printed.
3. Select the checks you want to print.
4. QuickBooks Pro will indicate the next check number on the system. Check to make sure this number matches the number on your first blank check. Also verify that you are using checks for the correct bank account.

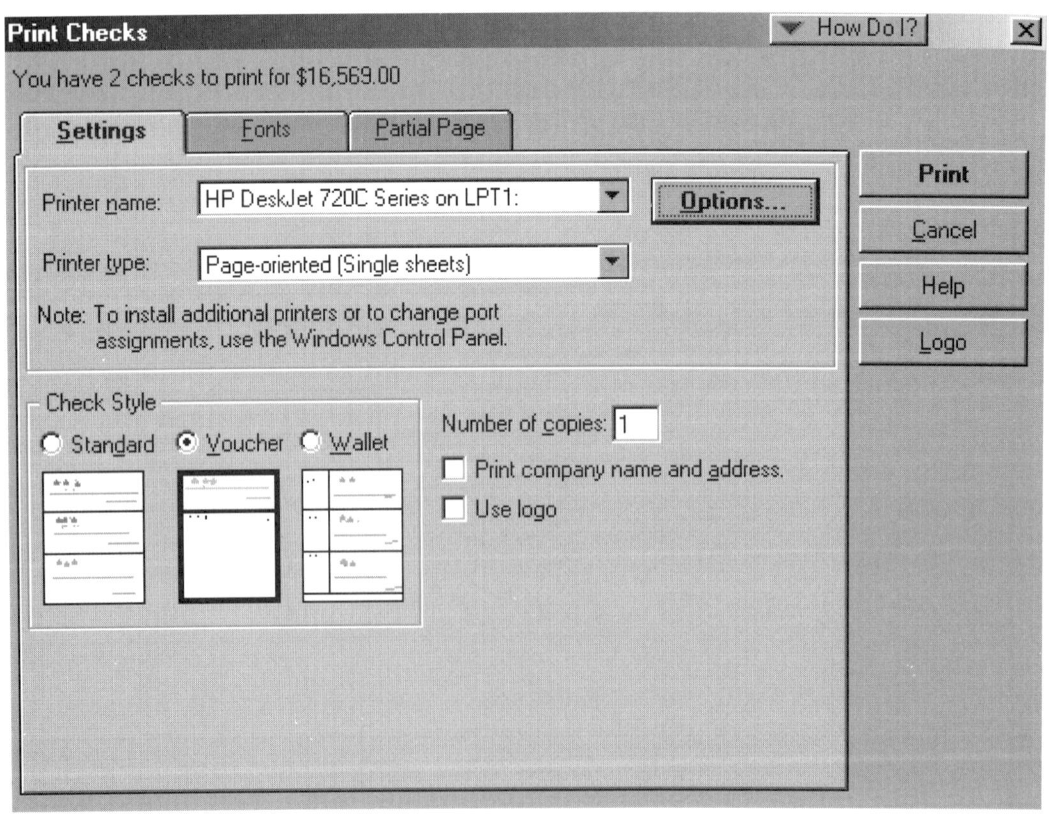

EXHIBIT 18.3 Print Checks window.

5. When you have verified the account number and check number, click **OK** to return to the Print Checks window.
6. The window will show the number of checks you have selected to print and the total dollar amount of the checks at the top of the screen.
7. Select the printer. Select **Page-oriented** for laser or single-sheet checks or **Continuous-feed** for continuous-feed checks.
8. Select the check style—either **3 checks to a page**, **voucher** (my preference), or **wallet** (I definitely don't recommend this).
9. Indicate the number of copies. If you have a voucher-style check you don't need extra copies.
10. Indicate if you want to print your company name and/or logo on the check. You cannot modify the check layout and usually won't print your company information, unless you don't use preprinted checks. (Preprinted checks are the most common.)
11. Click on **Print**.

QuickBooks Pro will print each of the checks. The window then will ask: **Did checks print OK? Remember to sign your checks!** If the answer is no, enter the first incorrect check number in the field provided. Using this feature QuickBooks Pro will void the check number and allow you to reprint the checks.

Tracking transactions

Once a bill is paid the system puts a paid stamp on it. If your report says a bill was entered on 9/14/00 and you want to see if it was paid, double-click on the bill in the report. You will see the paid stamp. To see when the bill was paid, click on the History button and the system will show you the transaction history for that bill, including vendor, date received, date due, bill amount, and so on. Toward the bottom of the window you will see the other transactions related to the bill, such as the purchase order if one was issued and the bill payment.

You can highlight any of these items—purchase order, bill, bill payment—and click the Go To button to see the initial transaction. Highlighting the purchase order will take you to the purchase order transaction. The purchase order, if all items were received, will be marked "Received in Full." If all items were not received, the purchase order will remain open. Highlighting a payment will bring up the actual payment window with all the information. The History feature lets you see the flow of all the related transactions. In addition, you can print the information if you need a printed copy of the transaction history.

Transaction histories can come in handy if you have a disagreement with your vendor about an outstanding check or bill due. If you get a statement at the end of the month from a vendor saying you still have outstanding bills and you don't agree, you need to research what is in your system and compare it to their statement to find the difference.

You might not have received or entered a bill, or the discrepancy could be due to a timing difference. You might have you mailed a check, but the vendor didn't post the check before preparing the statement. Sit down and match up the unpaid bills in your system to what the vendor says you owe according to their detailed statement. If the vendor shows a bill that you don't, request a copy of the bill and research it.

QuickBooks Pro's Vendor Balance feature offers an easy way to see the differences between what you have as an outstanding amount and what is on the vendor's statement.

1. Under **Reports** select **Vendors & Payables**, then **Vendor Balance Detail**, or under **Reports** select **Vendors & Payables**, then **Unpaid Bills Detail** and set the date to match the vendor's statement date.
2. Filter by the vendor's name. The window will show the transaction details for that vendor (see Exhibit 18.4).

LochLomond Development

10/09/00

Vendor Balance Detail

All Transactions

Type	Date	Num	Account	Amount	Balance
Stellar Counter Tops, Inc.					
Bill	06/27/2000	725-24	S.20P · Accounts ...	1,840.00	1,840.00
Bill	06/27/2000	705-25	S.20P · Accounts ...	1,741.00	3,581.00
Bill	07/24/2000	9	S.20P · Accounts ...	1,861.00	5,442.00
Bill	07/24/2000	10	S.20P · Accounts ...	1,741.00	7,183.00
Bill	08/16/2000	740-11	S.20P · Accounts ...	1,761.00	8,944.00
Bill	08/16/2000	12	S.20P · Accounts ...	1,741.00	10,685.00
Bill	08/24/2000		S.20P · Accounts ...	4,050.00	14,735.00
Bill	09/01/2000	13	S.20P · Accounts ...	1,741.00	16,476.00
Bill	09/01/2000	14	S.20P · Accounts ...	1,585.00	18,061.00
Bill Pmt -Check	10/09/2000	1366	S.20P · Accounts ...	(8,944.00)	9,117.00
Item Receipt	01/09/2001	3431...	S.20P · Accounts ...	4,040.00	13,157.00
Total Stellar Counter Tops, Inc.				13,157.00	13,157.00
TOTAL				**13,157.00**	**13,157.00**

EXHIBIT 18.4 Vendor Balance Detail window.

Keeping backup

From a paper flow standpoint I suggest you set up separate filing locations for new bills, entered bills, and the check support that will be permanently filed after the bill is paid. Set up a file or tray for bills that have not yet been entered in the system. Open the mail and put the new bills in the tray. Then several times a week enter the bills that are in the tray. Once the bills are entered, file them in an alphabetical file (I use an Expando file) to await payment.

Periodically run a report to see a list of the bills that have been entered but not paid. Highlight the bills to be paid on a given date. After the checks are prepared, remove the support from the alphabetical file and attach it to the check before the check is signed. Then whoever signs the check has the paperwork for reference purposes. I suggest that the bookkeeper or accountant print the checks and that the owner review and approve the bills, and sign the check. The owner or another staff member other than the accounting assistant should mail the checks.

I also recommend that you use voucher checks. A voucher check is a full page dedicated to paying one bill. The actual check is the top third of the page. The middle third is perforated and sent to the vendor with the check. The invoice number and any memo is printed on this portion. The bottom third stays with you and should be stapled to the top of the bill(s) from the vendor to create the backup.

After the checks are signed, file the check support either numerically by month or alphabetically by vendor. I prefer to file by vendor. If you keep job files you may want to copy the bill and voucher and put the copy in the job file. Don't file the original paperwork in the job file. This creates too fragmented of a filing system.

TIP Once you enter your purchase orders in QuickBooks Pro, it is relatively simple to enter bills and pay them. As far as QuickBooks Pro is concerned, you can put 60 jobs on a bill, but if you use job folders ask your vendor to send you separate bills for each job based on your individual purchase orders. Then you can file copies of them in your individual job files without having to make multiple copies and note which items are for which jobs. Even with the extra pieces of purchase order paper your life will be simpler.

Setting up reminders

In order to take advantage of all early payment discounts you may want to use the Reminders feature, shown in Exhibit 18.5, to remind you about when bills are due. To activate this feature, check the **Show Summary** preference for Bills to Pay in the Reminders preferences.

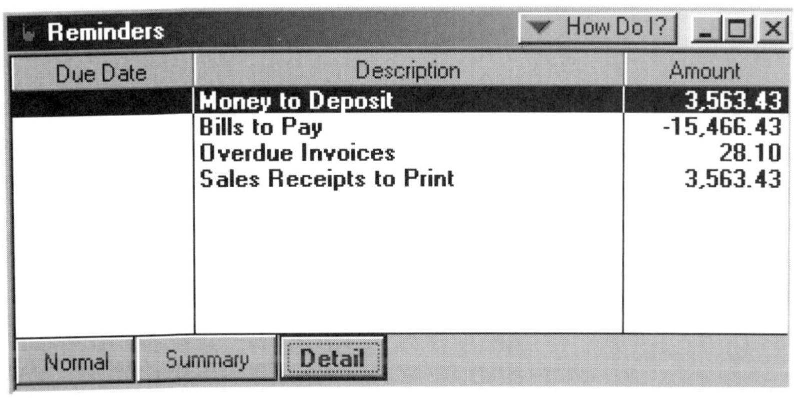

EXHIBIT 18.5 Reminders window.

Writing checks

Warning: Writing a check does not pay a bill! You cannot pay a bill using the Write Checks feature. If you have entered a bill the only way you can complete the payment process and reduce payables is to use the Pay Bills feature. If you use Write Checks you will enter the expense twice and still have the payable on your books!

So, you may ask, when do you use this feature? You use Write Checks instead of Enter Bills to enter historical information when you are playing catch-up or when you need to record and assign checks that were written manually. A third reason to use Write Checks is if you don't use accounts payable. Checks entered through Write Checks do not flow through accounts payable.

To write a check in these situations, follow these steps.

1. Under **Banking**, select **Write Checks** or in the **Banking Navigator** click on the **Checks** icon.
2. In the Write Checks window (see Exhibit 18.6) enter all applicable information, which includes:
   - **Account**—Enter the correct bank account.
   - **Pay to the order of**—Enter the payee name. QuickBooks Pro automatically will fill in the address, depending on your setup.
   - **Date**—Enter the check date.

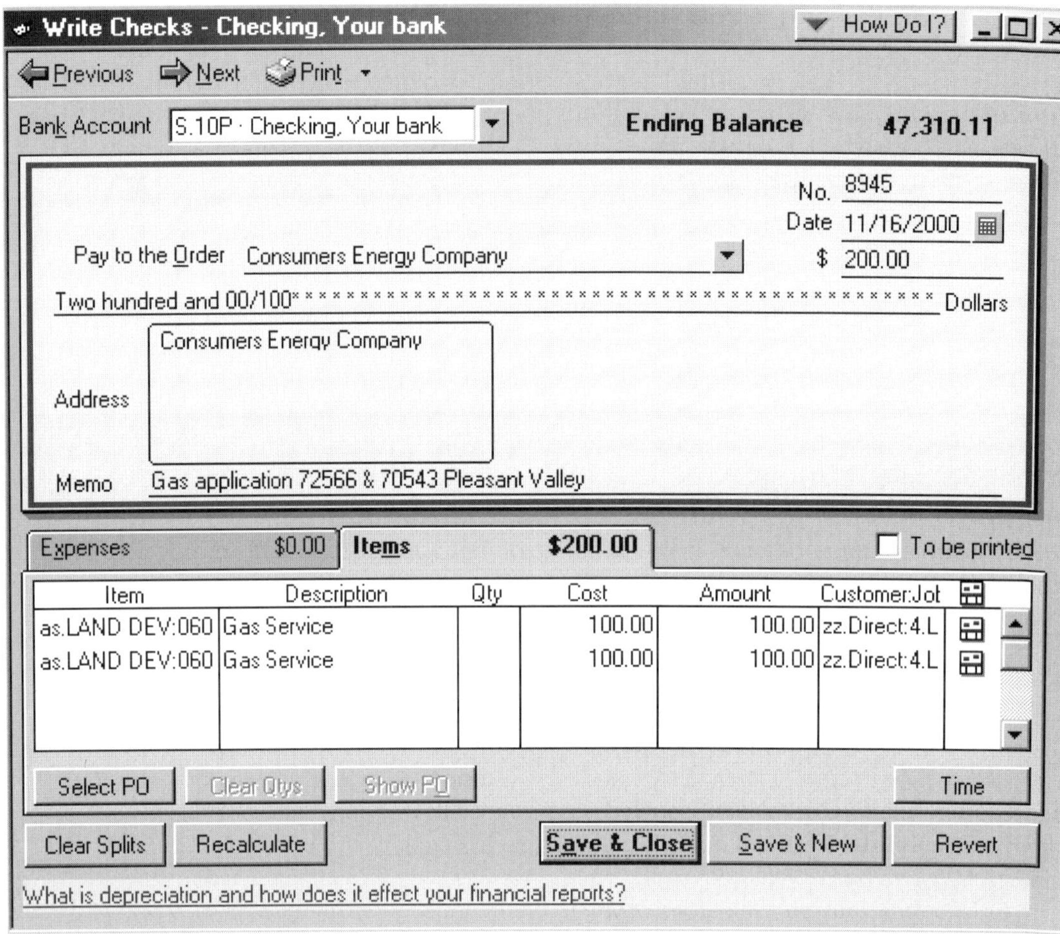

EXHIBIT 18.6 Write Checks window.

- **Check Number**—When recording checks that already have been issued, enter the check number.
- **Amount**—Enter the correct amount.

3. At the bottom of the window click on the **Items** tab or the **Expenses** tab to assign the merchandise or services paid for. Select the Items tab to record a direct cost using a cost of goods sold (COGS) item. If the item is not a direct cost select the Expenses tab and pick out the correct account from the Chart of Accounts.
4. Click the **To Be Printed** box to move the check to the print cue, if you choose.
5. Click **Save & New** to write another check or **Save & Close** to return to the previous window.
6. To print the checks, under **File** select **Print Forms** and follow the steps under the heading Printing checks above. Or, from the Write Checks window, click the Print button near the top. You can elect to print just the check you currently have on the screen, or to print all the checks in the batch.

TIP Intuit makes it easy to order checks. You may want to check alternative resources for checks. There can be large price differences—check several companies. I recommend that you make your letterhead fancy, but don't spend extra money on your checks. Do, however, be willing to pay for security features, such as checks that can't be copied on a color copier.

Preparing 1099s

You will need to print 1099s for each of your vendors who are individuals or unincorporated companies to whom you have paid more than $600 for services. The process of preparing and printing 1099s is a detail-oriented, time-consuming process. My advice is let your payroll service prepare the 1099s if they offer that service!

The next step is to go to File and select Preferences. Click on the Tax:1099 icon. Under the Company Preferences tab the system will ask: Do you file 1099-MISC forms? Click on Yes. The window will show a gray column that is labeled 1099 Category, a white column that is labeled Accounts, and another white column that is labeled Threshold. QuickBooks Pro will enter the current federal threshold. First, click on Box 7 and then click the account field. Click Selected Accounts. Click on accounts that might trigger 1099 status, such as:

- Trade contractors
- Other non-job-specific costs
- Uniforms and laundry
- Trade contractor with administrative labor
- Repairs and maintenance on office space
- Janitorial and office cleaning services
- Repairs and maintenance on office service
- Computer repairs and maintenance
- Software support
- Professional and service fees
- Other consulting services

Then click on Box 13A. Legal Service Proceeds and select any legal services accounts you have on your books. There is no threshold for reporting requirements for this line. Before you print the 1099s, create a report. In Reports, select Vendors and Payables, then 1099 Detail. For the date parameters, enter the last calendar year. The first time you run the report select All Vendors. The report will give you a list of vendors the system thinks you owe 1099s based on the 1099 preference information you entered during setup. Then run the report again selecting Only 1099 Vendors to see if you missed a 1099 vendor during your vendor setup.

NOTE If you entered a bill for the vendor in the last calendar year and didn't pay the bill during that calendar year, QuickBooks Pro will still include that bill on the report because it looks for bills, not payments.

Review and compare the two reports. Look for people who are not incorporated—trade contractors, cleaning people, accountant if not incorporated, and so on. Then highlight the ones subject to 1099s. If you find vendors that are not on your 1099 Vendor report, edit their vendor information to indicate they should receive 1099s, then rerun the report.

When the list is right for the timeframe, amount, and vendors, print the 1099 detail report.

To print the 1099s, follow these steps.

1. In **Vendors**, select **Print 1099s**, or in the **Vendors Navigator**, click on the **Print 1099s** icon.
2. If you are going to print 1099s and you have a laser or ink jet printer, you need to purchase special 1099 forms available at office supply stores or from your accountant. Be sure to buy the kind that is made to go through your printer—not the perforated kind. Don't wait until the last minute to do this. Be very careful that the forms you get are the ones that will work with your printer. The forms are expensive. They come with a 1096 transmittal form that has to accompany the 1099s. There is only one 1096 in the package. So be careful!

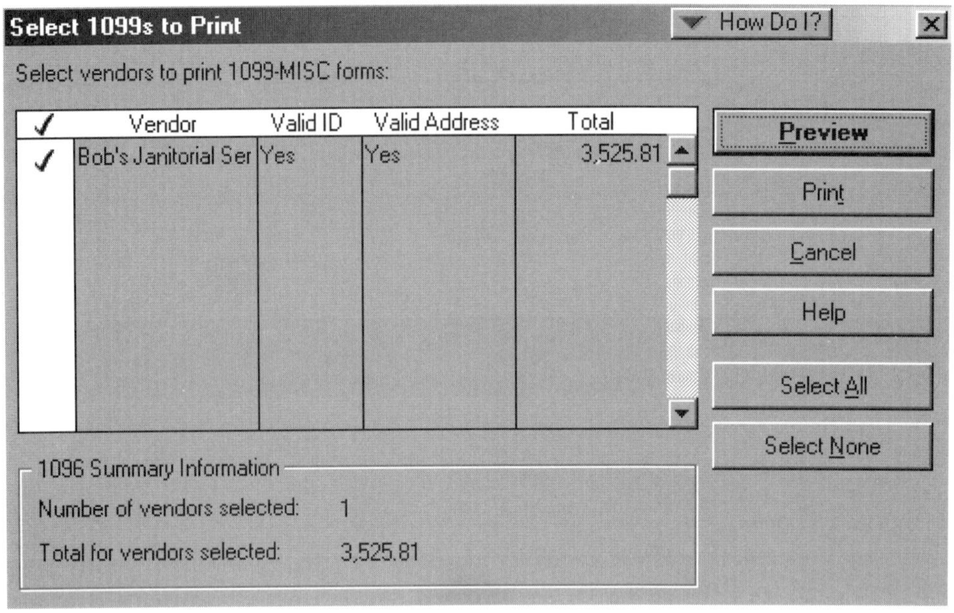

EXHIBIT 18.7 Printing 1099s window.

3. In the Print 1099 window specify a date range of the last calendar year.
4. Click **OK**.
5. Select the vendors. The system will show you the selected vendors with a valid social security number or employer identification number and address. The amount paid each vendor also will show on the window.
6. You can click on **Select All** or **Select None**, or you can pick and choose manually by placing a checkmark next to the vendors to whom you wish to send 1099s.
7. Go to **Print Preview**, **verify**, and then **Print**.
8. Select the screen settings, including printer, number of copies, and so on.
9. Click on the **Align** button to show the Fine Align Screen.
10. Click on **Print** to print a sample on blank paper. Don't load forms. Place your plain paper against the form and hold it up against the light to see if things are lining up correctly. If not, move the text in 1/100th inch increments up or down or left or right in the Fine Align Screen window. When you think you have it aligned correctly, print one 1099 on an extra form. You generally won't need all six pages that come with the 1099. If you don't need one set of those forms, practice with it.
11. When the sample aligns with the form, click on **Print**.
12. Indicate the number of copies and place the forms by sets in your printer. I print three vendors at a time because that is the number of 1099s that will print on one page. The sheet that goes to the federal government won't be perforated. The others will be perforated so you can split them apart and send them to the vendor.

Be prepared to spend the time to get the 1099s done. Don't wait until the night of January 30 because they technically must be mailed on January 31. You don't send the transmittal form to the federal government, however, until February 28. This gives you time to correct the 1099s you send out. The 1096 transmittal form cannot be prepared in QuickBooks Pro, but instead must be filled out or typed manually.

CHAPTER 19

Creating an invoice

You would think that creating an invoice shouldn't be a big deal. In fact, the creation of an invoice is fairly simple in QuickBooks Pro 2000, especially if you already have entered an estimate for the job.

Creating an invoice creates a transaction that affects both income and accounts receivable as of the date of the invoice. The date of the invoice can be very important, as it also is critical to keep your income and expenses relative to one another.

If you have created an estimate you can, if you choose, convert your estimate into an invoice. This helps you create an invoice that matches the items you used on the estimate, which in turn will result in accurate and easy-to-read reporting.

Creating a simple invoice

To start this discussion let's look at how you create a simple invoice. Then I'll show you how you modify the simple invoice to create time-and-material, fixed-price, and percentage of completion invoices.

1. Under **Customers** select **Create Invoices** or in the **Customers Navigator** click on **Invoices** (see Exhibit 19.1).
2. Select the appropriate template from the Form Template list. For help customizing your invoice templates, click on the **How Do I?** Help button and select **Reformat** and **Print**.
3. Edit the following fields, as applicable.
   - **Customer:Job**—Select a customer or job from the list. QuickBooks Pro will fill in the customer information.
   - **Bill to**—The name and address will appear in this field. You can change the information or accept it as is.
   - **Date**—Type the date of the invoice. This is the date on which the system will record the income.
   - **Invoice #**—QuickBooks Pro assigns an invoice number. You can change the number or accept it as is.

TIP I recommend you start numbering your invoices with the number 100 or 1000 so it doesn't look like you are a new company. QuickBooks Pro then will increase the number by one for each invoice. I recommend sequential numbering—and so does the IRS!

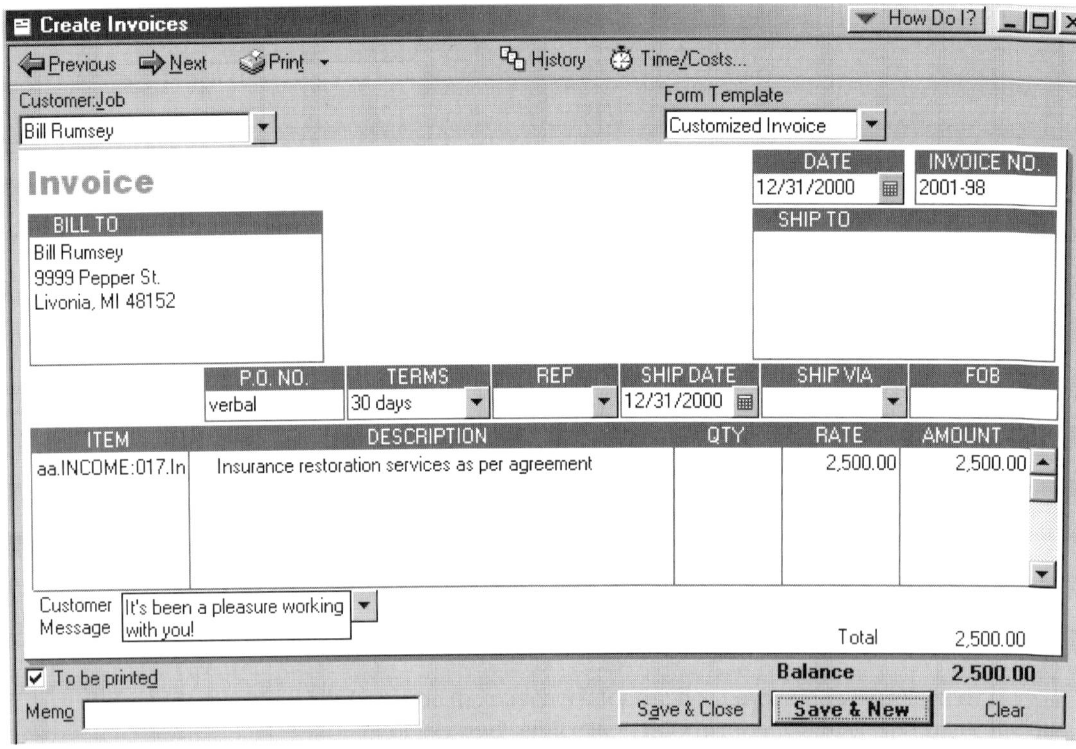

EXHIBIT 19.1 Create Invoice window set up for simple invoice.

- **Purchase Order #**—Enter the customer's purchase order number.
- **Terms**—This is entered automatically from the customer or job information.

> **TIP** In your Item List create incremental items for your fixed-price invoices, such as Progress Payment A, Progress Payment B, and so on. Make your final item Progress Payment F for final. If you create invoices for three payments, begin with item Progress Payment D and end with Progress Payment F. Then when you see the actual vs. estimate report, you will know where you are in relation to the final invoice.

- **Items**—Enter an item number or select from the Item List. The description for the item will fill in automatically. You may accept or modify the description as you wish.
- **Quantity**—Enter a quantity, if applicable.
- **Rate**—After you select an item the default price for the item appears in this field. You can change the information or accept it as is.
- **Amount**—This amount is calculated automatically based on information in the QTY and Rate fields.
- **T**—A T appears automatically if the item is taxable. In order to calculate and add tax, however, the customer also must be taxable.
- **Customer Message**—Select a message from the pull-down list. Use this message to convey marketing information, send a thank you, or alert the customer to late fees for late payments.
- **Tax**—Review the sales tax item to be sure you have the correct item or group.

- **Customer is Taxable**—After you select a customer or job the tax information for the customer appears. You can change the information or accept it as is.

4. To print a copy of the invoice, click **Print**.
5. When you are satisfied with the invoice, click **Save & New** to save the current invoice and create another, or click **Save & Close** to save the current invoice and return to the previous window.

Converting an estimate to an invoice

We have described how to create basic estimates and invoices. In this section, we will provide procedures for:

- Selecting costs attributable to the job, and copying the information to the invoice, including items such as roofing materials, or painting by a trade contractor and building permit fees.
- Selecting and copying time assigned to the job to an invoice using QuickBooks Pro's Time Tracking feature.

QuickBooks Pro makes it easy to convert an estimate to an invoice. You will have some decisions to make, however. If your estimate was for a fixed price you may need to issue progress invoices per your contract. If your estimate was for time and materials you may need to create progress invoices for separate line items. Whether you create a straight or progress invoice, the process for creating an invoice from an estimate is the same. See the options below for the different ways to create progress invoices.

> **NOTE** Before we begin, a word about job costing. Accurate job costing is critical to determining your company's overall profitability. It is important to be vigilant when tracking employees' time, trade contractors' time, materials, and other costs. In some cases these costs can be billed to the customer; in other instances costs can't be billed, but must be tracked.

Creating a progress invoice from an estimate

To create a progress invoice from an estimate, follow these steps.

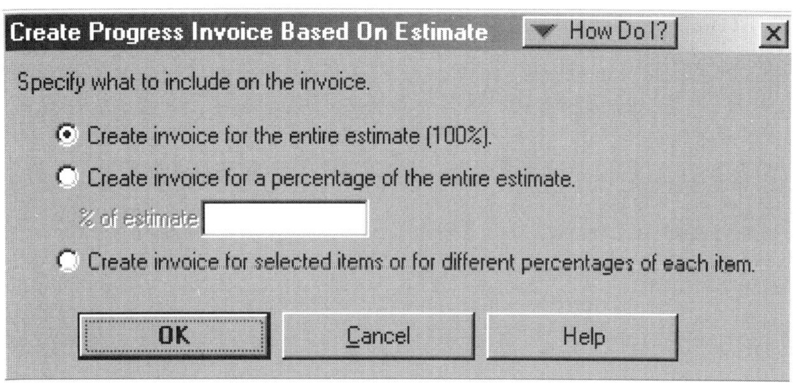

EXHIBIT 19.2a Create Progress Invoice Based on Estimate window.

1. Open the estimate and click **Create Invoice**. In this window you will have three choices:
   - Create Invoice for the Entire Estimate (100%) (Exhibit 19.2a)
   - Create Invoice for a Percentage of the Entire Estimate (Exhibit 19.2b)
   - Create invoice for selected items or for difference percentages of each item (Exhibit 19.2c)

You also can use the **Progress Detail button to** Show Quantity and Rate and Percentage of your invoiced estimate line item. This option is discussed after these three options.

Entire estimate

If you select the first option, the Create Invoice window will show the estimated amount and prior amount invoiced. There also will be an Amount column on the window. If you decide that you want to change the amounts, you also could back up and choose the third choice listed above. The total amount is the dollar amount that will be recognized as income as of the date of the invoice. Click Save & Close to exit.

Percentage of entire estimate

If you select the second option: Create an invoice for a percentage of the entire estimate, the system will ask you for the percentage that you want to invoice. If you enter 75% QuickBooks Pro will enter and compute 75% of each line item. The system will warn you if any of the items will be billed at more than 100 percent based on previous invoices and ask you if you want to continue. Use this option to create fixed-price progress invoices from your estimate. Click Save & Close to exit.

Invoice for Selected Items

The third option allows you to invoice for different percentages of each line item. Selecting this type of invoicing method brings up the **Specify Invoice Amounts for Items on Estimate** window. This window shows the estimated amount, the prior amount invoiced, and an amount column for the current invoice. If you click the **Show Quantity and Rate** box, you also will see the estimated quantity, estimated rate, prior quantity invoiced, prior rate invoiced, and current quantity and rate being invoiced. You also can choose to click the **Show Percentage** box, which shows the prior percentage invoiced, the current percentage invoiced, and the total percentage invoiced. You can decide which column to use to drive the invoicing price and percentage separately for each line item. Use this option to create percentage of completion invoices.

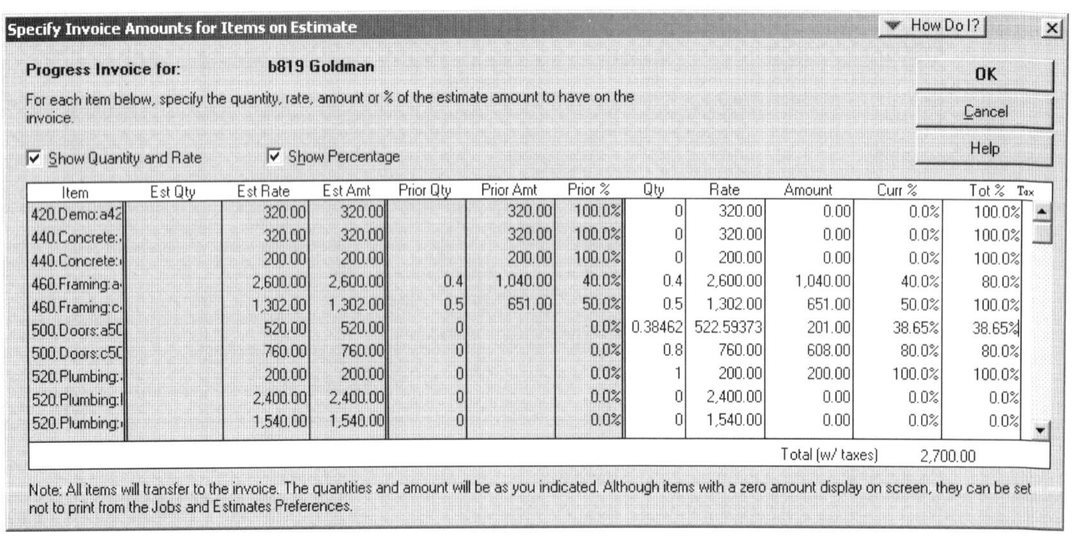

Item	Est Qty	Est Rate	Est Amt	Prior Qty	Prior Amt	Prior %	Qty	Rate	Amount	Curr %	Tot %	Tax
420.Demo:a42		320.00	320.00		320.00	100.0%	0	320.00	0.00	0.0%	100.0%	
440.Concrete:		320.00	320.00		320.00	100.0%	0	320.00	0.00	0.0%	100.0%	
440.Concrete:		200.00	200.00		200.00	100.0%	0	200.00	0.00	0.0%	100.0%	
460.Framing:a		2,600.00	2,600.00	0.4	1,040.00	40.0%	0.4	2,600.00	1,040.00	40.0%	80.0%	
460.Framing:c		1,302.00	1,302.00	0.5	651.00	50.0%	0.5	1,302.00	651.00	50.0%	100.0%	
500.Doors:a50		520.00	520.00	0		0.0%	0.38462	522.59373	201.00	38.65%	38.65%	
500.Doors:c50		760.00	760.00	0		0.0%	0.8	760.00	608.00	80.0%	80.0%	
520.Plumbing:		200.00	200.00	0		0.0%	1	200.00	200.00	100.0%	100.0%	
520.Plumbing:l		2,400.00	2,400.00	0		0.0%	0	2,400.00	0.00	0.0%	0.0%	
520.Plumbing:		1,540.00	1,540.00	0		0.0%	0	1,540.00	0.00	0.0%	0.0%	

EXHIBIT 19.2b Specify Invoice Amounts for Items on Estimate window.

Quantity, rate, and percentage

Regardless of the invoicing method you choose, you can select the **Progress Detail** button near the top of the Create Invoice window. This brings up the **Specify Invoice Amounts for Items on Estimate** window, which you see when you choose the third option above.

I prefer the third option because you can see what you have invoiced, including the total estimated amount. This gives you more control over the information before it is entered on the invoice.

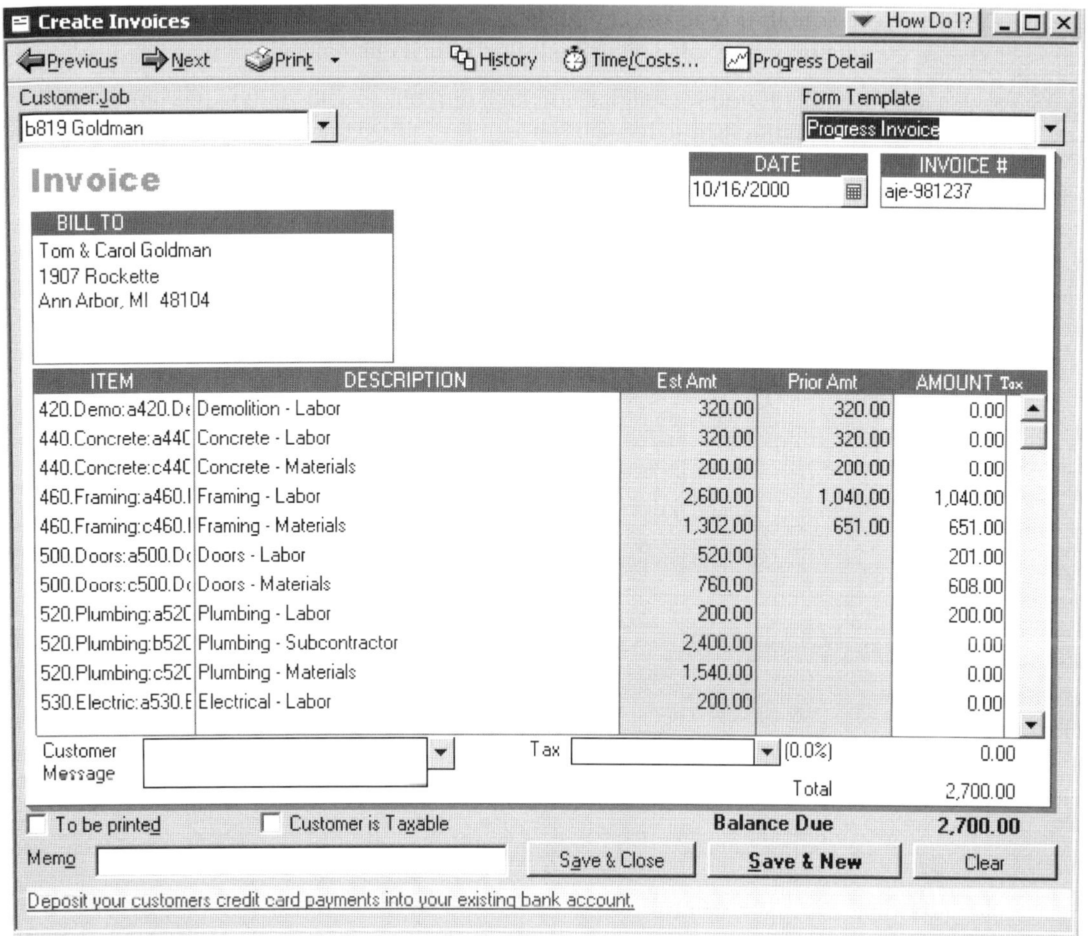

EXHIBIT 19.2c Create Invoices window set up for Progress Invoice.

Creating a time-and-cost invoice

To create a time-and-cost invoice, follow these steps.

1. In the **Create Invoice** window, with the customer or job information filled in, click on the **Time/Costs** button to the right of the History button.
2. The window will display Time and Costs assigned to the customer or job.
3. Under the Items, Expenses, and/or Time tabs select the line items you want to put on the invoice. The line items that appear on the Items and Expenses tab include costs from vendor bills, checks, or credit card charges for materials attributed to the customer or job.

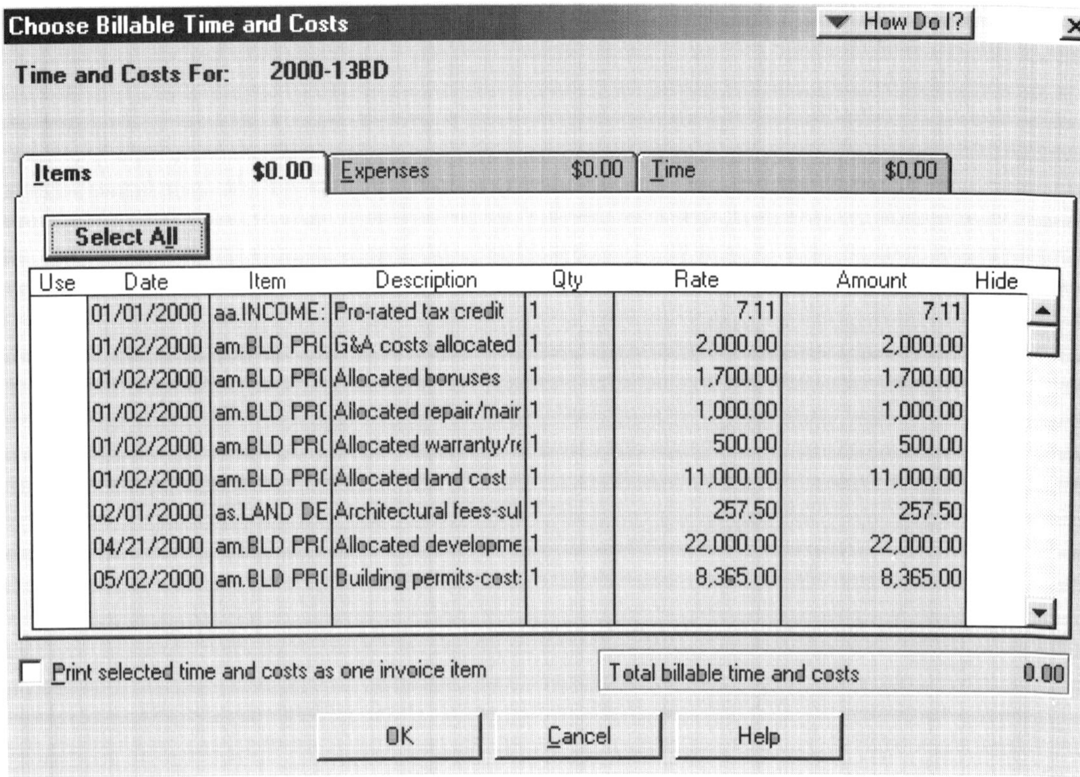

Use	Date	Item	Description	Qty	Rate	Amount	Hide
	01/01/2000	aa.INCOME:	Pro-rated tax credit	1	7.11	7.11	
	01/02/2000	am.BLD PR(	G&A costs allocated	1	2,000.00	2,000.00	
	01/02/2000	am.BLD PR(	Allocated bonuses	1	1,700.00	1,700.00	
	01/02/2000	am.BLD PR(	Allocated repair/mair	1	1,000.00	1,000.00	
	01/02/2000	am.BLD PR(	Allocated warranty/re	1	500.00	500.00	
	01/02/2000	am.BLD PR(	Allocated land cost	1	11,000.00	11,000.00	
	02/01/2000	as.LAND DE	Architectural fees-sul	1	257.50	257.50	
	04/21/2000	am.BLD PR(	Allocated developme	1	22,000.00	22,000.00	
	05/02/2000	am.BLD PR(	Building permits-cost:	1	8,365.00	8,365.00	

EXHIBIT 19.3 Choose Billable Time and Costs window.

You may click on the Select All button to select all the line items or you may select individual line items. Select line items by placing a checkmark next to them. (If you assign direct costs to the cost of goods sold (COGS) items there should not be anything under the Expenses tab.)

4. Click **OK** to add the selected items to the invoice.
5. If you know the percentage you want to mark up each line item, but not the dollar amount, you can compute the invoice price in the amount field on your invoice by using QuickMath. See Exhibit 19.4 for a QuickMath example.
6. Click **Save & Close** to exit.

NOTE This is why you have to enter bills and timesheet information before you can create a time-and-cost invoice! QuickBooks Pro will accumulate all time and costs, excluding payroll because you see that in the time reports, attributed to a job and display them under the Time/Costs button when you create the invoice. You then have a complete list from which to choose for invoicing purposes.

There are several ways you can set up the line items on your invoice.

- **Variation #1.** To show all of the time charges first go to the **Time** tab and click on all of your time charges first. Then click **OK**. Then go to the **Items** tab and click on those items. Again, click **OK**.
- **Variation #2.** To associate time with an item, select the item in the **Time** tab and click **OK**. Then select the item from the Items tab and click **OK**. Continue alternating between the Time and Items tabs until all the information is on the invoice.

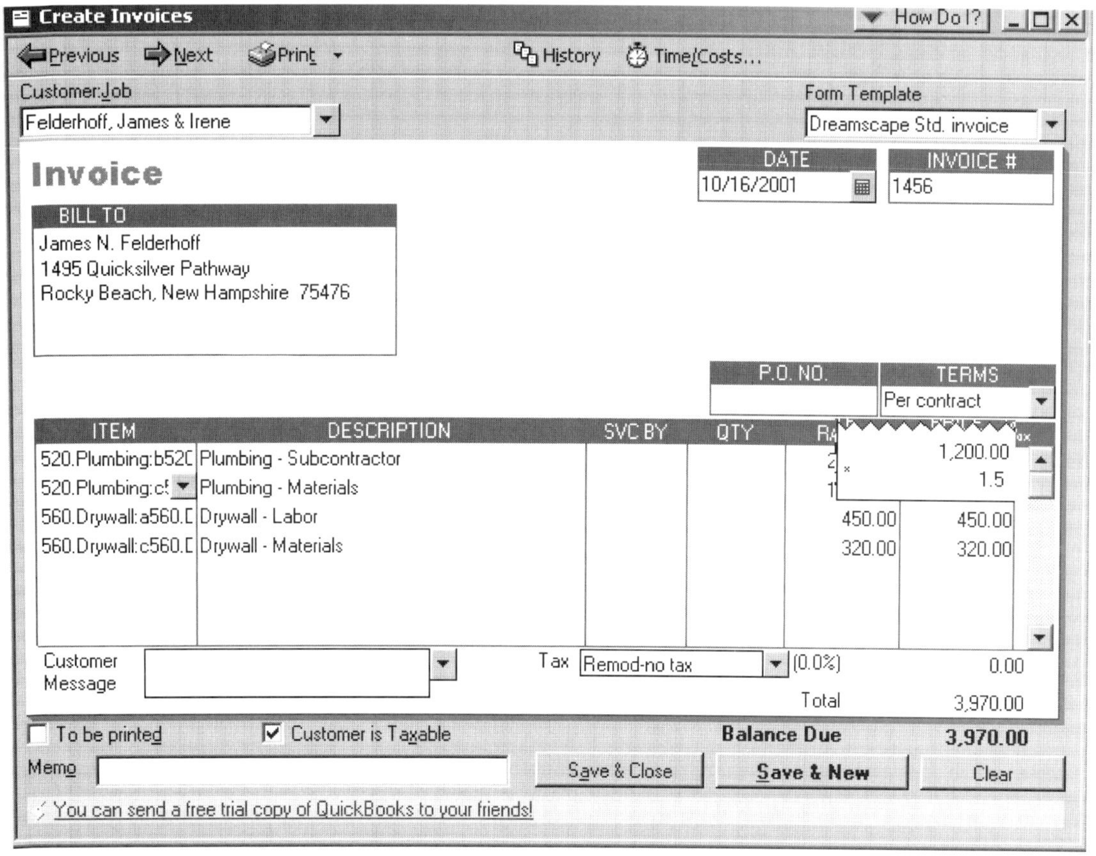

EXHIBIT 19.4 QuickMath example.

- **Variation #3.** If you want to charge for several items within one tab without providing the detailed line items you can combine them into one line item on the invoice. Highlight the items and click on the **Print Selected Time and Costs as One Invoice Item** button under the tabs. QuickBooks Pro makes the items a reimbursable group. The selected items will appear on the window so you can see the details, but if you go to Print Preview the line item on the invoice will read Total Reimbursable Expenses with a combined total amount.

To change the wording on the invoice, go to the invoice and change the description to say whatever you want—Labor, 16 hours, and so on. Check the Print Preview window to see your changes.

Recognizing income issues

Remodelers, custom and spec home builders, even handyman services encounter specific issues related to income—both payments and deposits. The primary concern is matching expenses to income as they overlap the accounting periods. Starting with the simplest and moving to the most complex, let's look at some of the issues.

Handyman service

What you see is what you get with this type of service—income and expenses. You do the job, invoice the customer, and receive payment. Payments may be received the day work is performed or shortly thereafter. There is a good match between expenses and income on a day-to-day basis.

Remodelers and custom home builders

Since these businesses engage in defined service arrangements that take place over time they experience more of a challenge when it comes to matching income to expense, particularly when draws are required before the work is done. This means that there is income on the books before there are any expenses. At other times, invoices may be mailed well after expenses are incurred. With the two circumstances—income with no matching expense or expense and no matching income—the challenge is to bring the income on your books into agreement with the expense. You have two options to accomplish this accounting feat—the completed contract method or the percentage of completion method.

Completed contract

Under the completed contract method you would not show income or expense on your formal financial statements or your taxes until the contract is complete. At the end of the year all expense and income amounts for contracts not yet completed are moved to the balance sheet using journal entries. No income or expense for the job is recognized until the contract is completed.

If you have elected this method for tax purposes, the good news is that you defer paying taxes on the job until it is complete. The bad news is that it can make your income appear very erratic from a tax standpoint. In the years you have high income you will be in a high tax bracket. Another limitation is that you have to be relatively small to use the completed contract method. Once you reach a certain dollar volume the IRS requires you to use the percentage of completion method.

See your tax advisor for more details. Once you have chosen a method with the IRS, it can be troublesome to change. So, if that is your tax method you probably will want to stick with it for tax purposes. For accounting, measurement, and management purposes, however, I prefer the accrual-based percentage of completion approach.

Percentage of completion

The percentage of completion method is very close to the accrual concept. Using this method you match your income and expense to the percentage of the job that is complete.

For example, if you enter into a $1 million contract for a custom home, $1 million would be considered 100 percent of the income you are going to receive. Your estimate shows that you expect to incur costs of $800,000 dollars or 80 percent of the total income. At the end of an accounting period your accountant or tax advisor looks at your books and sees $600,000 in costs and total customer invoices for $500,000—an obvious mismatch. You have incurred 75 percent of estimated cost, but only invoiced for 50 percent of income. With this method the expense influences the percentage considered complete so, unless you have other evidence to the contrary, the job is considered 75 percent complete. This means that you must recognize another $250,000 more income, even though you have not yet invoiced that amount.

Another way to look at your percentage of completion scenario, however, is that you may have incurred more expense than you anticipated, so you might actually be only 60 percent of the way through the job. This is why you need to see the estimate and actual vs. anticipated cost reports to acknowledge how far through the job you are. If you have incurred 75 percent of expense, but are only 65 percent of the way through the job, then you would recognize only 65 percent of $1 million as income during the current period, or another $150,000 (rather than $250,000). If you had chosen percentage of completion for tax purposes, you also would have to support why you aren't currently recognizing the more obvious choice of an additional $250,000 by documenting cost overruns, and so on.

To keep your books accurate for internal job-cost reports, make a journal entry at the end of the year to show the correct, estimated additional income. Then use another journal entry to reverse it the first of the following year. This won't affect your invoicing system. It also gives you a better match between your real income and expenses. Otherwise it would look like you had a $100,000 loss ($600,000 – $500,000 = $100,000).

Another important element of the percentage of completion method to remember is that you must continue to add the impact of change orders to your estimated income and expense for the job.

For mid-size companies, I recommend the percentage of completion method for book purposes because, as in accrual-based accounting, it gives you a better measurement system regarding the health of your business. If you also have chosen it for tax purposes, it may very well smooth out your income from a tax standpoint so you don't have those large mountains and valleys in your income. This can help you manage your cash flow and anticipate your taxes. If you have chosen the completed contract method for tax purposes, however, your books still can be percentage of completion if you wish.

Spec home builders and developers

By their very nature, spec home builders and developers operate on a version of the completed contract method, only they don't have a contract! They are, in effect, building assets that they will sell at a future time. In spite of the usual approach of putting all costs in progress straight into the balance sheet, I suggest that all spec home builders and developers initially put all of their expenses into the QuickBooks Pro job-cost system as they accumulate them and link the costs to the COGS accounts.

At the end of an accounting period (month, quarter, year) convert all costs attached to unsold units into assets by posting journal entries to clearly visible cost-offset accounts. For example, at the end of the accounting period you have $1,750,000 in COGS expenditures for your current development and no units have yet been sold. Offset, or take the amount from the COGS accounts and enter it into the balance sheet Work in Process (WIP) asset account using a journal entry at the end of the accounting period. This way the costs are temporarily shown as an asset instead of an expense. On the first day of the new accounting period, use another journal entry to take the costs out of WIP assets and put them back into the COGS accounts. You do this so that on a day-to-day basis the job cost reports agree with the general ledger and additionally, at the end of the accounting periods, so they agree with Generally Accepted Accounting Principles (GAAP).

As a spec home builder or developer, you will not show true income until you close on your houses and lots. You can't recognize, for financial statement purposes, any income until the sales are finalized. In between, however, you can link customer deposits and invoices for extras to the Item List and job-cost system, so that on most days you can see immediately what is going on with the development and the individual jobs. Here, too, the goal is to maintain the integrity of the job-cost information in QuickBooks Pro by keeping the items reports and the general ledger in agreement for the jobs. At the same time, by matching income and expenses on a periodic basis, you keep your bankers, tax advisors, and everyone else happy.

Now, let's look at customer down payments. On December 25th a customer gives you a $15,000 deposit. The deposit actually is a liability because the money is due back to the customer if the sale doesn't go through. It is not yet income because you haven't closed on the house. But for tracking purposes, I would have you put the deposit into an income account so that you can see the $15,000 deposit. On December 31st create an entry to move the deposit and any miscellaneous invoices to the balance sheet as a liability. Then on January 1st create another journal entry to move the deposit and any miscellaneous invoices back to the income account.

Now let's say that sometime in the middle of February the customer closes on the house. Create an invoice for the gross sales price of the house, which will create income in the system. Then, after you get the cash, which will be reduced by the initial $15,000 customer deposit, at the bank closing, you reduce the deposit "income" by entering it as a payment against the invoice. You now have all of the income and life-to-date costs for that home so your income statement is accurate for that particular unit.

Generating statements

Sometimes customers need to be reminded that they owe you money, or they may request a listing of their account activity. Statements provide a quick and easy way to inform customers of their overdue balances, account activity, or open credits.

There is a difference between invoices and statements. Invoices are entered on the system once because they put income on the system. If you need a copy of the invoice, go back and print it. Write copy across the top.

A statement is a history of the invoices and payments received for a customer and the outstanding balance. In most cases, reminder statements serve as helpful account management tools for home builders and remodelers. QuickBooks Pro's statements are based on customer account activity and balances. Although you can enter statement charges, which I don't recommend, and modify the format of the customer statement, it is not a transaction form—you cannot edit the information as it is presented on a statement.

TIP Run an accounts receivable aging report before you create the criteria for the statements. Then you can select the customers you need to send statements to.

To generate a statement, follow these steps.

1. Under **Customers**, select **Create Statements** or in the **Customers Navigator** click on **Statements** (see Exhibit 19.5).
2. Enter the period you want included in the statement as numbers, such as 8/1/00–9/30/00.
3. Enter the ending date or date of the statement.
4. Select the customers you want to prepare a statement for from the options: **All Customers**, **Selected Customers**, **One Customer**, or **Customers of Type**.
5. Select whether you want to print one statement for each customer or print one statement for each job. If one customer has three jobs, printing one statement for the customer will show the jobs in order with associated activity, including age of invoices, and a grand total at the bottom. If you do one statement for each job, the customer will get three statements.
6. Check the box that says **Don't Print Statements with a Zero Balance** if you have looked at a recent accounts receivable report. If you haven't, don't check this box.
7. Select **Finance Charges** to include a finance charge. Click on the **How Do I?** Help button if you have any questions.

TIP Statements are a great place to put collection-related messages because the statements show the age of the balances. Be creative and get your message across with memos such as, Avoid finance charges by paying outstanding balances, or Outstanding balances over 30 days are subject to finance charges.

8. When you are ready to print click **OK** to go to the Print Statements window.

TIP I recommend that you preview the statements before you print them. You will save yourself a lot of paper if you make sure the statements are accurate before they appear on paper.

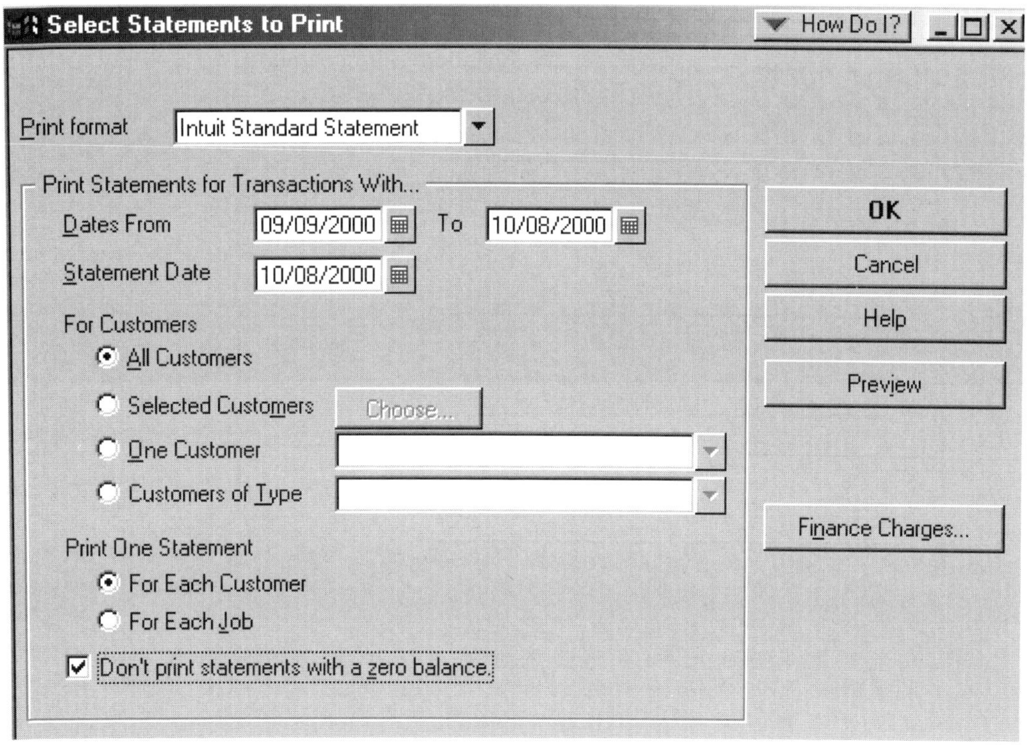

EXHIBIT 19.5 Select Statements to Print window.

9. Select the printer, type of paper or form, and number of copies. I suggest you print two copies—one for the customer and one for your files.
10. Click **Print**.
11. Click **OK** when you are done.

TIP See the steps for printing vendor labels in chapter 16 and modify them to print customer labels to send your statements.

CHAPTER 20

Receiving payments

Now that you have invoiced your customers you are ready to receive their payments. I recommend that you always enter payments as you receive them. When you use the Receive Payment process, QuickBooks Pro 2000 offers you two choices on how you will deposit the monies to your bank account.

What type of money collector are you? Are you the type of owner who runs to the bank with every customer's check? Then deposit the check directly to your bank account in QuickBooks Pro. Or do you like to wait and gather up all of your payments to make the big deposit? If you plan to keep your deposits in a "drawer," then classify the payments as Undeposited funds until you actually put them in the bank.

If you operate on a cash basis, you will not want to use the QuickBooks Pro Receive Payments function. You should be using the Cash Sale feature, and not the invoicing process. If you have created an invoice, however, you must use the Receive Payments function to record the invoice payment in order to relieve the receivable.

Receiving payments against invoices

To receive payments against invoices, follow these steps.

1. In **Customers** click on **Receive Payments**, or on the **Customers Navigator**, click on the **Receive Payments** icon.
2. In the Receive Payments window (see Exhibit 20.1), complete the following fields, as applicable. Many of these fields are self-explanatory. If you need help, click the **How Do I?** Help button and then select **Enter Payments**.
   - **Customer:Job**—Select the customer or job from the list.
   - **Date**—Type the date that you received the payment.
   - **Balance**—After you select a customer or job the account balance for the customer or job appears in this field. You cannot change this value.
   - **Amount**—Type the amount of the payment. You can receive a partial payment for an open invoice.
   - **Payment Method**—Select the method of payment.
   - **Check No.**—Type the number if paid by check.
   - **Existing Credits**—If this customer has a credit it will appear in this field. You can elect to apply a credit to any open invoice by placing a check in the **Apply Existing Credits** box.
   - **Outstanding Invoices/Statement Charges**—After you select a customer or job this portion of the window will show all open invoices for the customer or job.

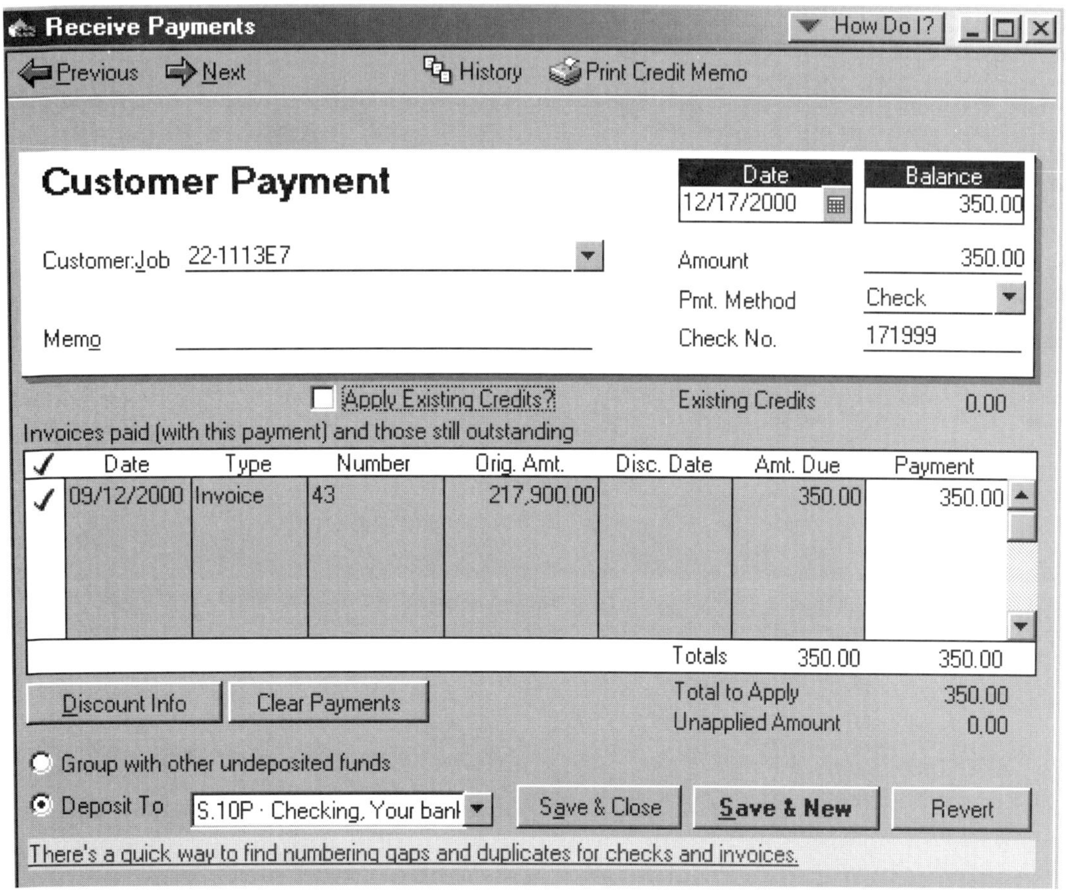

EXHIBIT 20.1 Receive Payments window.

- You apply a partial or full payment to any open invoice by placing a checkmark next to the invoice. You may select more than one invoice, and you can adjust the amount in the payment field.
- **Discount Info**—If you offer early payment discounts click this button to access the Discount Information dialog box.
- **Group with other Undeposited Funds**—Select this option if you have several checks to deposit, or if you do not plan to deposit the checks into the bank immediately. In your accounting system, the payments sit in the Undeposited Funds account until you indicate that you are ready to make the deposit. Select **Deposit To** *only* if you just have one check and you plan to actually deposit the payment in the bank immediately.

NOTE

Once you select Deposit To and the correct account number QuickBooks Pro will post the amount of the payment. If this is the only check you will be depositing in the bank, that is fine. If you have several checks that you are taking to the bank and you do that with each payment, however, the system will post each deposit separately to your account. This makes it difficult to reconcile your account when your bank statement comes. The second drawback to this option is that if you receive the check on Friday and don't put it into the bank until Tuesday the system will show a deposit date of Friday while the bank's deposit date will be Tuesday.

3. When you have finished making changes click **Save & New** to save the changes and apply another payment, or click **Save & Close** to save the changes and return to the previous window.

Making a deposit

When you are ready to go to the bank select the payments you want to deposit.

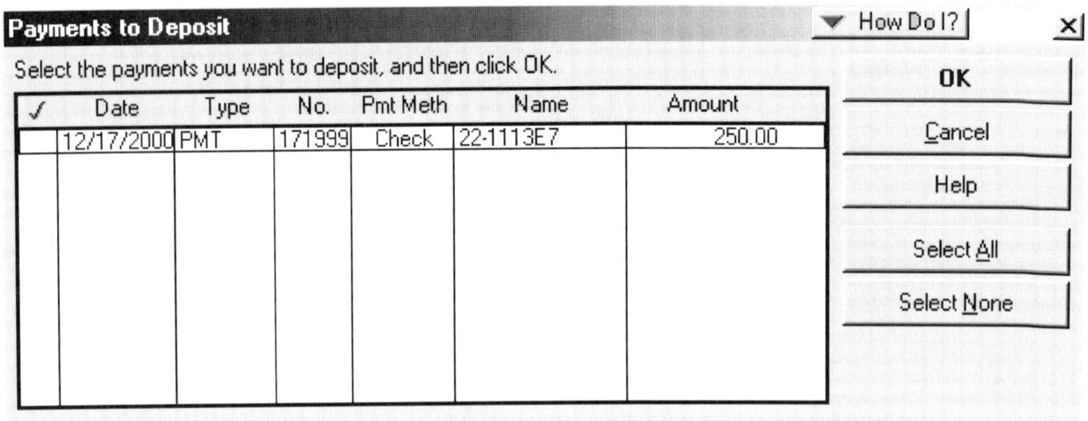

EXHIBIT 20.2a Payments to Deposit window.

1. In **Banking**, click on **Make Deposit** or in the **Banking Navigator** click on the **Deposits** icon.
2. Select the payments you want to deposit by clicking on them.
3. When you have selected the payments you want to deposit click **OK**.
4. Select the account into which you are going to deposit the payments.
5. Enter the date of the deposit.
6. The Received From field will pick up the information from the payments you are depositing. QuickBooks Pro will total your deposit.
7. Click **Save & Close**. QuickBooks Pro has appropriately adjusted your bank balance.
8. Go to the bank and make the deposit.

NOTE If you want to add other amounts you have received to this deposit, enter the receipts in this window as an additional deposit. Enter the name the money was received from, the income or expense account the dollars should be assigned to, and add an explanatory memo, such as rebate on software. Enter the check number, payment method, and the amount. QuickBooks Pro will add it to the deposit (see Exhibit 20.2b).

Warning—If you receive a refund or rebate check that relates to a job you must enter it as a special transaction because the Make Deposit window does not give you the option of selecting an item. The system workaround that I've developed is to create the equivalent of a journal entry, using the one QuickBooks Pro form that will let me access all of the parts of the system I need to use. The form? A check. Here's how to do it with what we refer to as a $0 check.

1. From the Banking Navigator, select **Checks**.
2. Choose any of your bank accounts as this transaction will not reduce the account.
3. Enter today's date, and nothing in the payee field.
4. Go to the **Expenses** tab, select the account where you will be depositing the check and enter the amount of the refund check in the Amount column. Do not select a customer:job in this tab.

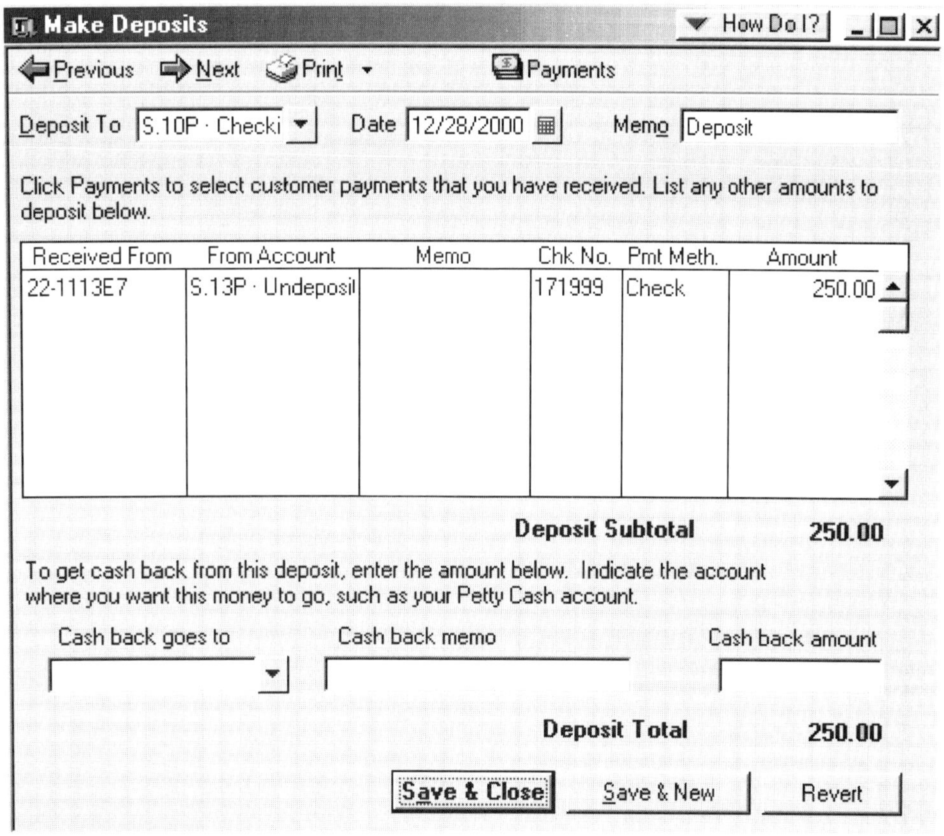

EXHIBIT 20.2b Make Deposits window.

5. Go to the **Items** tab, select the item(s) for which you are receiving a refund, enter the description, and be sure to enter the customer:job to which the refund relates.
6. Enter the amount of the refund as a *negative* number. This will make the amount of your check $0.
7. To make reconciliation easier, you should deposit this amount separately at your bank, as you will not be able to use the normal QBP Undeposited Funds or receive payments feature.
8. Click **OK** and close.

You have just reduced the items cost for a particular job, and increased your bank account by the amount of the refund.

Issuing refunds and credits

Your customer called, upset because the wrong colored countertops were installed in the kitchen. You know that you can't get a new countertop in the time you have left to finish the job, so you offer a $300 credit if the customer will live with the color.

When you have to issue a credit to a customer, use a credit memo. The credit memo reduces the amount that the customer owes you.

In accounting terminology, a refund is separate from a credit memo. While a credit memo reduces the customer's account balance, a refund is issued, usually in the form of a check, when you owe the customer money as a result of a credit to them or to return an overpayment.

Creating credit memos and refund checks are relatively easy tasks. One thing to keep in mind, however, is that you must issue a credit memo before you can write a refund check.

Creating credit memos

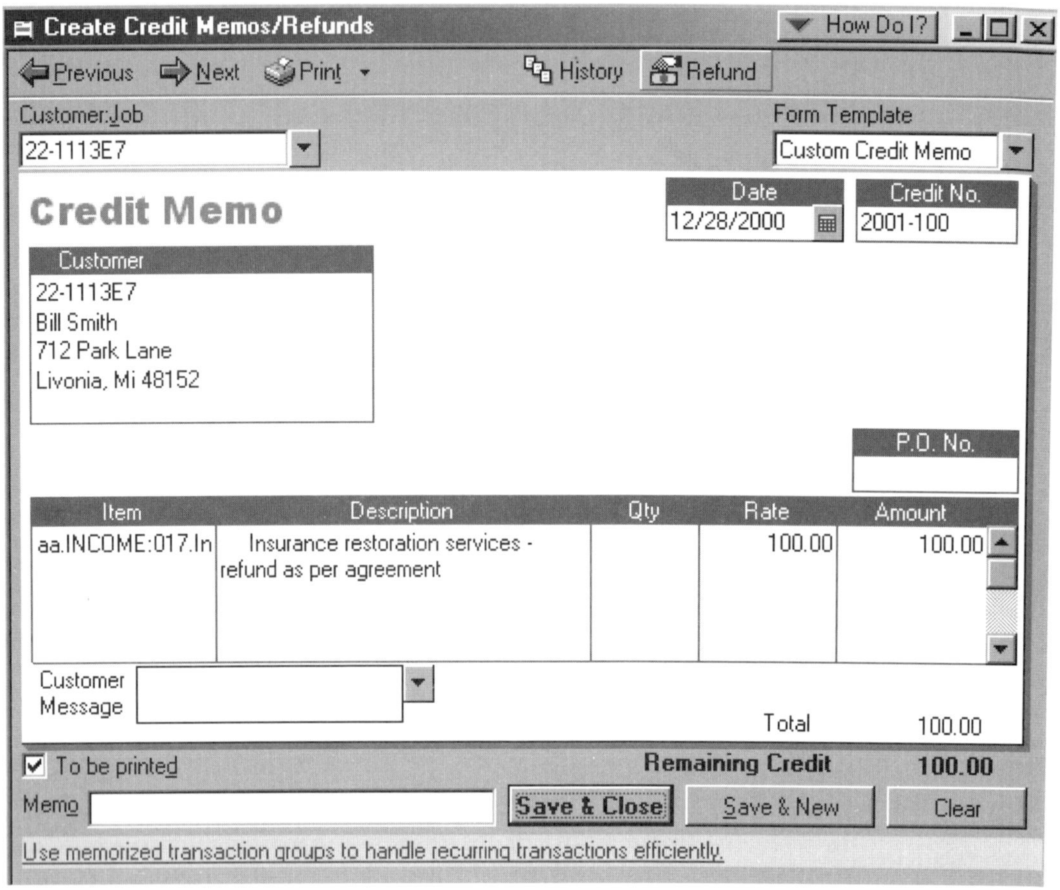

EXHIBIT 20.3 Create Credit Memos/Refunds window.

1. In **Customers** click on **Create Credit Memos/Refunds**, or in the **Customers Navigator**, click on the **Refunds and Credits** icon.
2. Select the appropriate credit memo form from the Form Template drop-down list. Enter or edit the following fields, as applicable. Many of these fields are self-explanatory. If you need help, click the **How Do I?** Help button, and then select **Create Credit Memos**.
   - **Customer:Job**—Select a customer or job from the list.
   - **Name/Address**—After you select a customer the default name and address information for the customer appears in this field. You can change the information or accept it as is.
   - **Date**—Type the date of the credit memo, such as the date you offered your customer the $300 credit.
   - **Credit No.**—QuickBooks Pro assigns the Credit Memo number, increasing the number incrementally by one with each credit memo. You can change the number or accept it as is.
   - **Item**—Select the item name for which you are providing a credit from the list. Use the same item number you originally invoiced the item under.
   - **Description**—After you select an item the default description appears in this field. You can change the information or accept it as is.
   - **Quantity**—Type a quantity, if applicable.

- **Rate**—After you select an item, the default price for the item appears in this field. You can change the information or accept it as is.
- **Amount**—This amount is calculated automatically based on the information in the QTY and Rate fields.
- **T**—A T appears automatically if the item is taxable. In order to calculate and reduce the tax, however, the customer must also be taxable.
- **Customer Message**—Select a message from the list.
- **Tax**—Review the sales tax item to be sure you have the correct item or group.
- **Customer is Taxable**—After you select a customer, his or her tax information appears. You can change the information or accept it as is.

3. When you have finished making changes you can either issue a **Refund** check, or click **Save & Close** to return to the previous window.

Issuing a refund check

If you want to issue a check *do not close* the Credit Memo form. Leave it open and follow these steps.

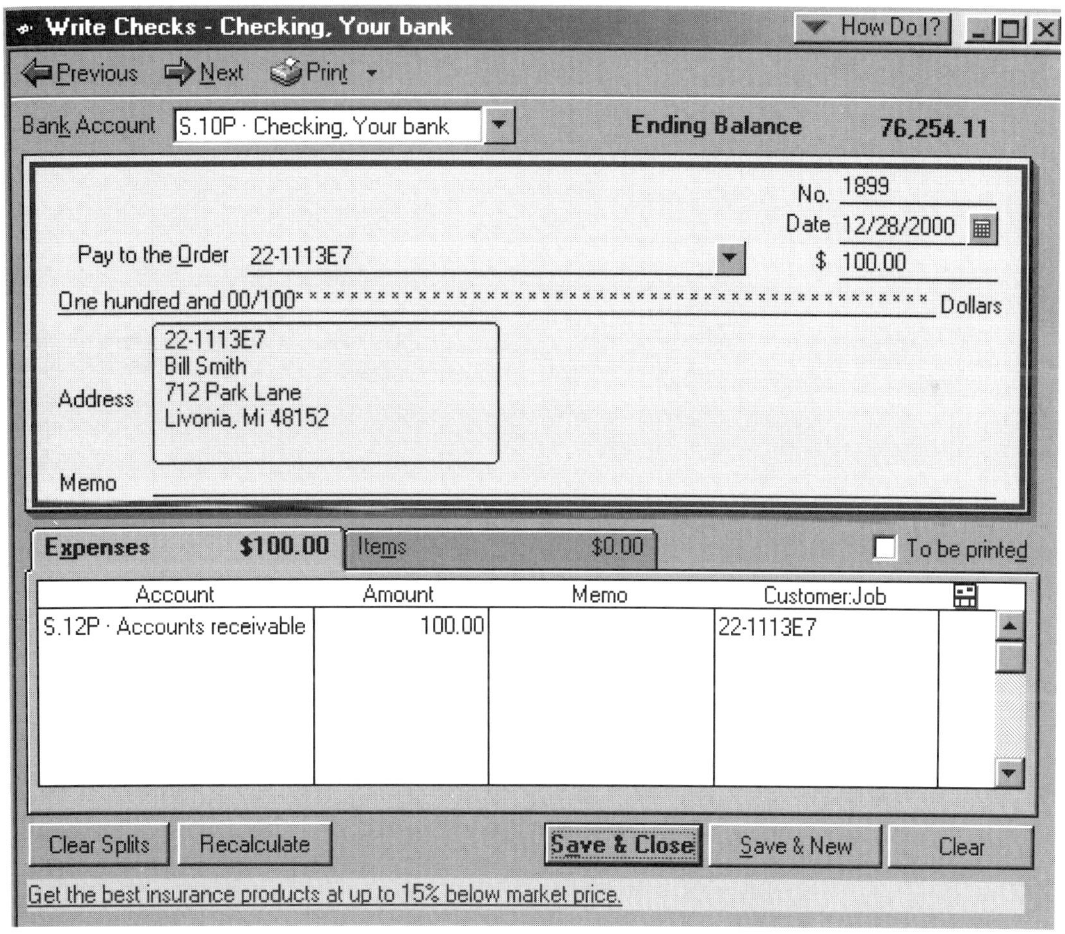

EXHIBIT 20.4 Writing a refund check using the Write Checks window.

1. With the applicable credit memo open, click the **Refund** button near the top.
2. QuickBooks Pro's AutoFill feature will fill in the check with the data from the Credit Memo form. You can change the date or check number as necessary.

3. When you have finished making changes, click **Save & Close** *two times* to save the changes to the refund check and to save the credit memo.

NOTE To issue a Refund check you must be in the applicable Credit Memo window.

Transferring funds

QuickBooks Pro makes it easy to transfer funds on the system between accounts.

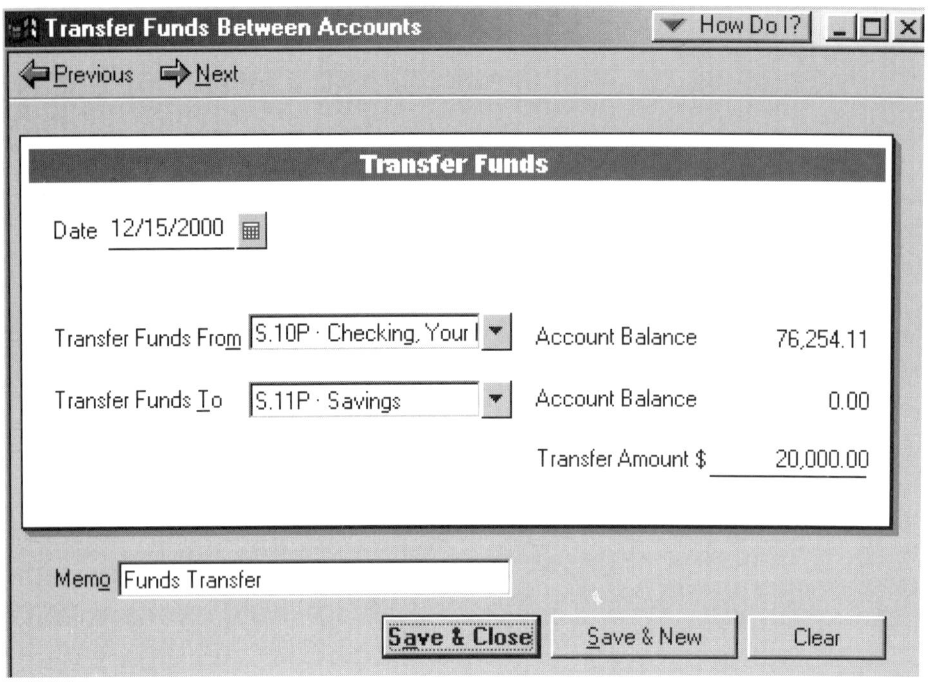

EXHIBIT 20.5 Transfer Funds Between Accounts window.

1. In **Banking**, select **Transfer Funds** or in the **Banking Navigator** click on the **Transfer Funds** icon.
2. Enter the date of the transfer.
3. In the **Transfer Funds From** field select the account from which you want to transfer funds. QuickBooks Pro will show you the account balance.
4. In the **Transfer Funds To** field select the account to receive the funds being transferred. QuickBooks Pro will show you the account balance.
5. Enter the amount to be transferred.
6. When you are satisfied with your transaction, click **Save & Close**.

If you actually write a check from your checking account to transfer the funds to your savings account, for example, you can go to the Check Register and enter the actual check number. You can access the Check Register two ways:

1. Under **List**, click on **Chart of Accounts**. Then select the correct checking account and double click on it.
2. In the **Banking Navigator** click on the **Check Register** icon and select the correct checking account.

CHAPTER 21

Recording credit card transactions

Why should you enter your monthly credit card payments and charges in QuickBooks Pro 2000? Why not just enter the monthly amount due directly into accounts payable?

The main reason is because you have multiple transactions with different vendors on different dates charged to your card. If you put the monthly payment in as an account payable, the vendor will be VISA, Master Card, American Express, and so on. Then when you create a report you will see a charge for drywall, but the vendor will be VISA. You won't have a record of the vendor from which you bought the drywall. You have lost part of the trail in the accounting system. You also will have lost transaction dates, and very well may not be realizing expense in the proper period.

The QuickBooks Pro Credit Card feature lets you enter individual transactions with the correct vendor and date. If you carry an outstanding balance on the card you also can enter the finance charge when you reconcile the account within the system.

I suggest that you collect the slips throughout the month from employees, owners, and so on. Write who made the charge, what it was for, and the customer or job, if relevant. You could put all the slips into an envelope until you receive the statement, or, if you need to see the information right away to invoice customers, you may want to enter the charges as soon as you receive the slips. If you use this method, reconcile the charges in the system against the statement when the statement comes in. Attach the slips to the statement as you reconcile the statement.

TIP

There are two tips I recommend to keep your Vendor List from ballooning out of control and to reduce the number of individual monthly transactions.

If you use your credit cards for meals, instead of entering every restaurant as a separate vendor I recommend setting up a vendor called restaurants. In the memo section enter the actual dates and amounts of each transaction because for tax purposes you have to document where you were, whom you were with, and the purpose of the meeting. This trick allows you to record the information you need for IRS meals and entertainment documentation without cluttering up your Vendor List.

Likewise, consolidate your gasoline charges for the month into one entry and date the entry the last day of the month under a vendor called gasoline. If you need the details in the future you can go back and pull the charge statements. If your statement is from the 15th of one month to the 14th of the next, for example, enter the charges through the end of the first month as one entry under gasoline on the last day of the month. Then add together the charges for the next month and enter them under gasoline. Date this entry the closing date of the credit card statement.

TIP
Continued

Entering receipts for frequent charges this way keeps you from being overwhelmed, while still giving you the information you need for tax and accounting reports.

Entering credit card purchases

To enter credit card purchases, follow these steps.

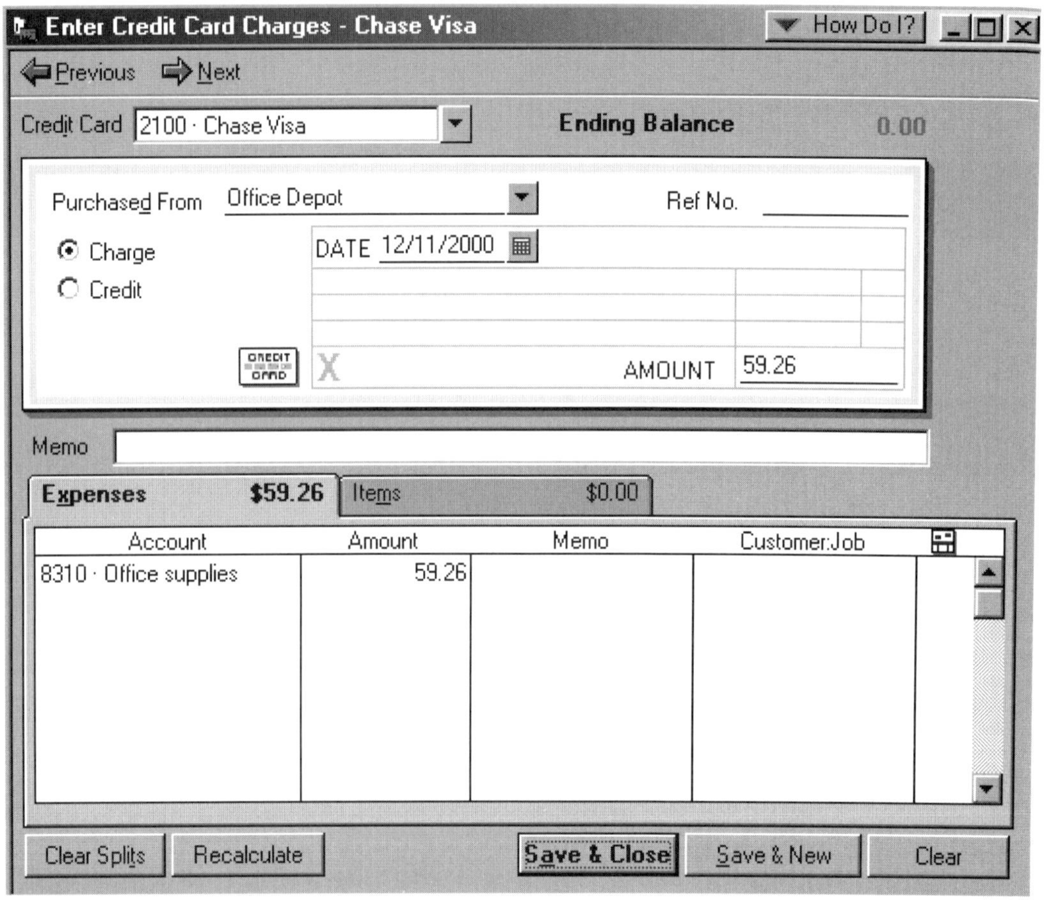

EXHIBIT 21.1 Enter Credit Card Charges window.

1. In **Banking** select **Enter Credit Card Charges** or on the **Banking Navigator**, click on the **Credit Card** icon.
2. Select the correct credit card account.
3. Enter the following information for each purchase, as applicable. Most of the information is self-explanatory. If you need help, click on the **How Do I?** Help button and select **Credit Cards**.
   - **Vendor**—Enter the actual name of the vendor, such as the store name, the lumber company's name, and so on.
   - **Reference number**—I usually don't enter one because the numbers on the credit card statement are long and complicated and you can refer back to the statement, if necessary.
   - **Charge or credit**—Select whether the transaction is a charge or a credit.
   - **Date**—Enter the transaction date on the statement.

> **NOTE** For a cash-basis business, you still will want to use the Credit Card feature to record your purchases. The date you put something on a charge card is the same as cash to the IRS, so be careful to use the actual transaction date.

- **Amount**—Enter the amount of the purchase.
- **Memo**—Enter a description of the purchase, such as tools for Bob, drywall, and so on.

4. Select the **Items** or **Expenses** tab at the bottom of window, just like you do when entering a bill. If it is a direct expense use a cost of goods sold (COGS) item and enter it under the Items tab. Otherwise, under the Expenses tab assign the purchase to an another account in the Chart of Accounts.
5. Modify the description if necessary and enter the quantity. Adjust the cost of the item and/or the amount, if necessary, and add the customer or job, if applicable.
6. Click on **Save & Close** or **Save & New** to enter another purchase. QuickBooks Pro adjusts the balance of your credit card appropriately.

> **NOTE** If you send out a purchase order and then pay for the materials with a credit card, QuickBooks Pro will tell you that there is an open purchase order for the vendor and ask you if you want to apply the credit card purchase against the purchase order. If you say yes the process is the same as that for entering bills against a purchase order.

Recording your monthly payment

To enter your monthly payment, follow these steps.

1. In **Vendors**, select **Enter Bills** or in the **Vendors Navigator**, click on the **Enter Bills** icon (see Exhibit 21.2).
2. In the Vendor field enter the charge card vendor, such as Master Card, VISA, or American Express.
3. In the Amount field enter the amount you are going to pay, either the full amount, or a partial payment.
4. In the Due Date field enter the date the payment is due.
5. At the bottom of the window select the correct charge card account. Do not enter a customer or job.
6. Click on **Save & Close**. At this point the credit card payment becomes another bill waiting to be paid.

Making the most of the Credit Card feature

If you have a credit line with the local lumber company or another vendor set up a credit card account for the line of credit. Use it the same way you would a credit card. The vendor would be the lumber company; the balance would be the amount of the credit you have used. This is a good way to keep track of individual transactions and your outstanding balance. Reconcile the lumber company's statement just like you would your bank or credit card statements (see chapter 23).

Likewise, if you have reimbursements due to the owner or an employee for travel or another type of purchase, set up a credit card account for the individual with the owner's or employee's name as the credit card name. If you put amounts due to the owner or employee in accounts payable, your job-cost reports will show the owner or employee as the vendor and you will lose the true vendor detail. It also doesn't look good if you are audited.

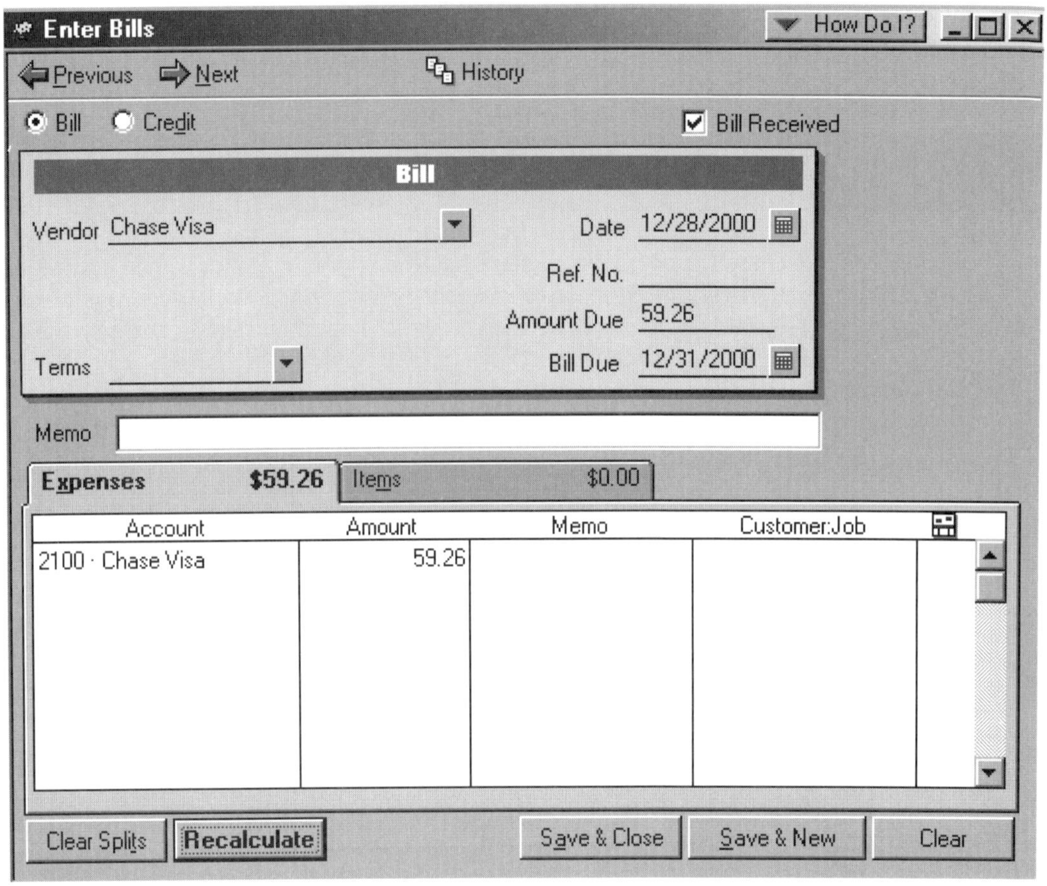

EXHIBIT 21.2 Entering a credit card bill in the Enter Bills window.

Separating owner's credit cards

I recommend that you segregate shareholders' credit cards from company credit cards. Many owners or shareholders use their personal credit cards to help finance the company. For tax reasons you also should designate specific credit cards for company business. Put only company business on that card so that the debt is shifted to the credit card. For tax purposes you can write off all the interest as if the credit card was owned by the business.

There also are legal considerations for designating a card as the company card. When an owner or shareholder uses his or her credit card, it is the same as if the owner or shareholder loaned the company money. The owner or shareholder is legally responsible for paying the credit card that is in his or her name. In certain loss situations the owner or shareholder must show enough basis or investment in the company to take the loss on his or her personal tax return. Segregating the account will show this. See your tax advisor for more details.

C H A P T E R 22

Accepting cash sales

In addition to the ability to create invoices, QuickBooks Pro 2000 has features to assist remodelers and home builders who make cash sales, including retail sales or handyman services. A cash sales receipt also may be the best place to enter a sale if your company does not use invoices. Entering all types of payments in this one-step process will let you keep track of sales in a detailed manner.

The cash sales window looks almost identical to the invoice window. Entering information on a cash sale, however, does not create a receivable. Instead, the amount of the cash sale posts directly to Undeposited Funds or your bank account depending on which account you specify. Your income also is increased.

Don't be confused! Cash doesn't mean green currency. Cash in this case means that you have received payment at the time of the service performed.

Entering a cash sale

1. In **Customers**, select **Enter Cash Sales** or in the **Customers Navigator** click on the **Cash Sales** icon (see Exhibit 22.1).
2. Select the appropriate template from the Template Form pull-down list. For help creating custom templates for receipts click on the **How Do I?** Help button and select **Print and format the cash sale form**.
3. Enter the following information for each sale, as applicable. Most of the information is self-explanatory. If you need help, click on the **How Do I?** Help button and select **Enter Cash Sales**.
   - **Customer:Job**—Enter the customer or job.
   - **Date**—Enter the date of the sale.
   - **Sales Number**—QuickBooks Pro will automatically increase the number by one starting with the number your designate.
   - **Sold To**—This automatically will fill in based on the information you entered when you set up the customer. You may modify this information if you wish.
   - **Check Number**—If the customer paid with a check enter the check number.
   - **Payment Method**—Indicate how the customer paid—cash, check, credit card, and so on.
   - **Item**—Select the correct item from the Item List. Modify the item description if you wish.
   - **Quantity**—Enter the quantity. QuickBooks Pro will calculate the amount due.
   - **Rate**—This is the default price assigned to the item. You may change the price if you wish.
   - **Amount (Price)**—This is calculated automatically based on the information in the quantity and rate fields.

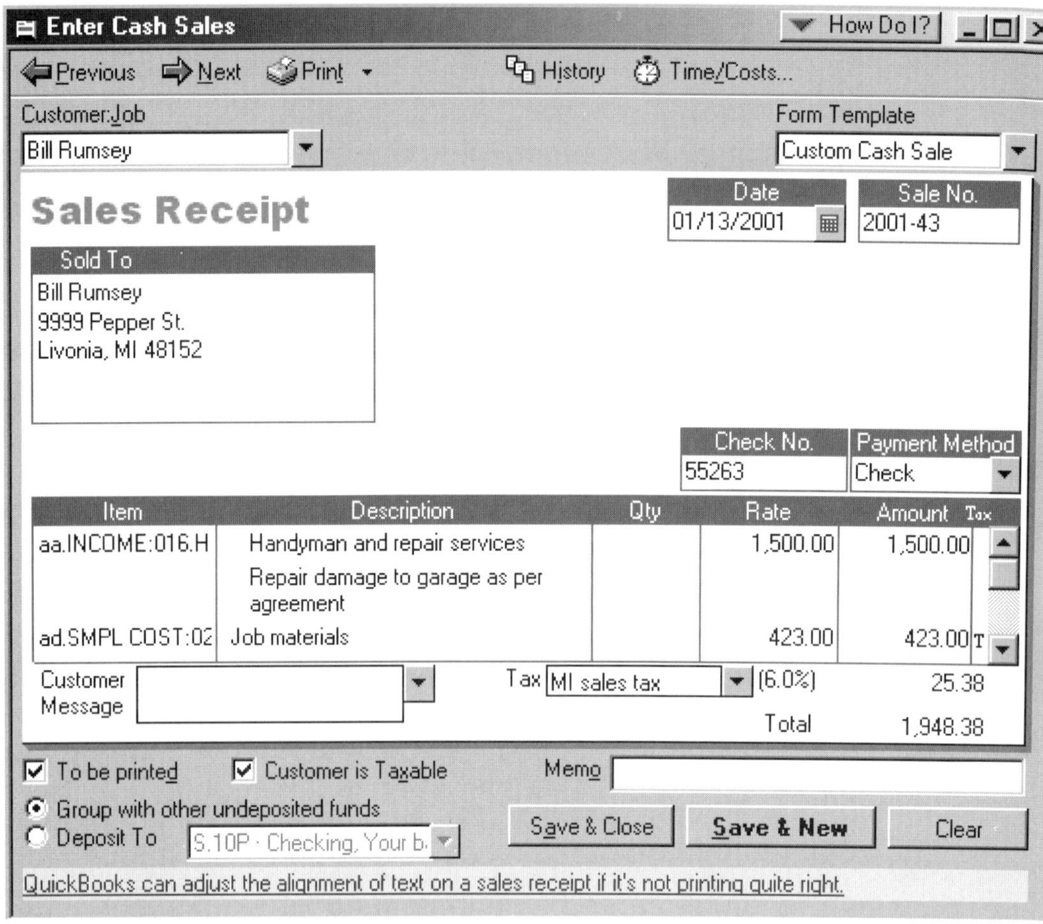

EXHIBIT 22.1 Enter Cash Sales window.

4. Decide if the item is taxable. If the item is taxable, select the appropriate sales tax item at the bottom of the window.

NOTE For the system to calculate and apply sales tax, the customer and the item must be designated as taxable.

---

5. Click on the **To Be Printed** box near the bottom to send the receipt to the print cue. Or click on the **Print** button near the top while the form is open and the system will print the sales receipt.
6. Click either on the **Group with Other Undeposited Funds** button or the **Deposit to** button (and select the appropriate bank account), depending on the timing of your actual bank deposit. QuickBooks Pro will handle the receipts in the same way it tracks payments entered through the Receive Payments feature.
7. Click **Save & New** or **Save & Close** to end the transaction.

A cash sales receipt will not show up on a statement because it does not create a receivable. If you want the transaction to show up on a statement showing receipts and invoices, create an invoice and receive the payment against the invoice.

Another way you could see the information is by running a sales by customer summary report or a sales by customer detail report. Both of these reports show cash sales receipts and invoices.

Accounting for sales taxes

The sales tax system in QuickBooks Pro is a subsystem like the accounts receivable and accounts payable subsystems. When you have a sale in which you are creating a sales tax payable, QuickBooks Pro keeps track of the tax liability. The sales tax liability report shows your tax liability for the period you define.

1. In **Reports**, select **Vendors & Payables** and then the **Sales Tax Liability** report or in the **Vendors Navigator** click on the **Report Finder** icon and then select **Sales Tax Liability** from the list.
2. Select the timeframe for the report, usually the prior month.
3. The system will generate a list of the different types of taxes owed, the amount due, and the sales dollars by sales tax item.

Making the most of the Sales Tax feature

Most states require that you submit a report that breaks out:

- Gross sales
- Sales to exempt buyers or resellers
- Taxable sales
- Type(s) of taxes collected
- Total amount due by tax, if more than one

To accumulate this information set up sales tax items that match the types of information you need to exclude or report, such as items called 0%-Reseller, 0%-School, or 0%-Finance charges. By keeping track of your nontaxable sales by sales item, you can get a sales report subdivided into each of your sales tax classifications. Talk to your tax advisor about the requirements for your state. The sales tax liability report will show all sales tax items, their gross sales, and their tax totals.

Paying sales taxes

When it is time to pay your sales tax, use the Pay Sales Tax feature.

1. In **Vendors**, select **Pay Sales Tax** or in the **Vendors Navigator** click on the **Pay Sales Tax** icon.
2. In the Pay Sales Tax window select the account from which you will pay the taxes (see Exhibit 22.2).
3. Enter the check date. In the **Show Sales Tax Through** field enter the date through which you are paying your sales taxes, usually the last day of the prior month. The system will show you the amount due for the different sales tax items.
4. Click next to the appropriate tax item(s).
5. QuickBooks Pro will enter the amount due. You can modify the amount if necessary.
6. The system will show your ending bank balance after the tax payment.
7. Click **Print** to print the check or click **OK** to close the window.

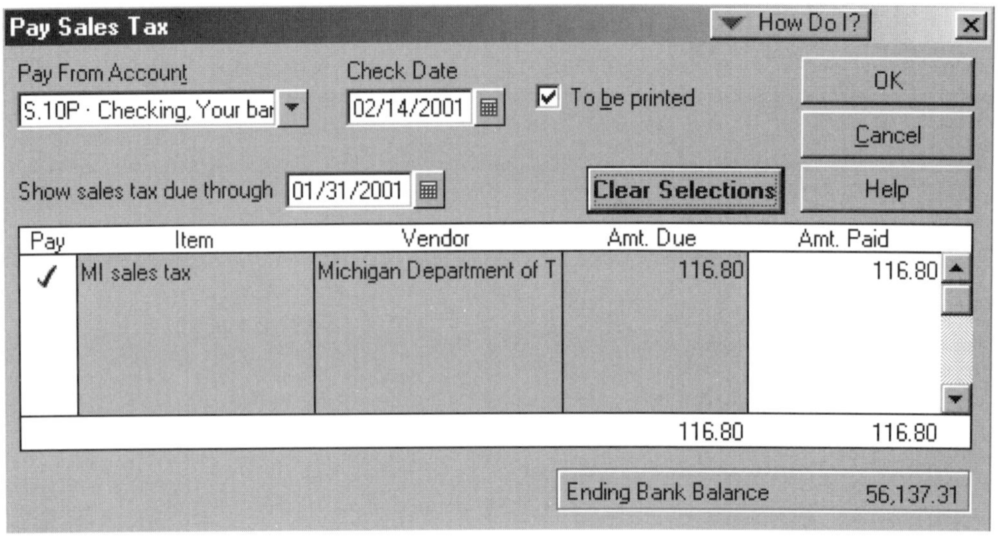

EXHIBIT 22.2 Pay Sales Tax window.

CHAPTER 23

Reconciling statements

QuickBooks Pro 2000 allows you to reconcile any account on the balance sheet—asset, liability, and equity accounts. You definitely will want to reconcile your bank, credit card, loan, and line of credit accounts each month when you receive the statement. The goal of the reconciliation process is to make sure everything shown on the statement has been recorded on your books and that any differences between your QuickBooks Pro balance and the statement balance can be explained (reconciled). There also are reasons to reconcile other accounts, such as customer deposits.

To begin the reconciliation process for a bank account, follow these steps.

1. In **Banking** select **Reconcile** or in the **Banking Navigator** click on the **Reconcile** icon.
2. Select the account you want to reconcile (Exhibit 23.1).
3. Enter the ending balance from the bank statement.
4. Enter the service charge, if there is one, and the bank statement date.
5. Select the bank service charge account from the Chart of Accounts.
6. Enter the interest earned, if any, and the bank statement date.
7. Enter the interest income account where any interest income will be posted.
8. The upper window will show all the deposits and other credits, and the bottom window will show checks and other withdrawals in the account that have not yet been cleared.

TIP Maximize the window so you can see as many transactions as possible.

9. Begin with deposits. Click off each deposit shown on the bank statement.

NOTE If you used the Group by Undeposited Funds feature, the deposit totals on the bank statement should match the deposit totals on the window.

TIP To keep your eyes on the Amount column, instead of making a checkmark in the far left column, cover the amount that is in the far right column with your cursor and click. The system will enter the checkmark in the far left column for you.

10. Next check off all checks and other withdrawals from your account that have cleared according to your bank statement.

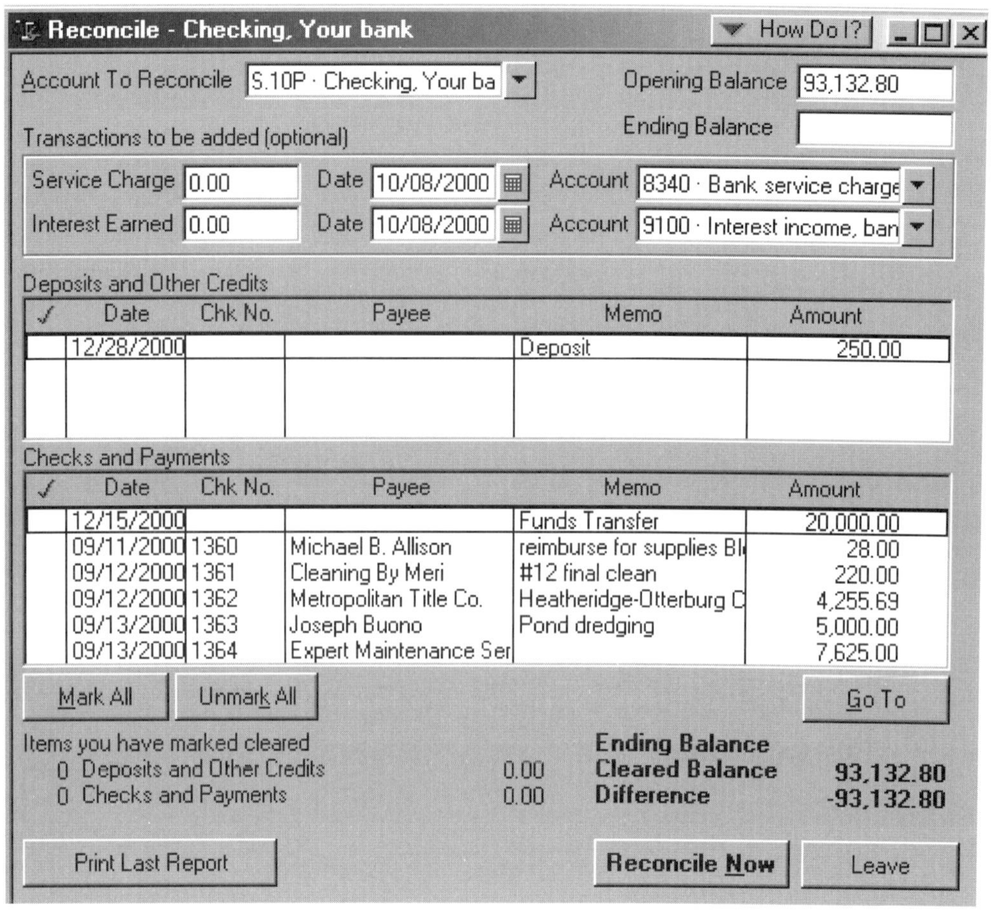

EXHIBIT 23.1 Reconcile window.

In the left-hand side of the window there are two fields that are helpful. One is the deposits and other credits, which many banks show as a total on their statements. The other field is the total for checks and other payments. Now is a good time to stop to make sure the totals match. If they don't match your bank statement, you need to find out the difference and research the missing amount.

Taking care of discrepancies

If the amount of a check in your Check Register and the bank statement disagree there are two ways to take care of the discrepancies.

For example, say the Check Register shows the amount of a check as $520.20 and the bank statement says the check was for $520.29. Highlight the transaction and click on the Go To button to go back to the transaction and change the check. Alternatively, you can create a check for $0.09. Instead of entering a check number, enter "adj bank" to note that the check was for a bank adjustment. Pick out the miscellaneous account for the $0.09 and make a notation that says the extra $0.09 was charged by the bank on check #1234. The $0.09 check will appear on the Reconcile window. Clear the original check and the check for $0.09 so the two together equal the amount that cleared on the bank statement. If the adjustment to the check is for a larger amount, you may want to select the account the original check posted to as your adjustment account.

Another common reconciliation discrepancy results from not recording all bank charges. For example, the statement shows a charge for reprinting checks that you have not recorded on your books.

In Banking, select Write Checks. Enter the vendor (the bank), the amount, and the appropriate expense for the check printing fee. Instead of giving the check a number, enter "bank chg" to indicate it was a bank charge. Click Save & Close. When the "check" appears on the Reconcile window, clear it with a checkmark.

TIP Create a transaction and memorize it for automatic withdrawals. Tell the system to post the transaction every month for the next 12 months. Then, the entry already is in the checking account as uncleared when the statement arrives.

When you reconcile you are working against the ending balance field in the lower right-hand corner. The Reconcile window shows the ending balance, cleared balance, and difference. The cleared balance changes each time you clear a check or deposit. The goal is to get the difference to $0. Once you get the difference to $0, click on Reconcile Now.

The Reconciliation Complete window (see Exhibit 23.2) will appear and the system will ask you: What type of reconciliation report would you like to print? Select the third choice—A full reconciliation report. Enter the bank statement closing date and click OK. The report will show the transactions that have cleared and those that are still outstanding. Between now and the next time you reconcile the account you can reprint this report if you wish. Once you complete another reconciliation, however, this report is lost, replaced by the more current report.

After you click OK, QuickBooks Pro will take you to the Print Lists window. Select the printer, orientation, number of copies, and so on. A window will pop up asking if the report has printed correctly. Before you say yes, look at the report. The front page is a summary with the date printed, name of your reconciled account, and the statement ending date. The report also will show the previous balance and the cleared balance. The cleared balance should match the ending balance on the bank statement.

The uncleared transactions are the total of transactions on the system that haven't cleared yet, including deposits and checks, as of the date of the report. The account balance as of the date of the report should be your bank balance on QuickBooks Pro as of today.

NOTE If at midstream you need to leave the reconciliation process, click on the Leave button at the bottom of the window. The system will let you leave the reconciliation process in midstream and go do other things in the system. When you come back to the reconciliation process you will have to enter the interest earned, service charge, and ending balance again.

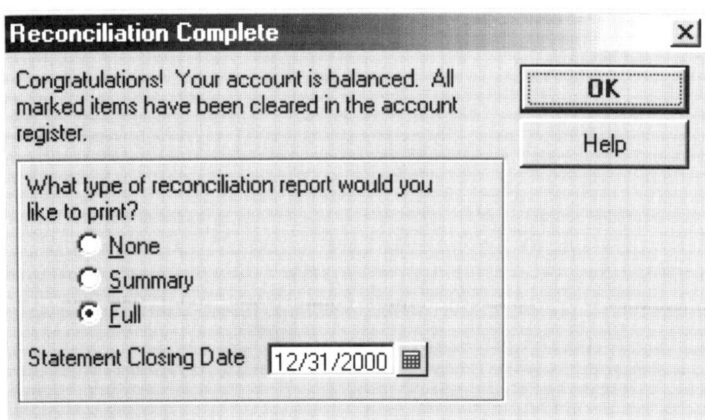

EXHIBIT 23.2 Reconciliation Complete window.

Clearing old transactions

The next step is for the owner to look at the uncleared transactions. If a check that is more than six months old shows as uncleared, the person may have gotten it and never cashed it, or the check may be lost. Depending on the circumstances, you can either bring the account up to date by not reissuing the check, or you can issue a replacement check.

If you are not going to reissue the check, don't void it. You probably have issued financial statements, job reports, and possibly even tax reports that included the effect of that check. Voiding the check will alter all these transactions and create havoc in your books.

Clear a check

Instead, research the check so you know what it was issued for initially. Look at it in the Check Register or view the original transaction. If the check wasn't written for materials that were assigned to a cost of goods sold (COGS) item, create a journal entry. The entry will increase your cash by the amount of the check by debiting your checking account and should credit the account the original check was issued against. Date the journal entry as of the date of the reconciliation. Then go to the Check Register and manually clear the old check and the journal entry "deposit." This removes the old check, increases your checking account balance, and leaves your system intact.

Replacement checks

If you need to issue a replacement check, go to the Banking Navigator and click on the Checks icon. Enter the vendor and the check amount. Click on the Expenses tab and select the bank account you are reconciling. In the memo field note that you are replacing the check and include the original check number. Print the check using the next check number and give it to the vendor. Go to the Check Register and manually clear the original check. Then find the new check. It will be posted on two lines in the account. One line will show it as a payment and the other will show it as a deposit. Go to the line that says deposit and manually clear the amount. You now have a new outstanding check in your system with the replacement check date. You haven't affected your income or expense and the old original check has gone away.

Old deposits and credits

You should not expect to see any deposits or credits remaining as uncleared, unless they were deposited on the day before the statement. If you see an older deposit it should raise questions. Find out why the deposit wasn't made.

Implementing owner safeguards

If, as the owner, you prepare the checks and have total control over the bank account, then you may not need the actual cancelled checks to be returned to you with the bank statement. If someone else has control over your checks, however, look at the cancelled checks and make sure the vendors are names that you recognize. This is a 10-minute control process that can help detect fraud. Place the cleared checks in numerical order so you can account for all the checks.

I suggest that the bank statement come to the owner unopened, or be mailed directly to the owner's home. The owner should remove and review the cancelled checks, and then give the statement to the person doing the reconciliation. The statement and reconciliation report then should be given back to the owner to make sure they match.

Reconciliation should be done within a day or two after you receive the bank statement so you can track down any problems right away.

While reviewing the reconciliation report and the bank statement, I recommend that the owner look at all of the cleared transactions. Look at both the checks and deposits to make sure they all seem reasonable and that the vendors all look familiar, especially if someone else keeps the books.

These steps don't take a long time, but they do help to ensure that you know everything that is recorded on your books.

Reconciling other statements

Use the same reconciliation process and owner review steps for savings accounts and other accounts, including credit card statements. The only difference with the credit card reconciliation process is that you can't print a reconciliation report. The owner should see the statement and look at the QuickBooks Pro reconciliation window.

When you select a credit card to reconcile, the Reconcile window changes a bit. You can't enter the service charge or interest in this window. Instead the window has a line for finance charges. If there is a late payment charge or annual fee, go out and create a credit card charge for the transaction.

When you reconcile other balance sheet accounts, you are really just clearing transactions for informational purposes. If you put customer deposits through a liability account, for example, you can keep an eye on what is outstanding by reconciling the account. When you pay back the deposit or apply it, the account balance will increase or decrease. Check off transactions as they clear so that when you look at the account you see only the uncleared items that are outstanding.

SECTION FOUR

EMPLOYEES AND PAYROLL

The chapters in this section discuss the QuickBooks Pro 2000 functions related to payroll. Payroll is a detail-oriented, exacting process that requires great care. While QuickBooks Pro is a forgiving system if you make an error, your employees may not be, and the governments to which you must pay employee deductions and company contributions definitely are not!

Employees are those individuals considered a part of your company who you compensate with a paycheck issued by your company. A paycheck is different than a payment on a trade contractor's invoice. You withhold taxes from your employees' wages and assume the liability for the employee withholdings and deductions as well as the contributions your company makes on behalf of the employee for a variety of taxes and benefits. Talk to your CPA or tax advisor to make sure that you have clearly distinguished your trade contractors from your employees.

I recommend that you seriously consider contracting with a payroll company such as ADP or Paychex to handle the processing of your payroll checks and the liability payments to the many entities for which you withhold and collect monies. These companies perform these functions very well. All you have to do, theoretically, is have the money in the bank. Paying a payroll service to submit the paperwork and payments on time could save you money in the long run—especially if you miss a filing deadline and are charged a substantial penalty.

So I've convinced you to use a payroll service. Now why, you ask, do you need to know about the Payroll features in QuickBooks Pro? There are two short answers: job costing and a check-and-balance on the payroll service.

There is a Payroll Item List that works much like your Item List. By creating different payroll items to be used on paychecks and liability checks, you can create a simple paycheck or you can do some pretty powerful stuff. You can assign payroll burden to your jobs and customers—not just employee pay, but the benefits and employee contributions you pay, as well as a portion of the costs of your administrative employee, if you choose. The ability to assign all your employee-related expenses to your different jobs gives you a true job cost, which in turn will help with your budgeting and estimating processes.

The second reason for using payroll in QuickBooks Pro is that the payroll service still must get the payroll information from you. You can gather the time card information and compute it manually, which I don't recommend, or you can enter it in QuickBooks Pro. Like the other functions we have discussed, QuickBooks Pro simplifies data entry related to payroll, and lets you track sick and vacation leave, overtime, and other information you need. QuickBooks Pro reports also provide a way to check and verify the information provided to you by the payroll service and vice versa.

Even if you do not have employees at this time, I recommend you review these chapters so, in the future when you do hire your first employees, you understand the payroll process and all that it involves.

CHAPTER 24

Adding payroll items

The Payroll Item List operates in much the same way as the Item List. As with the Item List, you decide how detailed you want your job-cost reports and develop the payroll items you need to accomplish that detail level. You also will need to create payroll items for your company's special needs, such as deductions, contributions, local taxes, etc.

Again, like items on the Item List, you link the payroll items to accounts in the Chart of Accounts. For some items, like those that represent monies forwarded to an agency, you designate two accounts—one for the expense-debit side of the entry, and one for the liability-credit side of the entry. Other payroll items, such as administrative wages, only require that you assign one account for the expense side of the entry.

The first way to categorize your employees is to divide them between hourly and salary. These are the two basic compensation categories. Under each of these categories are five primary types of employees:

- Direct labor
- Supervisor
- Administrative
- Sales and marketing
- Owner

Generally, your direct labor and supervisory employees are those most often associated with specific jobs. If you refer to Exhibit 8.2 in chapter 8 you will see the direct cost (COGS) account **S.520.Direct labor comp/taxes/burden** to which these payroll costs will be posted.

On the other hand, the payroll costs associated with administrative employees, sales and marketing employees, and the owners typically are not considered to be direct labor. Instead, they are considered an overhead expense and, therefore, are posted to expense accounts as opposed to direct cost (COGS) accounts. Exhibit 9.2 in chapter 9 includes the **S.730.Sales-comp/taxes/burden**, **S.802.Admin employment costs**, and **S.804.Owner's comp & benefits** accounts.

Finally, for each of these types of employees you create additional payroll items which, when used in payroll transactions, will assign correctly all payroll dollars to the appropriate accounts, such as:

- Regular compensation
- Overtime compensation
- Payroll taxes and other specific benefits
- Health and life insurance cost offsets
- Retirement benefits
- Union costs
- Allocated burden

Payroll items allow you to direct detailed costs for various types of employees to different kinds of cost accounts, such as direct labor, administrative, sales, or owner. Payroll items also allow you to assign, track, and view detailed compensation-related costs. You can see that to get a detailed break out of payroll costs on your income statement and job-cost reports you could possibly have a number of items on your Payroll Item List.

NOTE Payroll does not include trade contractors. Trade contractors are considered vendors. Talk with your tax advisor to make sure you are not treating any of your trade contractors as employees. The fines and consequences of treating a trade contractor as an employee or vice versa are serious. Be careful!

Naming payroll items

To keep the payroll item names easy to understand and remember I have developed a simple naming procedure to use as a prefix to the longer, more descriptive name. A side benefit of coding this way is that the payroll item names are easy to key in, a benefit when you have a large number of employees.

In my system, the first letter designates whether the employee is hourly (H) or salaried (S). The second letter(s) give the type of employee:

- A—Administrative
- D—Direct labor
- SA—Sales and marketing
- O—Owner
- SD—Supervisor, direct labor

The last letter indicates the type of pay:

- R—Regular pay
- S—Sick pay
- V—Vacation pay
- T—Training
- H—Holiday
- B—Bonus

Using this naming procedure, HDS would be used to track the amount of time incurred for, and the amount paid for hourly, direct labor, sick leave—or the cost of the sick leave benefit provided to an hourly employee whose time is normally attributed to specific jobs.

Tax-related items, however, should be listed by their actual name so they can be identified easily.

Selecting a QuickBooks Pro payroll service

QuickBooks Pro 2000 offers three types of payroll services, and you must select one before you can enter payroll information. The first choice is to figure your payroll manually and key the information into QuickBooks Pro. I do not recommend this method.

The second choice is to subscribe to the tax table service, which you pay for either monthly or annually. QuickBooks Pro calls this their Basic Payroll Service. The service provides tax tables, accessible only through the Internet, that QuickBooks Pro uses to calculate the taxes for your payroll. The system will remind you to look for tax table updates monthly, to be sure that you have the most current installed on your system.

If you want actual payroll services, like those offered by ADP or Paychex, you can use the Deluxe Payroll Service offered by Intuit. This QuickBooks Pro service feature will provide direct deposits, fill out the government tax forms, pay your taxes on time, issue W-2s, and generally perform all the services of an office-based payroll service only via the Internet. Intuit also makes the actual entries into your QuickBooks Pro file, so that all you need to do is enter your employees' timesheets or hours.

Creating a wage-related payroll item

To create a payroll item for wages, follow these steps.

1. In **List**, select **Payroll Item List**. Click on the **Payroll Item** button and then **New**, or in the **Employees Navigator**, click on the **Payroll Item** icon. Click on the **Payroll Item** button and then **New**.
2. QuickBooks Pro will ask you for the payroll item type. Click on **Wage**. Do not select Commission. QuickBooks Pro will not allow you to assign commission costs to jobs, so if you have commissioned employees, another workaround will be required.
3. Click on **Next**.
4. In the Wages window, click on **Salary** or **Hourly**.
5. Select **Regular Pay**, **Sick Pay**, or **Vacation Pay**. This is the only place where you can designate the item as sick pay or vacation pay.
6. Click on **Next**.
7. On the next screen enter your name for the payroll item. If you use the naming procedure above, an example of an item name prefix might be HDS—hourly, direct, sick pay.
8. Click on **Next**.
9. The next screen will ask you to **Enter the account for tracking this expense**. Enter the appropriate account, as discussed above, from the Chart of Accounts.
10. Click on **Finish**.

Repeat these steps until you have created wage payroll items for each type of employee and type of pay.

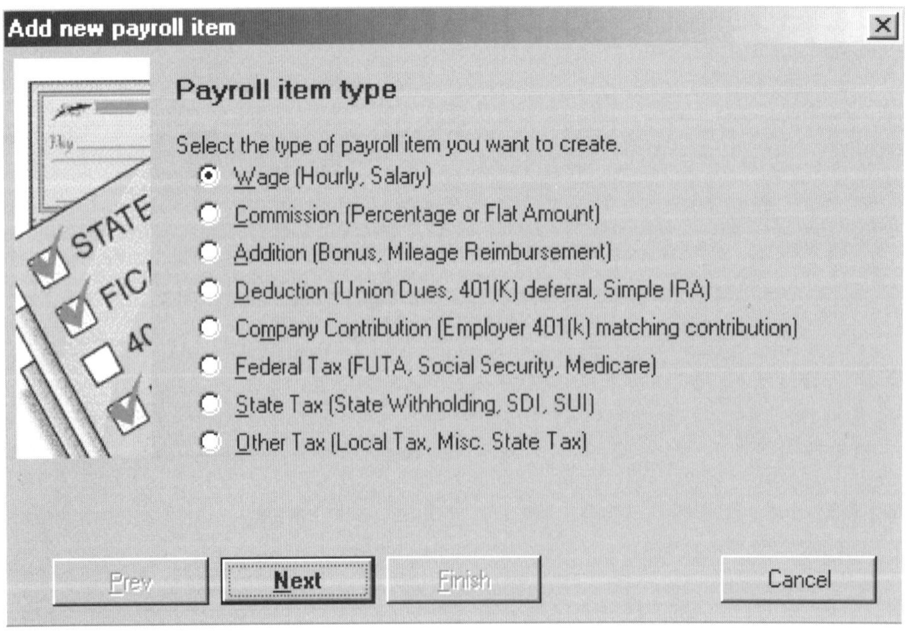

EXHIBIT 24.1 Add new payroll item window for wages.

Creating a deduction or contribution payroll item

You use the same steps for creating items that track employee deductions and company contributions.

1. In **List**, select **Payroll Item List**. Click the **Payroll Item** button and then **New**, or in the **Employees Navigator**, click on the **Payroll Item** icon. Click the **Payroll Item** button and then **New**.
2. Select the payroll item type, for example Deduction (Union Dues, 401K, simple IRA).
3. Enter the name of the contribution or deduction item. For company contributions, you also should check the **Track Expenses by Job** box.
4. Click on **Next**.
5. Enter the name of the agency to which the money will be paid. If your payroll service writes the check on your behalf, enter the name and address of the payroll service.
6. Enter the appropriate liability and expense accounts.
7. Click on **Next**.
8. Do not enter anything for Tax Tracking Type.
9. Click on **Next**.
10. The QuickBooks Pro default for how to calculate taxes is blank. If you ever have any questions, check with your tax advisor. This can have big implications.
11. Click on **Next**.
12. Click to have QuickBooks Pro calculate based on quantity.
13. Click on **Next**.
14. Enter the percentage if it applies to most employees. If there is not a general percentage or flat rate, leave it blank here and enter the amount for each individual employee in their individual Employee Payroll Info tab. You also can enter an annual limit to stop a contribution or deduction at a certain maximum without having to carefully monitor it.
15. Click on **Finish**.

Create a payroll item for each deduction and contribution you make on behalf of your employees. If your company makes a matching contribution, create two items. One should be an employee deduction and the other a company contribution.

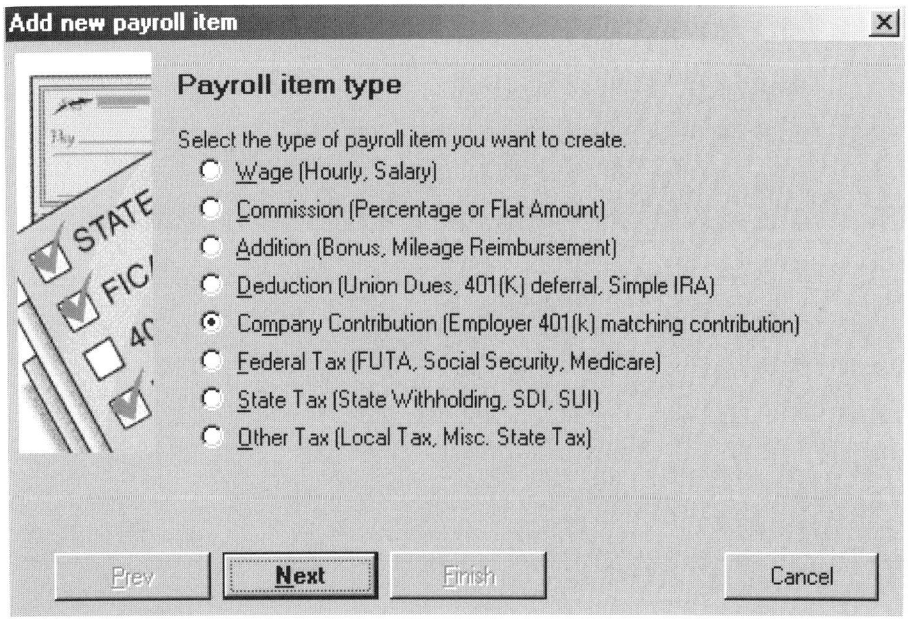

EXHIBIT 24.2 Add new payroll item window for company contribution.

Creating a payroll tax item

The same procedures also apply to tax items. QuickBooks Pro will ask for some additional information, however, so follow these steps.

1. In **List**, select **Payroll Item List**. Click on the **Payroll Item** button and then **New**, or in the **Employees Navigator**, click on the **Payroll Item** icon. Click on the **Payroll Item** button and then **New**.
2. Select the payroll item type, for example Federal Tax (FUTA, Social Security, Medicare).
3. Enter the name of the payroll tax item.
4. Click on **Next**.
5. Enter the name of the agency to which the money will be paid. If your payroll service writes the check to the government for taxes and other liabilities, enter the name and address of the payroll service.
6. Enter the appropriate liability and expense accounts.
7. Click on **Next**.
8. Some tax items will ask you to enter the tax rate. Be sure to enter the correct rate for your taxes! If you ever have any questions, check with your tax advisor. Wrong rates can have big implications.
9. Do not enter anything for Tax Tracking Type until you check with your tax advisor.
10. Click on **Next**.
11. QuickBooks Pro has selected its default for how to calculate taxes. Leave it this way unless you have specific circumstances that would change the status of these taxes. If you decide to make changes, be sure to read the window very carefully! If you ever have any questions, check with your tax advisor. This can have big implications.
12. Click on **Next**.
13. Select the appropriate pre-tax deductions. QuickBooks Pro has selected its default. Leave it this way unless you have specific circumstances that would change the status of these taxes. If you decide to make changes be sure to read the window very care-

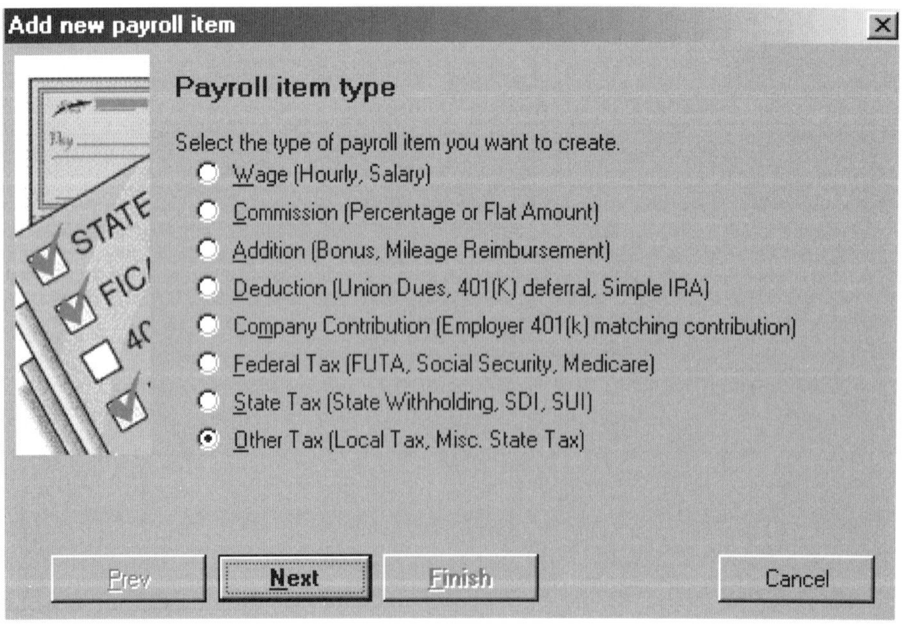

EXHIBIT 24.3 Add new payroll tax item window for other tax.

fully! If you ever have any questions, check with your tax advisor. This can have big implications.

14. Click on **Finish**.

Create an item for each type of tax you withhold and/or pay on behalf of your employees.

> **NOTE** QuickBooks Pro automatically will compute and post your federal withholdings and employer taxes, as well as most state withholdings and employer taxes, based on the specific information you enter in the Employees List within each employee's three *taxes* windows. Check with your tax advisor to make sure you have created items for all the appropriate taxes and that the percentages and limits you are using are correct.

Including special payroll items

Two additional items you should create are called Round Gross Pay Up, and Round Gross Pay Down. When you begin to assign paychecks to jobs, you will find that the more jobs an employee works on in a week, the more likely there is to be a rounding difference on the employee's paychecks. This is due to the wages being weighted based on the amount of time the employee spent at a particular job. You may find that the employee's paycheck does not match your payroll service report by up to several cents. These items can be entered on the paycheck creation screen to either add or subtract cents from the paycheck to make it balance to the actual paycheck as shown on the payroll service report. Create the Round Gross Pay Up payroll item as an addition and the Round Gross Pay Down payroll item as a deduction. Select the Direct Labor wage account as the account for both. Set the amount for 1 cent and do not set a limit.

Although QuickBooks Pro allows you to add mileage reimbursement as an addition to pay, the feature doesn't allow you to assign the expense to a specific job. I recommend that you reimburse mileage, meals, telephone, and related reimbursements in a separate check. Don't muddy up the payroll. Have the owner or employees submit an expense report and pay them from that.

Union-related deductions, employer costs, and special requirements also may be required for your employees. These items can be set up in QuickBooks Pro, but it can take some extensive trial and error experimentation to be certain that you have them set up correctly. You also can create labor burden payroll allocation items that are driven by your payroll postings. These payroll items can be used in such a way that the amounts calculated are spread by job, similar to the way payroll taxes are allocated to various jobs. Contact your Certified QuickBooks Pro Advisor for assistance in these advanced areas.

CHAPTER 25

Developing an Employees List

Setting up your list of employees in QuickBooks Pro 2000 is relatively easy. The first step is to create a default template with a profile for your most common type of employee. If the majority of your employees are hourly, create the template for an hourly employee. You can modify the template for each individual employee, but the more preferences you can incorporate into the template the less keying you will have to do.

Creating a default employee template

To create the employee template, follow these steps.

1. In the **Edit** pull-down menu, click on **Preferences**.
2. Click on the **Payroll and Employees** icon.
3. On the Company Preferences tab, click the **Employee defaults** button.

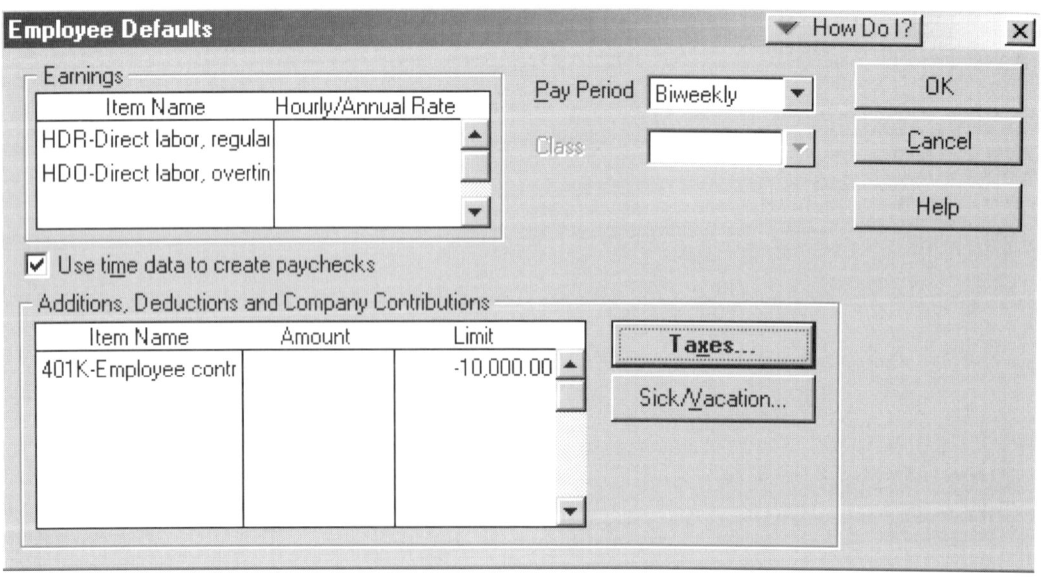

EXHIBIT 25.1 Employee Defaults window.

4. Under item name, select the most common wage-related items for your company. For example, HDR—direct labor, hourly, regular pay. Don't enter the amount. If you frequently pay overtime, enter the overtime item, such as HDO—direct labor, hourly, and overtime pay. Do not enter an amount.
5. In the Pay Period field enter weekly or bi-weekly.
6. Leave clicked the preference: **Use time data to create paychecks** if most employees complete a timesheet that is used as the basis for payroll.
7. Under Additions, Deductions, and Company Contributions add the items that apply to the majority of your employees from the pull-down list. Do not enter the Round Gross Payroll Down or Up items. If the items don't have a default percentage or amount, enter an amount or percentage, or leave it at 0.00 if the percentage is not correct for the majority of your employees.
8. Click on the **Taxes** button. Again, set the defaults so that they match the requirements of the majority of your employees. Enter filing status, state tax item(s), state where the employee works and lives, and so on (see Exhibit 25.3). In Other, enter local taxes, if applicable (see Exhibit 25.4).
9. On the Sick/Vacation tab (see Exhibit 25.5), select an accrual period, such as every paycheck, then select either a number of hours to accrue for every hour or paycheck, or an initial number of hours at the beginning of the year. Answer the questions—**What is the maximum number of hours they can accrue? Do you reset the hours each new year?** If you have a standard accrual, such as 80 hours per year, set this up as the default. You can always adjust this for the individual employee.
10. Click **OK** to save the employee default settings. You can use this template each time you want to add an employee.

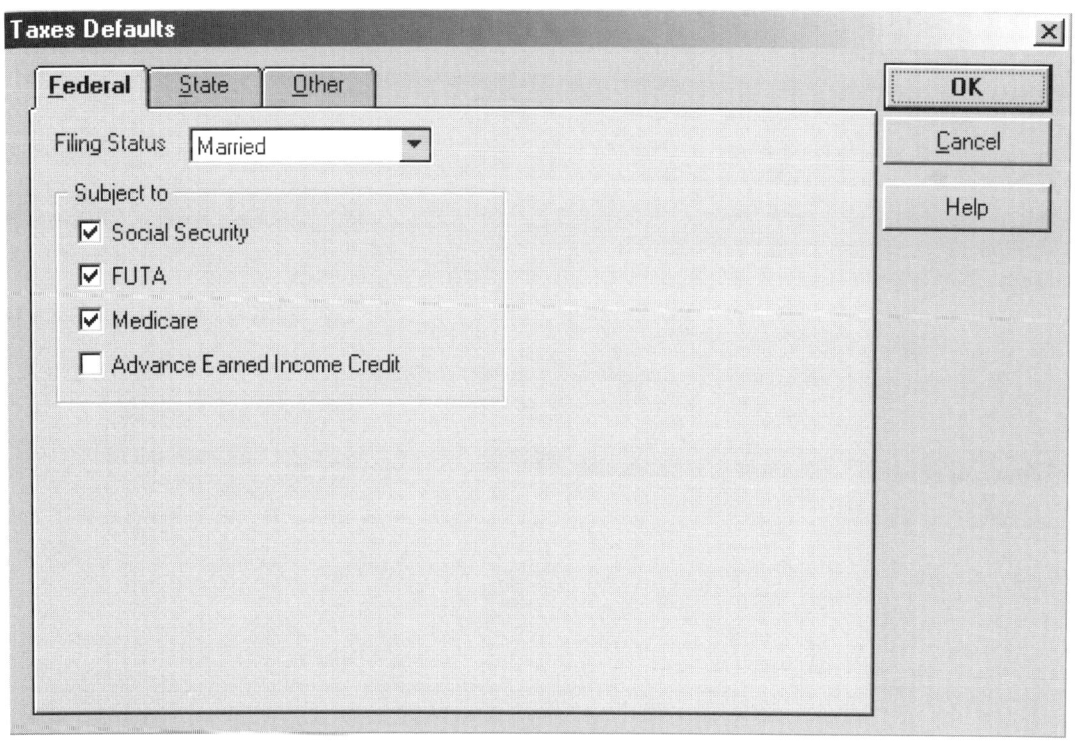

EXHIBIT 25.2 Taxes Defaults window with Federal tab.

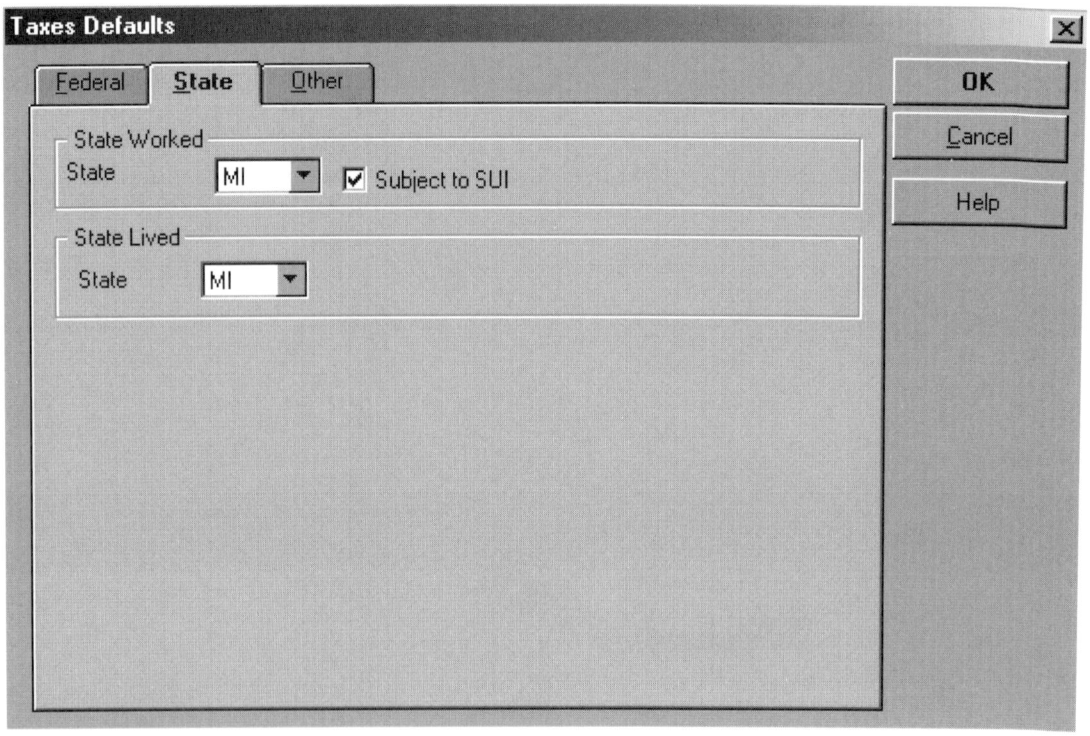

EXHIBIT 25.3 Taxes Defaults window with State tab.

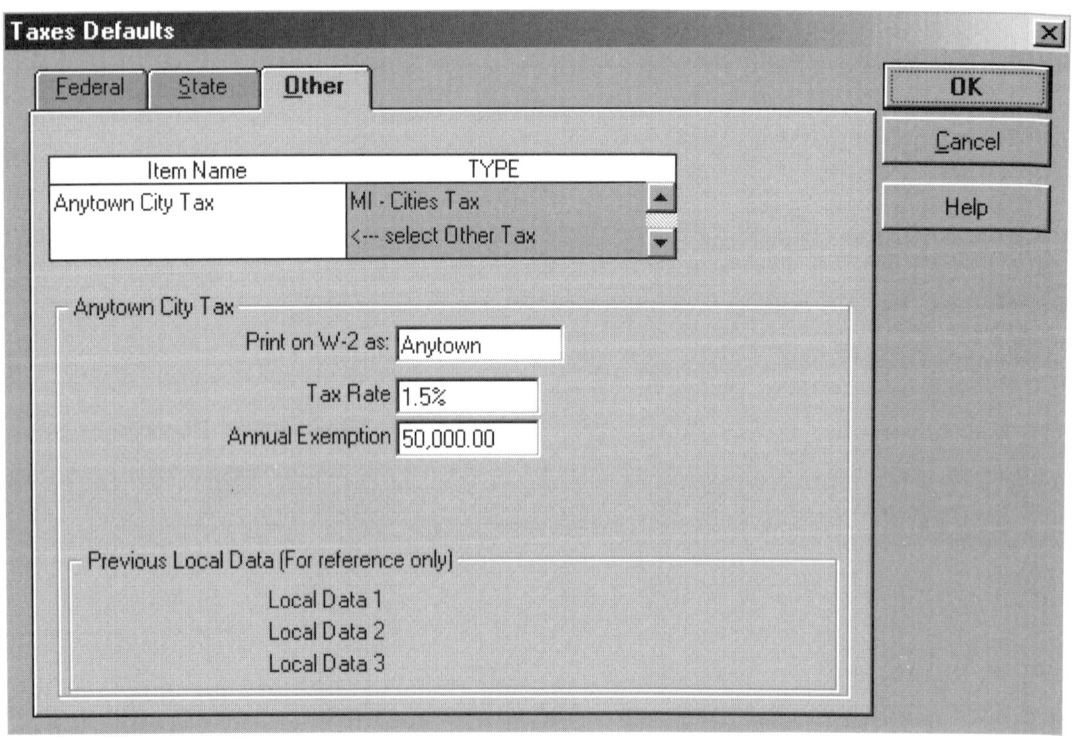

EXHIBIT 25.4 Taxes Defaults window with Other tab.

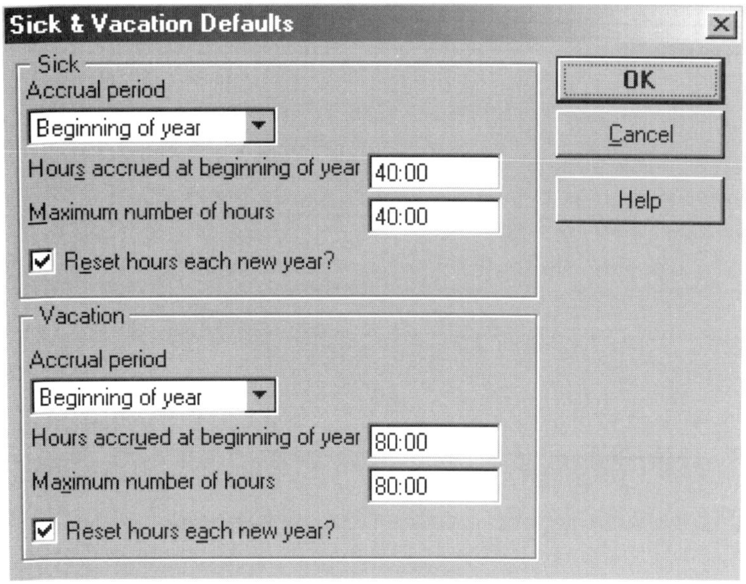

EXHIBIT 25.5 Sick & Vacation Defaults window.

Adding a new employee

To enter a new employee, follow these steps.

1. In **Lists**, select **Employees List**, then click the **Employee** button and **New**, or in the **Employees Navigator** click on the **Employees List** icon, then click the **Employee** button and **New**.

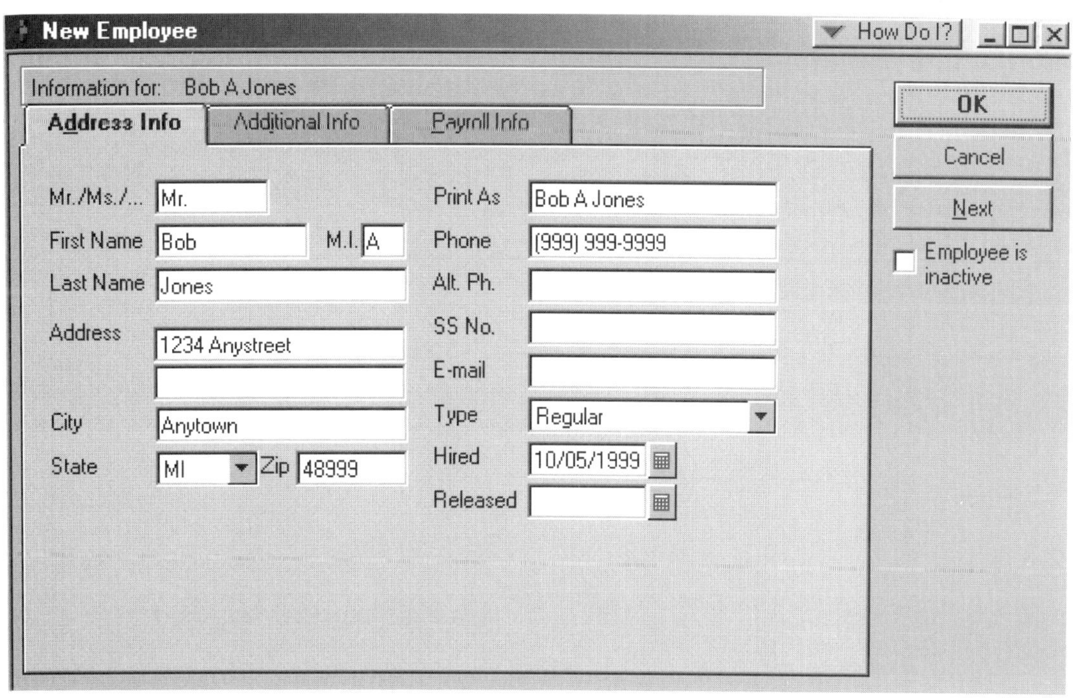

EXHIBIT 25.6 New Employee window.

2. Enter the employee's name, address, city, state, phone, alternative phone, social security number, and so on.
3. Enter all employees as **Regular**. Enter the hire date. If you let an employee go, do not enter the release date until the employee has received his or her last paycheck. QuickBooks Pro will not allow you to create paychecks if you attempt to prepare it after the employee's release date.
4. In the Additional Info tab enter the employee number if your company uses unique numbers to identify employees. You also can click on the **Define Fields** button to create custom fields that will show under this tab for all employees, such as birthday.
5. In the Payroll Info tab modify the default items that were set up earlier, as necessary, for this employee. Enter the regular and overtime pay rates. Make sure the pay period is correct.
6. Make sure the selection for the box **Use Time Data to Create Paychecks** is correct, based on the employee (only direct labor and supervisory personnel).
7. Verify that all deduction and/or contribution items are entered and that the percentages, rates, and limits are correct for the employee.
8. Click on the **Taxes** button. Under each of the tax tabs make sure the information is correct for the employee.
9. Click **OK**.
10. Click on the **Sick/Vacation** button. Modify the information as necessary to make it correct for this employee.
11. Click **OK** in this window and the next to save the employee information.

Enter each employee this way. If you have the information in a prior QuickBooks Pro file you can import the old information by exporting the list to an .iif file and importing it into this company file. You can import most lists in this manner. QuickBooks Pro will not let you import memorized transactions, reports, or customized forms. Click on the How Do I? Help button and select Importing Data for more information.

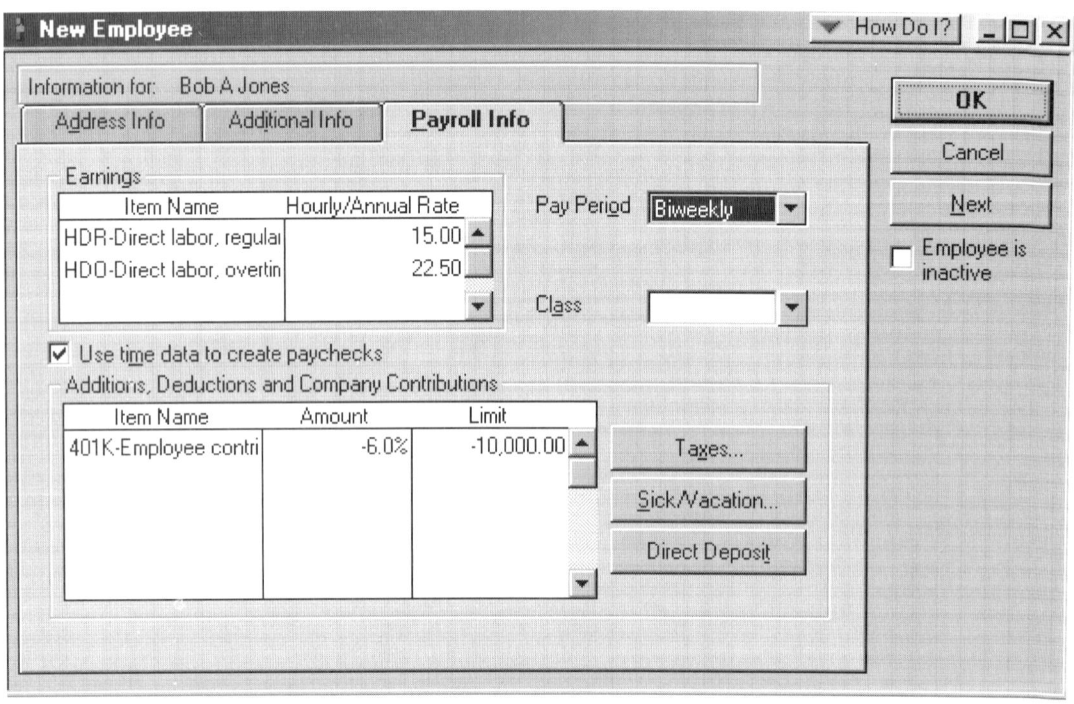

EXHIBIT 25.7 New Employee window with Payroll Info tab.

CHAPTER 26

Entering year-to-date opening payroll balances

If your QuickBooks Pro 2000 start date is not the first of the calendar year, then you will need to make year-to-date adjustments to make your payroll data correct. Because certain payroll items have annual limits, you must enter year-to-date information in order for these to calculate correctly. Your employees will need accurate year-to-date information for W-2s, payroll taxes must compute correctly for job-cost reports, and you will need to record your historical liability payments.

Entering year-to-date compensation totals

To enter year-to-date compensation totals, follow these steps.

1. Under **Employees** select **Enter YTD amounts**, or on the **Employees** Navigator, click on the **YTD amounts** icon.
2. QuickBooks Pro will ask you: **When should YTD summaries affect accounts?** Use the day prior to your QuickBooks Pro start date for both fields, because you will be entering any payrolls after the start date in a standard, detailed manner. For example, if your start date is April 1, enter March 31.
3. Click **Next.**
4. Enter the earliest QuickBooks Pro payroll date. This should be the first day of the pay period for which you will issue paychecks. It also should be your QuickBooks Pro start date, such as April 1. Make sure the year is correct.
5. Click **Next.**
6. Select the first employee from the pull-down list and click on **Enter Summary**.
7. Enter the period for the summary, which would be from the first of the year to the day before the QuickBooks Pro start date.
8. Click the **Accounts Affected** box. Click on **Do not affect accounts.** If you have followed my method for setting up your books, your account balances already are accurate. This process is designed to bring your payroll item subsystem balances into line with your liability accounts.
9. Enter the information requested, including gross wages and withholdings by item. If you have a payroll service, you should have access to reports with this information, or you may need to retrieve the information from your old system. You also may have to research certain deductions or reductions to pay such as cell-phone, tools reimbursements, or loan repayments.

10. When you have balanced this employee's year-to-date net pay in QuickBooks Pro to your source documents, click **OK**.
11. Continue with the next employee, and repeat until finished.
12. Then prepare a Payroll Summary Report through the setup date, which in our example would be March 31. Check your totals against your source documents. To access this report, go to Reports on the menu, select Employees & Payroll, and then select Payroll Summary.
13. If you see discrepancies between the reports and your source information, drill down on the data that you believe is incorrect, and fix it. When you are satisfied with the results, select **Next**.

Entering year-to-date liability payments

To enter year-to-date liability payments, follow these steps.

1. After you are finished entering employee data, select **Next** to enter historical liability payments.
2. Enter the date prior to your QuickBooks Pro start date, which in our example is March 31, as the Payment Date and Period Ending Date.
3. From the pull-down list select each liability item that has been paid against current-year earnings this calendar year and the amount that you have paid year-to-date.
4. Click the **Accounts Affected** box. Click on **Do not affect accounts.**
5. When you are satisfied, click **Done**.
6. To check to see that your payroll liability items balances match the amounts you posted to your accounts in the opening journal entry, single click on the **Pay Liabilities** icon on the Navigator bar, enter the day prior to your start up date in both fields, and review the results.
7. When you have confirmed that the Item List and your liability accounts agree, your year-to-date opening payroll balances are entered correctly. Congratulations!

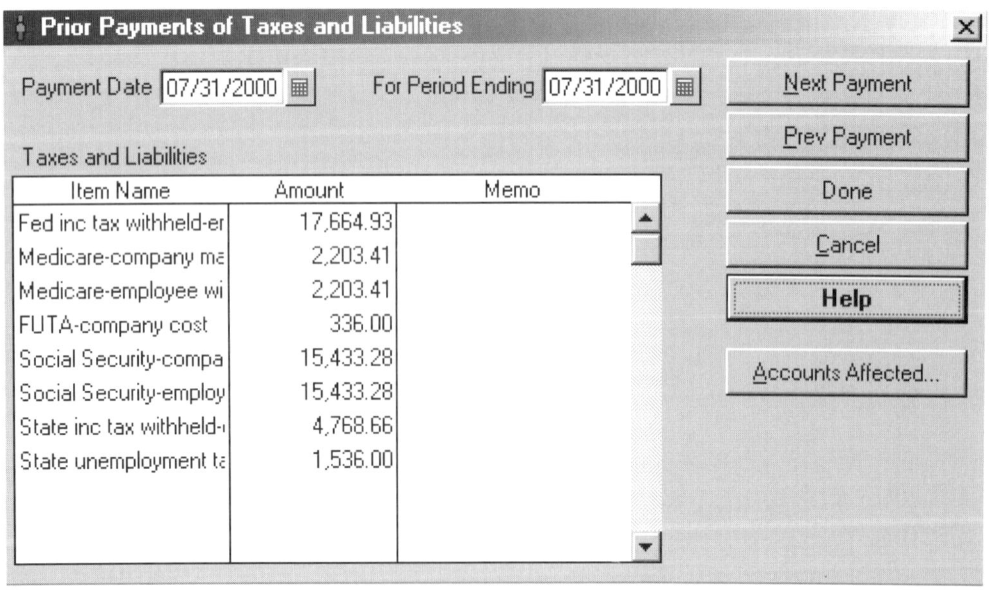

EXHIBIT 26.1 Prior Payments of Taxes and Liabilities window.

CHAPTER 27

Using timesheets

The best way to enter employee time that should be assigned to specific jobs is with the Timesheet feature in QuickBooks Pro 2000. The timesheet provides you with an easy-to-use template for entering employee hours by item using the Item List and then assigning those costs to customers and/or jobs using the Customer:Job List.

Follow these steps to enter an employee's time into the timesheet.

1. In the **Employees** menu, select **Time Tracking**, and then **Use Weekly Timesheet**, or in the **Employees Navigator**, click on the **Use Weekly Timesheet** icon.
2. Select the employee's name from the pull-down list.
3. The timesheet for the current week will pop up in the window. Check to see that the timesheet period is correct. If it's not, click the **Set Date** button near the bottom of the screen.

TIP Maximize the screen so you can see things more clearly.

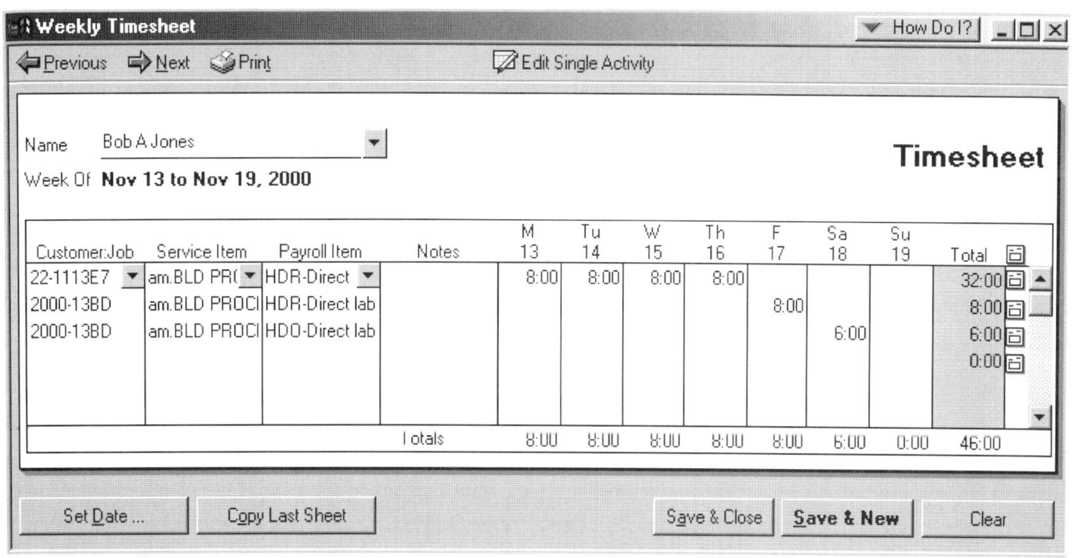

EXHIBIT 27.1 Weekly Timesheet window.

4. If the employee worked on the same jobs as last week, click on the **Copy Last Sheet** button. QuickBooks Pro will enter the data from the previous pay period. You may accept the information as is or modify it.
5. Enter the customer or job in the far-left column by typing in the name or selecting it from the pull-down list.
6. Enter the service item from the Item List by entering the item or selecting it from the pull-down list.
7. Enter the payroll item if QuickBooks Pro has not filled this column automatically based on the employee's default payroll items.

TIP Use the tab button to quickly navigate through the timesheet.

8. You can enter notes in the column to keep track of what people were doing with their time. When you are invoicing and accessing the Time/Costs feature, you can set Time/Costs to transfer the notes from the timesheet to the invoice if you wish. The notes can be further modified as desired on the face of the invoice.
9. Enter the time spent on various service items and various jobs for each day. Be sure to choose an overtime payroll item, if necessary.
10. QuickBooks Pro will calculate both daily and weekly hour totals on the timesheet, so you can keep an eye on when the employee has reached an overtime situation.
11. The small Invoice icon to the right of each line item represents that this time is billable to a job. If you leave the button alone, QuickBooks Pro will recognize the hours as billable for invoicing purposes. When invoicing the customer, you will be able to click on the Time/Costs button and select this line item to charge the customer. If you don't want the time to be billable, but still want it to appear in job-cost reports, then click on the small Invoice icon and an "X" will appear. This means that the time is not billable.
12. Click **Save & New** to create a new timesheet or **Save & Close** to exit.

By adding overtime payroll items you can easily track overtime pay by customer or job and service for your job-cost reports.

If you have indicated that a line item is billable, but have not included a customer or job, QuickBooks Pro will not let you save the timesheet until you enter a customer or job. For proofing, balancing, payroll tax assignments and re-assignments, and job-cost purposes, assign all time to a direct job or to one of the special company jobs.

CHAPTER 28

Paying your employees

When you pay an employee, a lot of things happen from an accounting standpoint. You create an expense for the gross wages you are paying your employees. You create liabilities on your books for the taxes and other deductions you withhold from your employees. You also create both expenses and liabilities for the taxes and contributions your company pays and/or matches.

In addition, all of the above costs can be assigned to jobs, as you've learned in our discussion of payroll items and timesheets. You then need to remit to the proper agencies the liabilities you have incurred as a result of this process. You, in a fiduciary capacity, have the obligation of passing the employees' withholdings along to the appropriate entities, as well as the unemployment and matching funds you are required to submit. Some of these taxes have annual limits, which you set up during the creation of the payroll tax items.

TIP If you click on the **Report all Payroll Taxes by Customer:Job and Service Item** box in the Payroll and Employees preference window, QuickBooks Pro 2000 will assign all payroll costs, including payroll taxes, to jobs. The system will assign the costs based on the time assigned to the jobs on the employee's timesheet.

NOTE You will have to select and set up a QuickBooks Pro payroll service before you pay your employees. See chapter 24 for more information.

Creating paychecks

With the year-to-date totals you recorded in the system in chapter 26, you now are ready to create paychecks for your employees using information you entered from the timesheets, as described in chapter 27. Follow these steps.

1. In **Employees**, select **Pay Employees**, or in the **Employees Navigator**, click on the **Pay Employees** icon.
2. Select the bank account from which the funds will be taken.
3. Click on the **To Be Printed** box.
4. Enter the date that should go on the payroll checks.
5. Click on the button **Enter hours and tax amounts before creating paychecks**.
6. Enter the appropriate **Pay Period Ends** date.

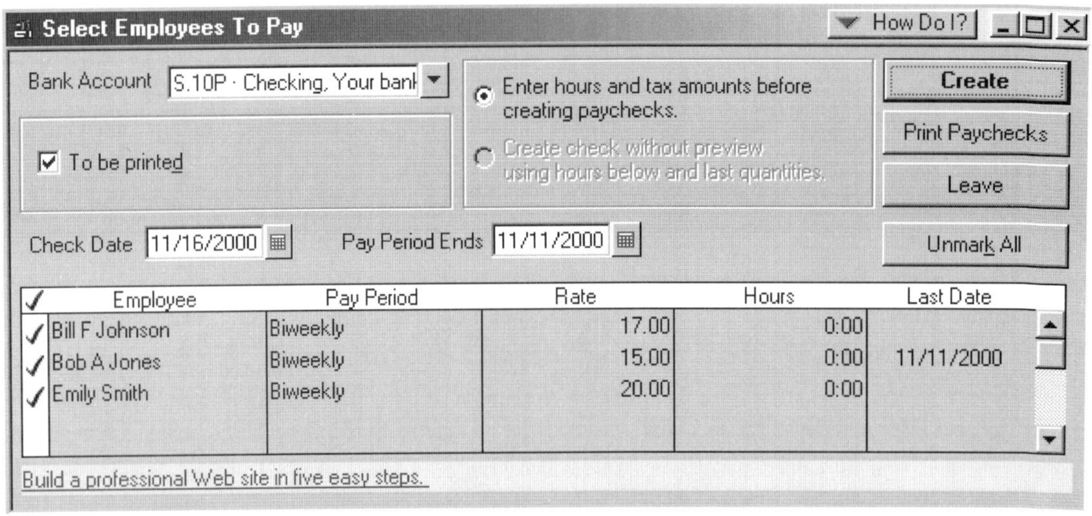

EXHIBIT 28.1 Select Employees to Pay window.

7. Select all of the employees you want to pay in the window.
8. Click **Create.**
9. QuickBooks Pro will show you the paycheck for the first employee you selected. The payroll item, hours, customer or job, and service item information will have automatically filled in from the timesheet for that pay period ending date. The rates will have automatically filled in from the individual employee template that you set up when you entered the employee. The taxes, contributions, deductions, and additions will all be calculated based on their setup.

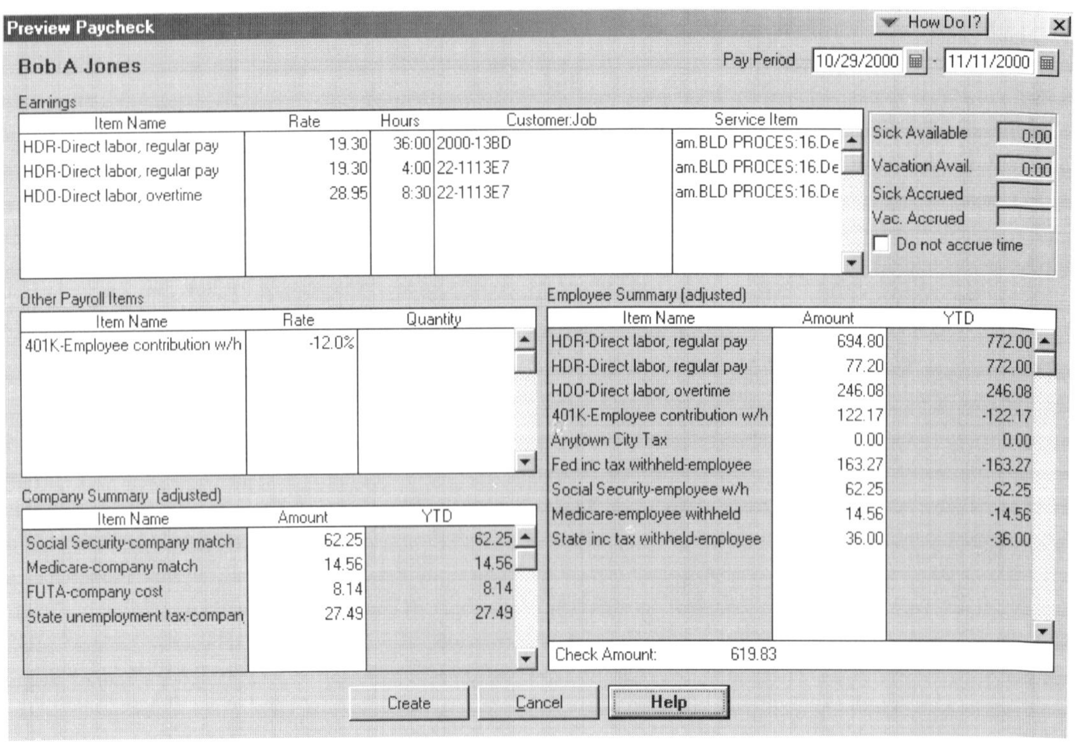

EXHIBIT 28.2 Preview Paychecks window.

10. If you have an outside payroll service, verify the net pay of the check against the payroll service report. Make sure the 401K and other withholdings match to the penny. Check that the company-paid Medicare and Social Security taxes balance to the employees' Medicare and Social Security deductions. Confirm the company's portion of total remaining payroll taxes against the payroll service report. If the employee's direct gross pay segments do not add up to the gross on the payroll service report, add or subtract a rounding payroll item (up to a few cents!) if necessary.

NOTE If the limit has been reached for a specific payroll item, you will not see a deduction for the current period. The year-to-date amount will remain constant for the rest of the calendar year.

11. After you have verified the information, click **Create** to create the paycheck and move on to the next employee's paycheck.
12. When you have verified all paychecks, click **Leave** on the **Select Employee to Pay** window.

TIP If you need to print paychecks, you can do so while in this window by clicking on Print Paychecks.

The employee summary on the check stub will show the amount paid during the pay period for each wage-related item, as well as year-to-date totals for the wage items. The same information will be provided for the employee deduction items.

Printing paychecks

To print the paychecks, if you did not print them above, go to File, select Print Forms, and then select Paychecks. You will see the paychecks waiting to be printed. Select the paychecks you want to print and click OK. Follow the same procedures as for printing other forms. If you are using a laser printer, you'll want to use laser voucher checks. Keep the bottom portion of the check with the payroll summary information for your records.

If you have a payroll service that prints the checks, you still will need to record the check numbers in your Check Register. In Lists, select Chart of Accounts, and double-click on the appropriate bank account. You will see the payroll checks entered in your register with check numbers that read "To be printed." Click on the check number field and enter the correct check number (or DD for direct deposit checks) taken from the payroll service report. Click Record. Repeat this process for all of the paychecks. Now when you reconcile your bank account, the payroll checks will be easy to identify and clear, and the direct deposit checks can be quickly added together to match the withdrawals of funds from your bank.

Paying liabilities

I recommend that you pay your liabilities every pay period. That is one of the real benefits of using a payroll service. They will deduct the payroll tax liabilities, as well as the net wages amounts, from your account and then pay your tax liabilities on time. This keeps your accounting system very clean. If you don't pay your liabilities each pay period, it is sometimes tempting to pay other bills before the tax liabilities. The penalties can be quite severe if payment is not made on time. Whether you pay the liabilities yourself or you have a payroll service that remits them, you still need to enter the liability payments on QuickBooks Pro.

To pay your liabilities, follow these steps.

1. **In Employees,** select **Pay Payroll Liabilities**, or in the **Employees Navigator**, click on the **Pay Liabilities** icon.
2. Enter the date range for which you wish to see your liabilities. Make sure the date range is correct. Remember that the liability is created not on the period ending date, but on the date of the paycheck itself.
3. Click **OK**.
4. The window will show all of your liabilities by payroll item due for the period.
5. Click on **To Be Printed**.
6. Select the bank account from which you will pay the liabilities.
7. Enter the date of the liability payment.
8. Check **Review liability check.**
9. Check the payroll summary information against the information in the Pay Liabilities window. The two figures should match. If they don't go back to the previous report, and instead of Total Only, select By Employee. Print the report, and check each employee's wages and taxes against your payroll service report one more time.
10. When the two reports are in agreement, place a checkmark next to the liability items you are paying.
11. Click **Create** to create checks for all the entities you owe, or have paid, liabilities. The system will consolidate amounts due to duplicate entities into one check.

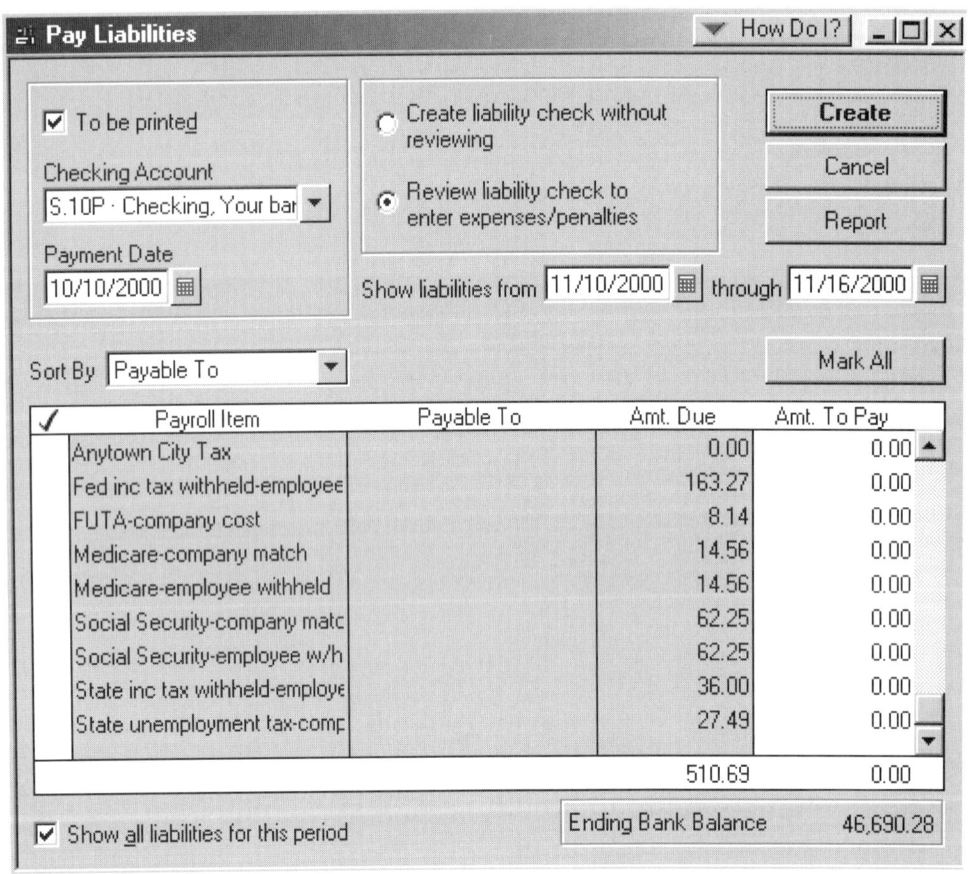

EXHIBIT 28.3 Pay Liabilities window.

TIP

To check the amounts due before you go further, create a report to check your liability balances. In Reports, select Employees and Payroll, and then select the Payroll Summary report. Choose as your date range the date of the paychecks for which you are remitting liabilities. Customize the report by clicking off the hours and rates, and choose to see Total Only. Print the report so you can compare the totals on the report with the amounts calculated by QuickBooks Pro for the liability checks.

12. Follow the same procedures to print these checks as you did for the paychecks above.
13. Creating the checks will clear out all the liabilities for the period.

To confirm that they all are paid, run a payroll liabilities report. In Reports, select Employees and Payroll, and then Payroll Liabilities. Enter the dates that match the period for which the liabilities were paid. You should see no outstanding liabilities.

To see how payroll affects your job-cost reports, create a profit and loss by job report. In Reports, select Company and Financial, and then Profit and Loss by Job. Click the Customize button and include source name, item, and item description. If you want to see the detail behind any amount on the report, double-click on the amount (you will see a tiny magnifying glass) and you will see the costs posted by individual transaction and item cost.

SECTION FIVE

THE FINAL WALK THROUGH

The lists are prepared; the information "buckets" are filled. Now it's time to move in and begin creating those financial management reports you are going to use to manage your business effectively and efficiently.

The chapters in this section take you on a final walk through the QuickBooks Pro system. Each of these topics will help you get the most out of your system, from customized reports to effective file management and security. The allocations and special transactions sections round out the accounting information you need to keep your financial books clean and consistent, and your CPA and tax advisor happy come year-end.

QuickBooks Pro has many features that can help you produce accurate and timely financial information. Click and try different things. Try to get your staff to try things. The only caution I offer is that you should always read what is on the screen before you start clicking away. If you look at the screen and carefully read what is there, you will find some wonderful features. Don't be afraid to try new things—carefully!

I'm reminded of the story of the salesman back in the forties who sold 56-volume Farming Encyclopedia sets. One day, while driving on a country road, he pulled over to make a pitch to an older farmer he saw working in the field. He hopped out of the car, waved the farmer over, and made his entire 15-minute presentation about the incomparable farming information and the great crops the farmer would get if he just bought the Farming Encyclopedia. When he concluded, the farmer scratched his chin for a moment, then drawled, "Son, that sure sounds great, but you know, I already don't farm half as good as I already know how."

One of the points of the story is that knowledge without action isn't much better than not having the knowledge. The second point I'd like to make from the story is that most of us are sitting on a secret reservoir of ideas and knowledge that we don't apply on a day-to-day basis because we don't take the time to truly focus on solving our problems. I'm not trying to convince you to buy an encyclopedia or change your operations on a wholesale basis. Instead I'm encouraging you to put your measurement system in place so that you can use the results to run your business as well as you already know how.

You are the business owner! You are the individual in control! You are the person in the best position to ask the questions and determine the answers! You have made the investment to design effective lists and build a robust accounting system. Now it's time to generate the reports you need and focus your energy on analyzing the results.

The true power of an accounting system lies not in the system or even in the information itself, but in the actions you take after you review and analyze your reports, so be sure to review them, understand them, and act!

CHAPTER 29

Enhancing and customizing your reports

When you developed your Chart of Accounts, Item List, and Customer:Job List I stressed the importance of planning the contents of each list so that your reports would provide meaningful management information for your business. Chapter 4 provided an overview of the basic reports—balance sheet, income statement, cost accounting by job, cost accounting by job variance, and cost accounting by job stage variance. Now let's look at how you can enhance these basic reports and create custom reports that will provide you with the information you need to make effective management decisions.

Enhancing the usefulness of your balance sheet

You may wish to briefly review the basic information about balance sheets presented in chapter 4. The balance sheet is a statement of your financial position on the date the report was created. It shows what you own (assets), what you owe (liabilities), and what's left over (equity).

The most important way to enhance the value and usefulness of your balance sheet is to thoughtfully review the information presented, understand the nature of the information included in each part of the balance sheet, and use the information to help manage your company.

Balance sheet overview

You also can use QuickBooks Pro 2000 to do some things that will help you in this process. For example, use the QuickBooks Pro selection buttons at the top of your onscreen reports to collapse or expand your report. Then click on the Customize button and check the "of Column" box to add a column for percentages of assets to use as a "measuring stick." Exhibit 29.1 shows a balance sheet for a remodeling business. You'll notice that the brevity of the report combined with percentages of assets reveals some basic information very quickly. For example:

- The books are handled on an accrual basis because the balance sheet includes accounts payable and accounts receivable accounts.
- The business' total current assets of $35,374.30 represent 78 percent of its total assets, and its total current liabilities of $33,313.90 are slightly less at 73.5 percent of its total assets. Therefore as of the date of the balance sheet it appears existing current liabilities can be met by existing current assets.
- There are no long-term liabilities in this balance sheet.
- There is a year-to-date net income of $1,445.84.

If you owned this company and reviewed this report you'd probably want to ask and obtain answers to the following questions:

Dreamscape Remodeling, Inc.
Balance Sheet - Summarized Version
As of March 31, 2001

	Mar 31, '01	% of Column
ASSETS		
Current Assets		
Checking/Savings		
T.100 · Checking account	944.41	2.08%
T.110 · Savings account	4,881.89	10.77%
T.119 · Cash on hand	200.00	0.44%
Total Checking/Savings	6,026.30	13.3%
Accounts Receivable		
T.120 · Accounts receivable, trade	26,848.00	59.23%
Total Accounts Receivable	26,848.00	59.23%
Other Current Assets		
T.130 · Other current assets	2,500.00	5.52%
Total Other Current Assets	2,500.00	5.52%
Total Current Assets	35,374.30	78.04%
Fixed Assets		
T.173 · Office FF&E, net book value	642.38	1.42%
T.174 · Computers/software-net book val	1,433.68	3.16%
T.175 · Vehicles-net book value	7,876.51	17.38%
Total Fixed Assets	9,952.57	21.96%
TOTAL ASSETS	**45,326.87**	**100.0%**
LIABILITIES & EQUITY		
Liabilities		
Current Liabilities		
Accounts Payable		
2000 · Accounts payable, trade	12,055.00	26.6%
Total Accounts Payable	12,055.00	26.6%
Credit Cards		
T.210 · Company credit cards	3,994.60	8.81%
T.215 · Shareholder cr cards	8,373.29	18.47%
Total Credit Cards	12,367.89	27.29%
Other Current Liabilities		
T.250 · Line(s) of credit	8,891.01	19.62%
Total Other Current Liabilities	8,891.01	19.62%
Total Current Liabilities	33,313.90	73.5%
Total Liabilities	33,313.90	73.5%
Equity		
3100 · Capital stock - 500 shares	500.00	1.1%
3900 · Retained Earnings	10,067.13	22.21%
Net Income	1,445.84	3.19%
Total Equity	12,012.97	26.5%
TOTAL LIABILITIES & EQUITY	**45,326.87**	**100.0%**

EXHIBIT 29.1 Summarized balance sheet.

- **Are all of our receivables entered?** If not, the year-to-date income will be understated.
- **How collectible are the receivables?** You probably would want to review an accounts receivable aging report. If there are exceptionally old balances, you may follow up with collection efforts or recognize the expense of bad debt.
- **Have all of our payables and credit purchases been entered?** Are the credit card purchases recorded through the date of the balance sheet? If not, the expenses are understated and the net income number is too high.
- **How current are the payables?** You probably would want to review an accounts payable aging report. If you have old outstanding balances due to vendors you could be losing discounts or jeopardizing important supplier relationships. It's a general rule that better service goes to those who pay promptly. Slow payment eventually will cause snags in your operations. If this is the case, you may need to investigate ways to speed up your cash flow, perhaps by revisiting the receivables situation.
- **Are the balances reasonable?** As you review the balance sheet, ask yourself if the balances you see appear to be reasonable, given what you know about your company.

Detailed version

You also should review a detailed version of the balance sheet. Exhibit 29.2 shows an expanded version of the same balance sheet with detailed information for the accounts. In this example I've kept the same account numbers and used the Customize button to choose Previous Period, $ Change, and % Change to see a snapshot of the balance sheet at the same time the previous quarter. In effect, by clicking three boxes on the computer screen, I've told the computer "genie" to compute and print all of the dollar and percentage changes since last quarter. Some of the interesting things to note and ask questions about in this expanded report include:

- The checking account balance is significantly lower than it was at this same time last year.
- The accounts receivable is significantly higher. It's time to find out who owes us money and how long the receivables have been outstanding since receivables mean we are financing someone else's operations.
- As of the end of last quarter $3,453 had been earned but not invoiced.
- The employee who owed us money paid us back.
- The additional depreciation expense within the fixed-asset categories are due to the new computer equipment we purchased last quarter.
- The payables are higher and the credit card debt has increased. Is it because there isn't enough cash due to the higher receivables? If some of the money owed was collected could the payables be lowered?
- The $4,500 that was a client deposit at this time last quarter was earned or returned.
- The taxes seem to be up-to-date, but are the current tax liabilities properly recorded? Has the tax advisor provided current information regarding estimated taxes?
- Principal in the amount of $2,320.62 has been paid the line of credit.
- Last year's income is now included in the current year's "retained earnings" account as it is supposed to be.

These are just some of the amazing things that you can learn, changes you'll be reminded of, and the kinds of questions you'll have and learn the answers to when you carefully review this type of balance sheet comparison.

Understanding the income statement

We've just finished examining how you can make your balance sheet a more useful management tool—now, we'll examine the income statement and how you can put it to work for you. This chapter addresses some of the ways you can:

- Enhance your income statements, also referred to as profit and loss reports by QuickBooks Pro, for your company as a whole

Dreamscape Remodeling, Inc.
Balance Sheet - Expanded Version
As of March 31, 2001

	Mar 31, '01	Dec 31, '00	$ Change	% Change
ASSETS				
Current Assets				
Checking/Savings				
T.100 · Checking account				
1002 · Regular cks/deposits/svc chgs	944.41	8,947.22	(8,002.81)	(89.45%)
Total T.100 · Checking account	944.41	8,947.22	(8,002.81)	(89.45%)
T.110 · Savings account				
1102 · Regular cks/deposits/svc chgs	4,881.89	4,831.10	50.79	1.05%
Total T.110 · Savings account	4,881.89	4,831.10	50.79	1.05%
T.119 · Cash on hand				
1190 · Petty cash	200.00	200.00	0.00	0.0%
Total T.119 · Cash on hand	200.00	200.00	0.00	0.0%
Total Checking/Savings	6,026.30	13,978.32	(7,952.02)	(56.89%)
Accounts Receivable				
T.120 · Accounts receivable, trade	26,848.00	14,226.15	12,621.85	88.72%
Total Accounts Receivable	26,848.00	14,226.15	12,621.85	88.72%
Other Current Assets				
T.122 · Accounts receivable, other				
1230 · Income earned - not yet billed	0.00	3,453.00	(3,453.00)	(100.0%)
Total T.122 · Accounts receivable, other	0.00	3,453.00	(3,453.00)	(100.0%)
T.130 · Other current assets				
1320 · Due from employees	0.00	422.83	(422.83)	(100.0%)
1360 · Refundable deposits	2,500.00	2,500.00	0.00	0.0%
Total T.130 · Other current assets	2,500.00	2,922.83	(422.83)	(14.47%)
Total Other Current Assets	2,500.00	6,375.83	(3,875.83)	(60.79%)
Total Current Assets	35,374.30	34,580.30	794.00	2.3%
Fixed Assets				
T.173 · Office FF&E, net book value				
1730 · Furn/fixt/equip, at cost	11,578.90	11,578.90	0.00	0.0%
1830 · (Less) Acc depr - FF&E	(10,936.52)	(10,523.00)	(413.52)	3.93%
Total T.173 · Office FF&E, net book value	642.38	1,055.90	(413.52)	(39.16%)
T.174 · Computers/software-net book val				
1740 · Computer h'ware/s'ware, at cost	5,031.87	3,573.89	1,457.98	40.8%
1840 · (Less) Acc dpr-cmptr h/s'ware	(3,598.19)	(3,573.89)	(24.30)	0.68%
Total T.174 · Computers/software-net book val	1,433.68	0.00	1,433.68	100.0%
T.175 · Vehicles-net book value				
1750 · Vehicles, at cost	23,578.42	23,578.42	0.00	0.0%
1850 · (Less) Accum dpr - vehicles	(15,701.91)	(14,523.00)	(1,178.91)	8.12%
Total T.175 · Vehicles-net book value	7,876.51	9,055.42	(1,178.91)	(13.02%)
Total Fixed Assets	9,952.57	10,111.32	(158.75)	(1.57%)
TOTAL ASSETS	45,326.87	44,691.62	635.25	1.42%

EXHIBIT 29.2 Expanded balance sheet.

	Mar 31, '01	Dec 31, '00	$ Change	% Change
LIABILITIES & EQUITY				
Liabilities				
Current Liabilities				
Accounts Payable				
2000 · Accounts payable, trade	12,055.00	4,691.25	7,363.75	156.97%
Total Accounts Payable	12,055.00	4,691.25	7,363.75	156.97%
Credit Cards				
T.210 · Company credit cards				
2100 · Corporate Optima	3,994.60	2,379.12	1,615.48	67.9%
2101 · American Express	0.00	342.00	(342.00)	(100.0%)
Total T.210 · Company credit cards	3,994.60	2,721.12	1,273.48	46.8%
T.215 · Shareholder cr cards				
2150 · Chase Gold VISA - M Marlowe	8,373.29	8,000.00	373.29	4.67%
Total T.215 · Shareholder cr cards	8,373.29	8,000.00	373.29	4.67%
Total Credit Cards	12,367.89	10,721.12	1,646.77	15.36%
Other Current Liabilities				
T.221 · Other accounts payable				
2020 · Client deposits (unearned inc)	0.00	4,500.00	(4,500.00)	(100.0%)
Total T.221 · Other accounts payable	0.00	4,500.00	(4,500.00)	(100.0%)
T.240 · Taxes payable				
T.241 · Payroll taxes due				
2416 · State unemployment taxes due	0.00	2,347.35	(2,347.35)	(100.0%)
Total T.241 · Payroll taxes due	0.00	2,347.35	(2,347.35)	(100.0%)
T.246 · Other taxes due				
2470 · Real est/prsnl property tax due	0.00	653.14	(653.14)	(100.0%)
Total T.246 · Other taxes due	0.00	653.14	(653.14)	(100.0%)
Total T.240 · Taxes payable	0.00	3,000.49	(3,000.49)	(100.0%)
T.250 · Line(s) of credit				
2200 · First Union Bank-line of credit	8,891.01	11,211.63	(2,320.62)	(20.7%)
Total T.250 · Line(s) of credit	8,891.01	11,211.63	(2,320.62)	(20.7%)
Total Other Current Liabilities	8,891.01	18,712.12	(9,821.11)	(52.49%)
Total Current Liabilities	33,313.90	34,124.49	(810.59)	(2.38%)
Total Liabilities	33,313.90	34,124.49	(810.59)	(2.38%)
Equity				
3100 · Capital stock - 500 shares	500.00	500.00	0.00	0.0%
3900 · Retained Earnings	10,067.13	2,057.00	8,010.13	389.41%
Net Income	1,445.84	8,010.13	(6,564.29)	(81.95%)
Total Equity	12,012.97	10,567.13	1,445.84	13.68%
TOTAL LIABILITIES & EQUITY	45,326.87	44,691.62	635.25	1.42%

EXHIBIT 29.2 Expanded balance sheet *continued.*

- Pose questions that you may want to ask while reviewing your results
- Suggest who may be in the best position to provide the answers for getting your company back on target

Gross profit

When you set up your Chart of Accounts in chapter 8 you created income accounts, including a special group of accounts for your direct costs or costs of goods sold (COGS)—the costs you incur only for jobs.

On the income statement, income accounts are listed first and then subtotaled. The COGS accounts are listed next and then subtotaled. QuickBooks Pro then generates a *gross profit* subtotal by subtracting the COGS subtotal from the income subtotal. In Exhibit 29.3, the gross profit is $45,903.55.

You can use this gross profit subtotal to:

- Analyze the profitability of your direct activities, such as construction or remodeling
- Create profit margin percentages
- Create budgets
- Help set your pricing guidelines or markup
- Compare your direct cost of doing business with that of other remodelers or custom home builders

Overhead costs

The section of the income statement that follows the gross profit subtotal is the company overhead group of accounts. These are the accounts that show the costs required to generally keep your company functioning.

These overhead accounts are grouped and subtotaled by business function, such as sales and marking expense, administrative employment costs, and occupancy expenses. The subtotal that follows these groupings is called *net ordinary income*. The *other income* and *other expense* groupings that follow the net ordinary income subtotal are the accounts I use for costs not directly related to operations.

Computing costs as a percentage of income

Another way to enhance the usefulness of your income statement is to compute and report your costs as a percentage of income. Percentages are useful because they provide you with a standardized measuring stick you can use to view the relationships between your income and the expenses you incur.

The most important relationship is between your income and direct costs. If your direct costs are 65 percent of your income, then you know that *on average* for every $100 of income your company earns, you will need to expend $65 to get the job done. On a larger scale, you also will know that on average for every $10,000 of income, you will need to spend $6,500 to get the job done.

NOTE This on-average cost assumes that all of your costs are variable or that they vary in direct proportion to income. Some direct costs, in fact, do *not* vary in proportion to income and therefore are known as fixed costs, such as a supervisor's salary, or semi-variable, meaning they may change but not directly in proportion to income. For example, volume purchase discounts are considered semi-variable because they can affect costs at certain levels of production. Keep this in mind when you are analyzing or budgeting your direct costs.

Dreamscape Remodeling, Inc.
Income Statement - Summarized Version
January through March, 2001

	Jan - Mar '01	% of Income
Ordinary Income/Expense		
Income		
T.450 · Design & consulting services	7,600.00	4.72%
T.460 · Construction services	153,332.00	95.28%
Total Income	160,932.00	100.0%
Cost of Goods Sold		
T.529 · Direct employment cost	55,068.27	34.22%
T.550 · Subcontractors	19,163.45	11.91%
T.560 · Building materials & supplies	31,781.17	19.75%
T.570 · Job-specific misc costs	8,692.50	5.4%
T.580 · Direct expense-not job specific	323.06	0.2%
Total COGS	115,028.45	71.48%
Gross Profit	45,903.55	28.52%
Expense		
T.710 · Sales & marketing expenses	5,324.58	3.31%
T.800 · Admin employment costs	21,647.00	13.45%
T.820 · Occupancy expense	3,071.00	1.91%
T.830 · Office expenses	3,774.05	2.35%
T.840 · Computer expense	24.30	0.02%
T.850 · General & admin expense	4,791.64	2.98%
T.860 · Vehicle expenses	2,805.32	1.74%
T.870 · Meals, travel & entertainment	1,104.46	0.69%
T.890 · Taxes & licenses	668.00	0.42%
Total Expense	43,210.35	26.85%
Net Ordinary Income	2,693.20	1.67%
Other Income/Expense		
Other Income		
9100 · Interest income	50.79	0.03%
Total Other Income	50.79	0.03%
Other Expense		
T.920 · Interest	710.17	0.44%
T.923 · Other financing expense	587.98	0.37%
Total Other Expense	1,298.15	0.81%
Net Other Income	(1,247.36)	(0.78%)
Net Income	**1,445.84**	**0.9%**

EXHIBIT 29.3 Summarized income statement.

In addition to covering your direct costs, you will need to charge enough to cover your company overhead and your salary (if it isn't already included in your overhead costs), as well as enough to create a profit for your company. The lower the direct cost percentage, the higher the gross margin or profit percentage and vice versa. Exhibit 29.4 shows this for three scenarios moving from scenario A (lowest gross profit) to scenario C (highest gross profit).

Create a customized report to see these percentages next to the accounts and subtotals as shown in Exhibit 29.5. Then become very familiar with these percentages as you use your reports and create your budgets and job estimates.

The next step is to review your overhead expense percentages. For example, if you have 35 percent of your income left to cover overhead, you need to make sure your overhead expenses don't exceed the available 35 percent. If they do, your business will show a loss. On the other hand, if your overhead plus your other income and expense accounts are less than the available 35 percent, you will show a profit.

NOTE

Costs that don't vary in relationship to sales volume, such as rent, owner's salary, or professional fees often are referred to as fixed costs. These costs stay relatively flat unless your sales volume changes dramatically and forces a change in your overhead activities. Then these costs have a tendency to increase in incremental levels. For example, you may decide that you need to hire an office manager because of increased volume. The office manager is paid a salary that is not related to your job volume, so your fixed cost is increased in incremental steps. Or you may decide that you need to move to a larger office space or create a showroom, so your previously fixed facilities costs increase, but then stay flat again. Keep this in mind when you are budgeting for changes in overhead costs.

Enhancing your income and expense reports

The most critical requirement for making your financial statements truly useful is for you to review, understand, investigate, and act upon the information you find in your financial statements. Directing your attention to unusual results, and focusing your attention on resolving the related issues, is referred to as *management by exception*, meaning you spend more time managing (exploring and/or modifying) activities or transactions that appear exceptional or unusual.

When reviewing the income and expense statement, begin by reviewing summarized and/or bottom line results. Then, on the basis of what you see, work your way backward into the various sections of the report and review the detailed information that produced the result. For example, you may start with a summarized report, which in QuickBooks Pro is referred to as the "collapsed" version of the income statement. If this was a report for your company, you probably would not be pleased to see that the net profit for the first three months of 1999 is just \$1,445.84 or less than 1 percent of your gross income.

Next, you might notice that your COGS percentage is 71.48 percent or 9.5 percent higher than the 62 percent you planned for this year. In addition, the report shows overhead of 26.85 percent of income versus the 22 percent you had anticipated.

You also can enhance the value of your income statement by comparing the current period's income and expense against a prior period's income and expense. You do this by reviewing an expanded version of the report such as the one in Exhibit 29.6. Typically you also would want to compare similar periods, for example the first quarter of this year against the first quarter of last year. Use QuickBooks Pro's customization features to add columns to your profit and loss report that calculate the dollar amount difference between this year's and last year's income and expense accounts and the percentage difference between the two years.

(text continues on page 238)

Dreamscape Remodeling, Inc.
Comparative Income Statement -Expanded Version
January through March, 2001

	Jan - Mar '01	Jan - Mar '00	$ Change	% Change
Ordinary Income/Expense				
Income				
T.450 · Design & consulting services				
4500 · Design services	7,600.00	2,360.00	5,240.00	222.03%
4510 · Consulting services	0.00	1,295.00	(1,295.00)	(100.0%)
Total T.450 · Design & consulting services	7,600.00	3,655.00	3,945.00	107.93%
T.460 · Construction services				
4600 · Remodeling - residential	130,437.00	134,554.25	(4,117.25)	(3.06%)
4610 · Remodeling-comm'l & industrial	22,000.00	10,000.00	12,000.00	120.0%
4620 · Handyman & repair services	895.00	2,897.00	(2,002.00)	(69.11%)
4640 · Subcontracted to other builders	0.00	1,235.00	(1,235.00)	(100.0%)
Total T.460 · Construction services	153,332.00	148,686.25	4,645.75	3.13%
Total Income	160,932.00	152,341.25	8,590.75	5.64%
Cost of Goods Sold				
T.529 · Direct employment cost				
T.530 · Direct labor cost				
5301 · DLR-Direct labor, regular comp	41,262.49	39,697.85	1,564.64	3.94%
5302 · DLO-Direct labor, overtime	0.00	2,542.73	(2,542.73)	(100.0%)
Total T.530 · Direct labor cost	41,262.49	42,240.58	(978.09)	(2.32%)
T.532 · Labor burden				
5320 · DLP-Dir labor, pd time off	3,000.00	1,189.75	1,810.25	152.15%
5324 · DLB-Dir labor, bonus/prof share	1,525.00	2,000.00	(475.00)	(23.75%)
5332 · D-FICA employer match	2,599.54	2,725.82	(126.28)	(4.63%)
5334 · D-Medicare employer match	557.04	613.31	(56.27)	(9.18%)
5336 · D-FUTA expense	330.10	363.43	(33.33)	(9.17%)
5338 · D-State unemployment insurance	1,237.10	1,037.62	199.48	19.23%
5342 · D-Workers' comp insurance	942.00	1,090.33	(148.33)	(13.6%)
5346 · D-Health insurance	3,615.00	3,549.00	66.00	1.86%
Total T.532 · Labor burden	13,805.78	12,569.26	1,236.52	9.84%
Total T.529 · Direct employment cost	55,068.27	54,809.84	258.43	0.47%
T.550 · Subcontractors				
5500 · Subcontractors	19,163.45	15,462.87	3,700.58	23.93%
Total T.550 · Subcontractors	19,163.45	15,462.87	3,700.58	23.93%
T.560 · Building materials & supplies				
5600 · Building materials & supplies	31,781.17	29,463.87	2,317.30	7.87%
Total T.560 · Building materials & supplies	31,781.17	29,463.87	2,317.30	7.87%
T.570 · Job-specific misc costs				
5710 · Site utilities	700.50	457.69	242.81	53.05%
5720 · Cnstrctn tools/equipmt rental	530.00	428.00	102.00	23.83%
5730 · Truck & auto expenses/mileage	1,367.09	842.56	524.53	62.25%
5790 · Other	6,094.91	6,043.62	51.29	0.85%
Total T.570 · Job-specific misc costs	8,692.50	7,771.87	920.63	11.85%
T.580 · Direct expense-not job specific				
5802 · Misc job supplies	179.64	281.31	(101.67)	(36.14%)
5804 · Cnstr equpt/vehicles-repair	143.42	487.28	(343.86)	(70.57%)
Total T.580 · Direct expense-not job specific	323.06	768.59	(445.53)	(57.97%)
Total COGS	115,028.45	108,277.04	6,751.41	6.24%
Gross Profit	45,903.55	44,064.21	1,839.34	4.17%

EXHIBIT 29.4 Comparative income statement.

	Jan - Mar '01	Jan - Mar '00	$ Change	% Change
Expense				
T.710 · Sales & marketing expenses				
T.711 · Marketing & business developmt				
7100 · Events (shows, tours, etc.)	1,239.00	895.00	344.00	38.44%
7110 · Newsletters	1,250.00	0.00	1,250.00	100.0%
7120 · Donations	0.00	250.00	(250.00)	(100.0%)
7130 · Sponsorships	0.00	300.00	(300.00)	(100.0%)
7160 · Gifts - ded'bl @ $25/person	0.00	125.00	(125.00)	(100.0%)
7165 · Gifts - non-ded'ble portion	0.00	42.00	(42.00)	(100.0%)
Total T.711 · Marketing & business developmt	2,489.00	1,612.00	877.00	54.4%
T.720 · Advertising expenses				
7210 · Yellow pages	679.00	440.00	239.00	54.32%
7220 · Print - nwsppr/mag/guides, etc.	2,156.58	365.00	1,791.58	490.84%
7230 · Specialty publications	0.00	225.00	(225.00)	(100.0%)
7250 · Site signs	0.00	255.00	(255.00)	(100.0%)
Total T.720 · Advertising expenses	2,835.58	1,285.00	1,550.58	120.67%
Total T.710 · Sales & marketing expenses	5,324.58	2,897.00	2,427.58	83.8%
T.800 · Admin employment costs				
T.801 · Administrative compensation				
8000 · ADC-Admin-company owner(s)	12,000.00	10,500.00	1,500.00	14.29%
8010 · ADA-Admin-office administration	6,000.00	5,400.00	600.00	11.11%
8040 · ADO-Admin overtime	0.00	379.00	(379.00)	(100.0%)
Total T.801 · Administrative compensation	18,000.00	16,279.00	1,721.00	10.57%
T.805 · Admin p/r taxes & benefits				
8100 · A-FICA employer match	1,188.00	976.74	211.26	21.63%
8110 · A-Medicare employer match	243.00	219.77	23.23	10.57%
8120 · A-FUTA expense	104.00	113.95	(9.95)	(8.73%)
8130 · A-State unemployment ins	462.00	439.53	22.47	5.11%
8150 · A-Workers' comp insurance	207.00	189.00	18.00	9.52%
8160 · A-Health insurance	1,443.00	1,247.00	196.00	15.72%
Total T.805 · Admin p/r taxes & benefits	3,647.00	3,185.99	461.01	14.47%
Total T.800 · Admin employment costs	21,647.00	19,464.99	2,182.01	11.21%
T.820 · Occupancy expense				
8210 · Rent/lease on office space	2,085.00	1,785.00	300.00	16.81%
T.823 · Utilities				
8231 · Gas	324.00	312.54	11.46	3.67%
8232 · Electric	362.00	389.00	(27.00)	(6.94%)
Total T.823 · Utilities	686.00	701.54	(15.54)	(2.22%)
8250 · Property/liability insurance	300.00	289.00	11.00	3.81%
Total T.820 · Occupancy expense	3,071.00	2,775.54	295.46	10.65%
T.830 · Office expenses				
T.831 · Communications				
8310 · Telephone	373.00	420.00	(47.00)	(11.19%)
8312 · Cell phones	478.63	283.72	194.91	68.7%
8314 · Pagers	425.00	389.42	35.58	9.14%
Total T.831 · Communications	1,276.63	1,093.14	183.49	16.79%
T.832 · Office equipment, general				
8320 · Equipment rental	936.00	0.00	936.00	100.0%
8340 · Deprec, furn/fixt/equip't	413.52	413.52	0.00	0.0%
Total T.832 · Office equipment, general	1,349.52	413.52	936.00	226.35%
8350 · Office supplies	647.00	472.00	175.00	37.08%
8360 · Printing & copying	87.50	367.31	(279.81)	(76.18%)
8370 · Postage & delivery	374.50	128.79	245.71	190.78%
8380 · Bank service charges	38.90	42.00	(3.10)	(7.38%)
Total T.830 · Office expenses	3,774.05	2,516.76	1,257.29	49.96%

EXHIBIT 29.4 Comparative income statement *continued.*

	Jan - Mar '01	Jan - Mar '00	$ Change	% Change
T.840 · Computer expense				
8470 · Computer depreciation	24.30	0.00	24.30	100.0%
Total T.840 · Computer expense	24.30	0.00	24.30	100.0%
T.850 · General & admin expense				
T.851 · Professional fees				
8510 · Accounting	1,568.00	1,425.00	143.00	10.04%
8512 · Payroll services	325.00	310.00	15.00	4.84%
8514 · Legal services	1,185.00	0.00	1,185.00	100.0%
8518 · Recruiting and hiring	117.53	0.00	117.53	100.0%
Total T.851 · Professional fees	3,195.53	1,735.00	1,460.53	84.18%
8530 · Dues	649.56	450.00	199.56	44.35%
8540 · Non-ded'ble dues/pen/contrib	38.70	0.00	38.70	100.0%
8550 · Training/education fees/mat'ls	750.00	542.00	208.00	38.38%
8560 · Books & subscriptions	157.85	98.95	58.90	59.53%
Total T.850 · General & admin expense	4,791.64	2,825.95	1,965.69	69.56%
T.860 · Vehicle expenses				
8600 · Mileage reimbursement	184.91	133.88	51.03	38.12%
8620 · Vehicle insurance	653.58	643.71	9.87	1.53%
8630 · Vehicle repairs & maintenance	1,070.36	550.00	520.36	94.61%
8640 · Vehicle gas & oil	298.97	248.97	50.00	20.08%
8690 · Depreciation, vehicles	597.50	597.50	0.00	0.0%
Total T.860 · Vehicle expenses	2,805.32	2,174.06	631.26	29.04%
T.870 · Meals, travel & entertainment				
8700 · Travel expense	227.00	483.00	(256.00)	(53.0%)
8710 · Lodging expense	398.78	425.63	(26.85)	(6.31%)
T.873 · Meals & entertainment				
8730 · 50% deductible	258.68	312.65	(53.97)	(17.26%)
8740 · 100% deductible	220.00	75.80	144.20	190.24%
Total T.873 · Meals & entertainment	478.68	388.45	90.23	23.23%
Total T.870 · Meals, travel & entertainment	1,104.46	1,297.08	(192.62)	(14.85%)
T.890 · Taxes & licenses				
8910 · Personal property taxes	668.00	642.00	26.00	4.05%
Total T.890 · Taxes & licenses	668.00	642.00	26.00	4.05%
Total Expense	43,210.35	34,593.38	8,616.97	24.91%
Net Ordinary Income	2,693.20	9,470.83	(6,777.63)	(71.56%)
Other Income/Expense				
Other Income				
9100 · Interest income	50.79	48.57	2.22	4.57%
Total Other Income	50.79	48.57	2.22	4.57%
Other Expense				
T.920 · Interest				
9210 · Interest, credit cards	530.79	476.38	54.41	11.42%
9215 · Interest, loans (non-direct)	179.38	280.29	(100.91)	(36.0%)
Total T.920 · Interest	710.17	756.67	(46.50)	(6.15%)
T.923 · Other financing expense				
9230 · Discounts lost & late fees	482.98	647.60	(164.62)	(25.42%)
9235 · Credit card fees	105.00	105.00	0.00	0.0%
Total T.923 · Other financing expense	587.98	752.60	(164.62)	(21.87%)
Total Other Expense	1,298.15	1,509.27	(211.12)	(13.99%)
Net Other Income	(1,247.36)	(1,460.70)	213.34	(14.61%)
Net Income	1,445.84	8,010.13	(6,564.29)	(81.95%)

EXHIBIT 29.4 Comparative income statement *continued.*

Dreamscape Remodeling, Inc.
Income Statement - Expanded Version
January through March 2001

	Jan - Mar '01	% of Income
Ordinary Income/Expense		
Income		
T.450 · Design & consulting services		
4500 · Design services	7,600.00	4.72%
Total T.450 · Design & consulting services	7,600.00	4.72%
T.460 · Construction services		
4600 · Remodeling - residential	130,437.00	81.05%
4610 · Remodeling-comm'l & industrial	22,000.00	13.67%
4620 · Handyman & repair services	895.00	0.56%
Total T.460 · Construction services	153,332.00	95.28%
Total Income	160,932.00	100.0%
Cost of Goods Sold		
T.529 · Direct employment cost		
T.530 · Direct labor cost		
5301 · DLR-Direct labor, regular comp	41,262.49	25.64%
Total T.530 · Direct labor cost	41,262.49	25.64%
T.532 · Labor burden		
5320 · DLP-Dir labor, pd time off	3,000.00	1.86%
5324 · DLB-Dir labor, bonus/prof share	1,525.00	0.95%
5332 · D-FICA employer match	2,599.54	1.62%
5334 · D-Medicare employer match	557.04	0.35%
5336 · D-FUTA expense	330.10	0.21%
5338 · D-State unemployment insurance	1,237.10	0.77%
5342 · D-Workers' comp insurance	942.00	0.59%
5346 · D-Health insurance	3,615.00	2.25%
Total T.532 · Labor burden	13,805.78	8.58%
Total T.529 · Direct employment cost	55,068.27	34.22%
T.550 · Subcontractors		
5500 · Subcontractors	19,163.45	11.91%
Total T.550 · Subcontractors	19,163.45	11.91%
T.560 · Building materials & supplies		
5600 · Building materials & supplies	31,781.17	19.75%
Total T.560 · Building materials & supplies	31,781.17	19.75%
T.570 · Job-specific misc costs		
5710 · Site utilities	700.50	0.44%
5720 · Cnstrctn tools/equipmt rental	530.00	0.33%
5730 · Truck & auto expenses/mileage	1,367.09	0.85%
5790 · Other	6,094.91	3.79%
Total T.570 · Job-specific misc costs	8,692.50	5.4%
T.580 · Direct expense-not job specific		
5802 · Misc job supplies	179.64	0.11%
5804 · Cnstr equpt/vehicles-repair	143.42	0.09%
Total T.580 · Direct expense-not job specific	323.06	0.2%
Total COGS	115,028.45	71.48%
Gross Profit	45,903.55	28.52%

EXHIBIT 29.5 Expanded income statement with percentages.

	Jan - Mar '01	% of Income
Expense		
T.710 · Sales & marketing expenses		
T.711 · Marketing & business developmt		
7100 · Events (shows, tours, etc.)	1,239.00	0.77%
7110 · Newsletters	1,250.00	0.78%
Total T.711 · Marketing & business developmt	2,489.00	1.55%
T.720 · Advertising expenses		
7210 · Yellow pages	679.00	0.42%
7220 · Print - nwsppr/mag/guides, etc.	2,156.58	1.34%
Total T.720 · Advertising expenses	2,835.58	1.76%
Total T.710 · Sales & marketing expenses	5,324.58	3.31%
T.800 · Admin employment costs		
T.801 · Administrative compensation		
8000 · ADC-Admin-company owner(s)	12,000.00	7.46%
8010 · ADA-Admin-office administration	6,000.00	3.73%
Total T.801 · Administrative compensation	18,000.00	11.19%
T.805 · Admin p/r taxes & benefits		
8100 · A-FICA employer match	1,188.00	0.74%
8110 · A-Medicare employer match	243.00	0.15%
8120 · A-FUTA expense	104.00	0.07%
8130 · A-State unemployment ins	462.00	0.29%
8150 · A-Workers' comp insurance	207.00	0.13%
8160 · A-Health insurance	1,443.00	0.9%
Total T.805 · Admin p/r taxes & benefits	3,647.00	2.27%
Total T.800 · Admin employment costs	21,647.00	13.45%
T.820 · Occupancy expense		
8210 · Rent/lease on office space	2,085.00	1.3%
T.823 · Utilities		
8231 · Gas	324.00	0.2%
8232 · Electric	362.00	0.23%
Total T.823 · Utilities	686.00	0.43%
8250 · Property/liability insurance	300.00	0.19%
Total T.820 · Occupancy expense	3,071.00	1.91%
T.830 · Office expenses		
T.831 · Communications		
8310 · Telephone	373.00	0.23%
8312 · Cell phones	478.63	0.3%
8314 · Pagers	425.00	0.26%
Total T.831 · Communications	1,276.63	0.79%
T.832 · Office equipment, general		
8320 · Equipment rental	936.00	0.58%
8340 · Deprec, furn/fixt/equip't	413.52	0.26%
Total T.832 · Office equipment, general	1,349.52	0.84%
8350 · Office supplies	647.00	0.4%
8360 · Printing & copying	87.50	0.05%
8370 · Postage & delivery	374.50	0.23%
8380 · Bank service charges	38.90	0.02%
Total T.830 · Office expenses	3,774.05	2.35%
T.840 · Computer expense		
8470 · Computer depreciation	24.30	0.02%
Total T.840 · Computer expense	24.30	0.02%
T.850 · General & admin expense		
T.851 · Professional fees		
8510 · Accounting	1,568.00	0.97%
8512 · Payroll services	325.00	0.2%
8514 · Legal services	1,185.00	0.74%
8518 · Recruiting and hiring	117.53	0.07%
Total T.851 · Professional fees	3,195.53	1.99%

EXHIBIT 29.5 Expanded income statement with percentages *continued.*

	Jan - Mar '01	% of Income
8530 · Dues	649.56	0.4%
8540 · Non-ded'ble dues/pen/contrib	38.70	0.02%
8550 · Training/education fees/mat'ls	750.00	0.47%
8560 · Books & subscriptions	157.85	0.1%
Total T.850 · General & admin expense	4,791.64	2.98%
T.860 · Vehicle expenses		
8600 · Mileage reimbursement	184.91	0.12%
8620 · Vehicle insurance	653.58	0.41%
8630 · Vehicle repairs & maintenance	1,070.36	0.67%
8640 · Vehicle gas & oil	298.97	0.19%
8690 · Depreciation, vehicles	597.50	0.37%
Total T.860 · Vehicle expenses	2,805.32	1.74%
T.870 · Meals, travel & entertainment		
8700 · Travel expense	227.00	0.14%
8710 · Lodging expense	398.78	0.25%
T.873 · Meals & entertainment		
8730 · 50% deductible	258.68	0.16%
8740 · 100% deductible	220.00	0.14%
Total T.873 · Meals & entertainment	478.68	0.3%
Total T.870 · Meals, travel & entertainment	1,104.46	0.69%
T.890 · Taxes & licenses		
8910 · Personal property taxes	668.00	0.42%
Total T.890 · Taxes & licenses	668.00	0.42%
Total Expense	43,210.35	26.85%
Net Ordinary Income	2,693.20	1.67%
Other Income/Expense		
Other Income		
9100 · Interest income	50.79	0.03%
Total Other Income	50.79	0.03%
Other Expense		
T.920 · Interest		
9210 · Interest, credit cards	530.79	0.33%
9215 · Interest, loans (non-direct)	179.38	0.11%
Total T.920 · Interest	710.17	0.44%
T.923 · Other financing expense		
9230 · Discounts lost & late fees	482.98	0.3%
9235 · Credit card fees	105.00	0.07%
Total T.923 · Other financing expense	587.98	0.37%
Total Other Expense	1,298.15	0.81%
Net Other Income	(1,247.36)	(0.78%)
Net Income	**1,445.84**	**0.9%**

EXHIBIT 29.5 Expanded income statement with percentages *continued.*

Final page only-detail does not agree to total.

Dreamscape Remodeling, Inc.
Profit & Loss by Job, with Percentages
January through March, 2001

	9002.Handyman (Winworth, Charles & Marge)		9008.Kitchen (Winworth, Charles & Marge)		Total Winworth, Charles & Marge		zx.Jobs-non specific		TOTAL	
	Mar 31, '01	% of Inc	Mar 31, '01	% of Inc	Mar 31, '01	% of Inc	Mar 31, '01	% of Inc	Mar 31, '01	% of Inc
Ordinary Income/Expense										
Income										
T.450 · Design & consulting services										
4500 · Design services									7,600.00	4.72%
Total T.450 · Design & consulting services									7,600.00	4.72%
T.460 · Construction services										
4600 · Remodeling - residential	1,437.00	100.0%	32,500.00	100.0%	33,937.00	100.0%			130,437.00	81.05%
4610 · Remodeling-comm'l & industrial									22,000.00	13.67%
4620 · Handyman & repair services									895.00	0.56%
Total T.460 · Construction services	1,437.00	100.0%	32,500.00	100.0%	33,937.00	100.0%			153,332.00	95.28%
Total Income	1,437.00	100.0%	32,500.00	100.0%	33,937.00	100.0%			160,932.00	100.0%
Cost of Goods Sold										
T.529 · Direct employment cost										
T.530 · Direct labor cost										
5301 · DLR-Direct labor, regular comp	285.86	19.89%	6,989.80	21.51%	7,275.66	21.44%			41,262.49	25.64%
Total T.530 · Direct labor cost	285.86	19.89%	6,989.80	21.51%	7,275.66	21.44%			41,262.49	25.64%
T.532 · Labor burden										
5356 · D-Labor brdn allocat'n to jobs	71.46	4.97%	1,747.00	5.38%	1,818.46	5.36%	(7,728.18)			
Total T.532 · Labor burden	71.46	4.97%	1,747.00	5.38%	1,818.46	5.36%	(7,728.18)			
Total T.529 · Direct employment cost	357.32	24.87%	8,736.80	26.88%	9,094.12	26.8%	(7,728.18)		41,262.49	25.64%
T.550 · Subcontractors										
5500 · Subcontractors	144.13	10.03%	3,189.50	9.81%	3,333.63	9.82%			19,163.45	11.91%
Total T.550 · Subcontractors	144.13	10.03%	3,189.50	9.81%	3,333.63	9.82%			19,163.45	11.91%
T.560 · Building materials & supplies										
5600 · Building materials & supplies	185.19	12.89%	8,198.25	25.23%	8,383.44	24.7%			31,781.17	19.75%
Total T.560 · Building materials & supplies	185.19	12.89%	8,198.25	25.23%	8,383.44	24.7%			31,781.17	19.75%
T.570 · Job-specific misc costs										
5710 · Site utilities									700.50	0.44%
5720 · Cnstrctn tools/equipmt rental									530.00	0.33%
5730 · Truck & auto expenses/mileage									188.18	0.12%
5790 · Other	47.41	3.3%			47.41	0.14%			6,094.91	3.79%
Total T.570 · Job-specific misc costs	47.41	3.3%			47.41	0.14%			7,513.59	4.67%
T.580 · Direct expense-not job specific										
5802 · Misc job supplies							179.64		179.64	0.11%
5804 · Cnstr equpt/vehicles-repair							143.42		143.42	0.09%
Total T.580 · Direct expense-not job specific							323.06		323.06	0.2%
Total COGS	734.05	51.08%	20,124.55	61.92%	20,858.60	61.46%	(7,405.12)		100,043.76	62.17%
Gross Profit	702.95	48.92%	12,375.45	38.08%	13,078.40	38.54%	7,405.12		60,888.24	37.84%
Net Ordinary Income	702.95	48.92%	12,375.45	38.08%	13,078.40	38.54%	7,405.12		60,888.24	37.84%
Net Income	702.95	48.92%	12,375.45	38.08%	13,078.40	38.54%	7,405.12		60,888.24	37.84%

EXHIBIT 29.6 Profit and loss by job report with percentages.

In this example you'll notice that even though this quarter's bottom line is considerably less than at the same time last year, the gross profit representing direct operations is higher. You also can find the detailed items that primarily account for the lower year-to-date net profit. Note that this account organization, subtotaling, and reporting format makes it relatively easy to see where the differences are, and what accounts would inspire you to conduct follow-up activities.

Using the information in this report you probably will want to review specific transactions to understand the details and/or decisions that created the current situations. You also may need to see the same type of detailed information for the prior year to understand the differences. The Drill Down feature in QuickBooks Pro makes it easy to view underlying transactions, which can be exceptionally helpful when you're truly using your financial reports as they should be used.

If these were your company's reports, the first several questions you'd probably want to ask yourself are:

- **Have I recognized all the income that is due to be recognized?** If your books are kept on a cash basis, and your checking and savings accounts are reconciled, you probably have. If your books are kept on an accrual basis, you'll want to be sure all jobs have been properly invoiced or entered in an income earned, not yet invoice account.
- **Given the income I have recognized, have all of my expenses been recorded?** Have expenditures that should actually be reductions to liabilities or increases to assets been recorded as expenses in error?
- **How do the details compare with my budget?**
- **How do these figures compare to last year at this time?**

As you review the numbers in comparison to those from the same time last year, you begin to provide yourself with a reference point. Use a highlighter to call attention to significant favorable variances, and use a different color on unfavorable variances. This helps keep favorable and unfavorable variances visible and clearly defined as it can be easy to confuse the two. Ask yourself questions such as:

- What's better? What's worse?
- Are any of the variances due to timing? Did an anticipated expense show up earlier or later than expected?

Enhancing your job-cost reports

Most of my clients get very enthused about their job-cost reports. Whether they see them in the summarized format, or in the detailed format shown in Exhibit 4.6 in chapter 4, the job-cost reports provide you with the opportunity get a real bottom-line payback in your day-to-day construction activities.

After you have constructed summarized job-cost reports for your company, as well as for the individual jobs that you want to review on a job-phase level, you can double-click your mouse to drill down into the underlying transaction detail. Then ask:

- What are the circumstances that account for positive changes and negative changes?
- Can anything be done now to correct or minimize any of the unfavorable results before they get worse?
- Which of the variances were controllable? What went wrong? For example, your lead carpenter didn't create a change order for a little change that became, in fact, a considerable expense. You found later that your client was astonished at the outrageous price you proposed for the change and decided that they shouldn't have to pay for the extra work because they figured it was just part of getting the job done. It can be useful, albeit painful, to actually determine the cost of each of the things that went wrong.
- Which of the unexpected costs were truly beyond your or your employees' control? Can they be controlled next time?
- What do you believe needs to change next time to resolve the problems you find? For example, would it help to:
    - Change policies or procedures?
    - Modify scheduling?
    - Locate more reliable trade contractors?
    - Create contingency plans?

- Obtain better information or perform more planning before starting on that part of the job?
- Correct an estimation calculation?
- Create clearer or better client contracts, or schedule more client meetings to discuss progress and issues before they get out of hand?
- Provide employees or trade contractors with more explicit information?
- Follow up more quickly?

- How and when will you implement the change?
- Who needs to know?
- How and when will you let them know?
- How will you follow up to see that the change is working?

Enhancing year-to-date budget vs. actual variance reports

When you prepare your annual budget for each of your company's income and expense accounts, enter the budgeted amount for each account on a monthly basis. Then prepare the budget vs. actual variance report on a year-to-date basis rather than the current month to get the most out the information. Month-to-month variances primarily reflect differences in the timing of budget and actual cost entries, while major variances will show over the long term.

Start at the bottom of your budget vs. actual variance report and review your net income variance. Look at the dollar difference and the percentage difference, and note whether the variance is favorable or unfavorable. Then work you way back up the report to figure out where the differences are. Examine the subtotals for each section of the report. First check your income total. Then look at its relationship to your direct costs or COGS accounts. For example, if your income is up or down by 10 percent, then on a simple, reactive level, you might expect your COGS to be higher or lower by approximately the same amount. If the direct cost variance is less than 10 percent it may be that some of your COGS costs do not vary in direct relationship to income, such as employee salary costs or vehicle maintenance costs.

Next look at your overhead and other income and expense subtotals. Since these accounts are of a more fixed nature, they should line up with your budget if your budgeting process was on target. After you have obtained the bird's-eye view of the results, you'll probably want to review significant dollar and percentage differences for individual accounts, and note areas that require follow up.

Creating custom reports

QuickBooks Pro offers a variety of default reports that you can access through the pull-down list under Reports. Reports that tie into QuickBooks Pro processes are discussed in the chapters related to the process. In addition, I have developed several custom reports for my clients that provide key information tailored to remodelers and custom home builders. Exhibit 29.7 provides a list of the custom reports, their purposes, and the modifications I made to the default report to achieve the customized report.

The customization process is similar for all reports. While the process is fairly simple, it does require planning. Once you have created a custom report, save it by having the system memorize it.

Before you customize a report, review the QuickBooks Pro default reports by selecting Report Finder in the Reports pull-down list.

Each customized report has as its basis a QuickBooks Pro default report. Select the default report that most closely resembles the report you want to create, or that includes the majority of the information you want in your custom report. Print out the default report and:

- Write down the additional information you need. For example, you may want to see only active accounts, or you may want to see a customer's current balance on the report.

Report Name	Purpose	Customized from QBPro default report (name)	Modifications from QBPro default report
REVIEW/PROOF/ BALANCE	Provides instant glance of income and gross profit from all income accounts	Profit & Loss Standard	• Select only **Design** and **Construction** income accounts • Do not show expense accounts
Missing Checks Report -Primary	Details missing checks from the primary checking account	Missing Checks	
A/R Aging Summary	Lists all open invoices by aging date	A/R Aging Summary	
A/P Aging Summary	Lists all open bills by aging date	A/P Aging Summary	
Gross Profit – by Job	Breaks down income and Cost of Goods Sold (COGS) and then provides a profitability percentage for each job		• Add **Percentage of Income** column to report
Balance Sheet	Displays Assets & Liabilities	Balance Sheet Summary	• Display **each** checking account • Show **Cash on Hand** account under "TotalChecking" • Show **Undeposited Funds** under "Other Current Assets" • Show **Sales Tax Payable** under "Other Current Liabilities"
Income Statement w%	Displays income and COGS total and as percentage of income	Profit & Loss Summary	• Add **Percentage of Income** column to report
Income Statement Budget vs. Actual	Compares budgeted to actual income	Profit & Loss Budget vs.Actual	• Show columns by **Total** only, rather than by month
Job Estimates by	Compares estimated	Job Estimates by	• Add **Percentage Difference**

EXHIBIT 29.7 Custom reports recommended for remodelers, developers, and custom home builders.

Report Name	Purpose	Customized from QBPro default report(name)	Modifications from QBPro default report
Actuals Summary	vs. actual income, expenses, and time for each job	Actuals Summary	columns to report
Job Profitability Detail for Job # ___-_______	Provides profitability information on each job	Job Profitability Summary	
Job Estimates vs. Actuals forJob # - _____-_______	Compares estimated vs.actual income and expense by job	Job Estimates vs. Actuals Summary	
Account Listing – All Accounts	Listing of **all** accounts that includes account number and current balance	Account Listing – All Accounts	• Do not show Description or Tax Line columns
Account Listing – Active Accounts	Listingof active accounts that includes account number and current balance	Account Listing – All Accounts	• Filter by **Active** accounts only • Do not show Description or Tax Line columns
Customer List – Basic Information	Displays most the useful information for each customer	Customer Contact List	• Show **Payroll Item**, **Type**, **Amount**, **Annual Limit**, **Expense Account**, **Liability Account**, **Allocate Cost**, and **Based on Quantity** columns

EXHIBIT 29.7 Custom reports recommended for remodelers, developers, and custom home builders *continued.*

- Cross off any information that you don't need. For example, you may not want to see a vendor's entire street address on the report. Plan how you want to see or filter the information. For example, do you want to see only active jobs?

The next step is to select a name and title for the new report. To minimize confusion, create a new name. Do not use the QuickBooks Pro default name. The report name appears in the title bar of the report window. You also need to create a report title that will make sense to the users. The report title appears at the top of the report.

The process for customizing and memorizing a report follows, using, as an example, a report called, Account Listing—Active Accounts. You can use the same basic steps to create other custom reports.

1. In **Reports** select **Report Finder** (see Exhibit 29.8).
2. Select a report that you want to customize. The reports in Report Finder are set up under general headings, such as Customers and Receivables or Accountant and Taxes, under which is the default report named Account Listing that we will use for this example.
3. Click **Display** to show the selected default report on the screen.
4. Click on the **Customize** button on the Account Listing report window (see Exhibit 29.9).

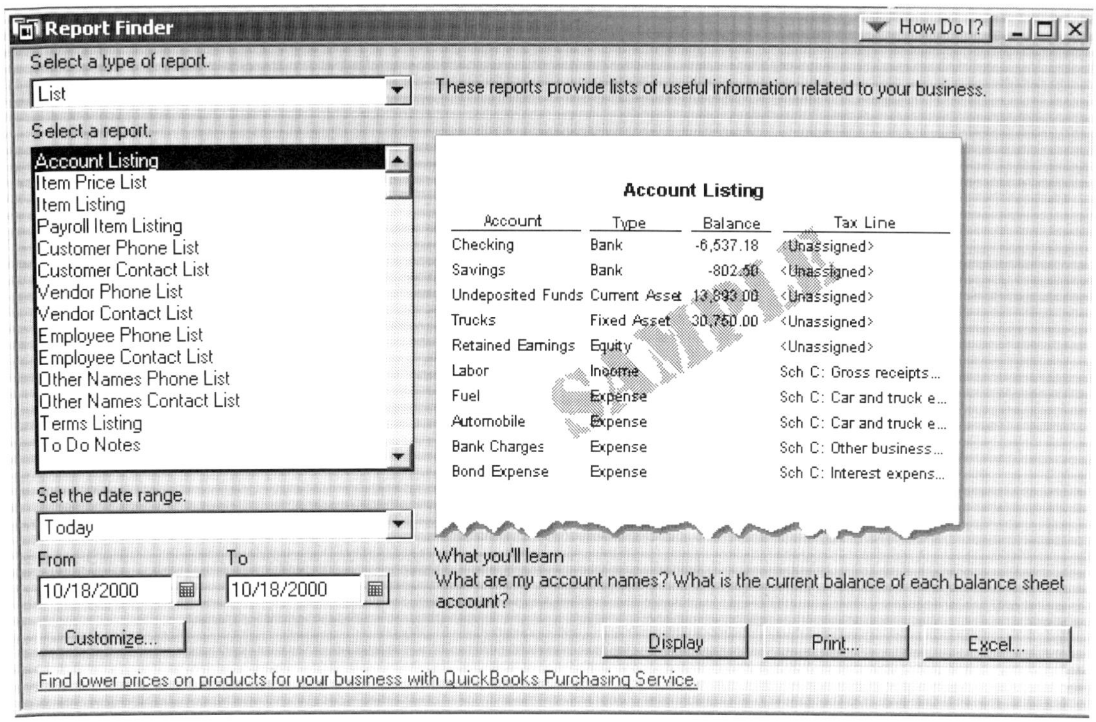

EXHIBIT 29.8 Report Finder window with Account Listing default report highlighted.

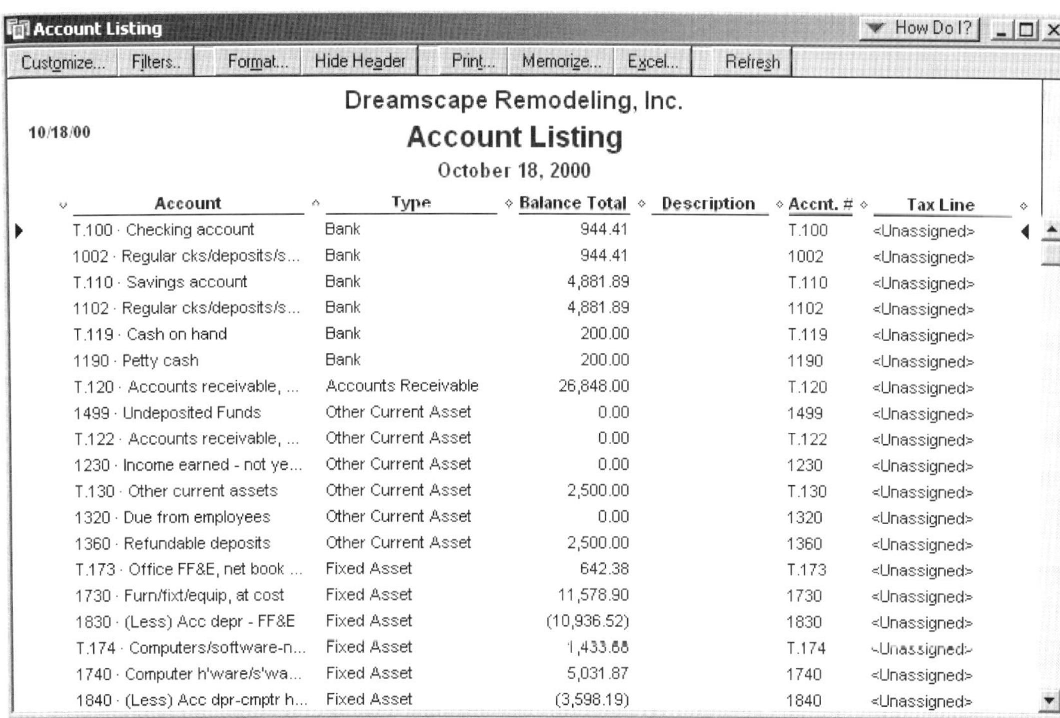

EXHIBIT 29.9 Account Listing report window.

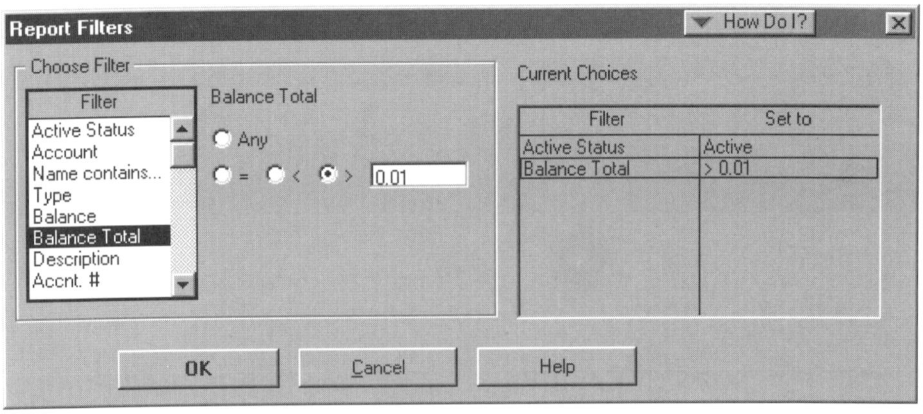

EXHIBIT 29.10 Report Filters window.

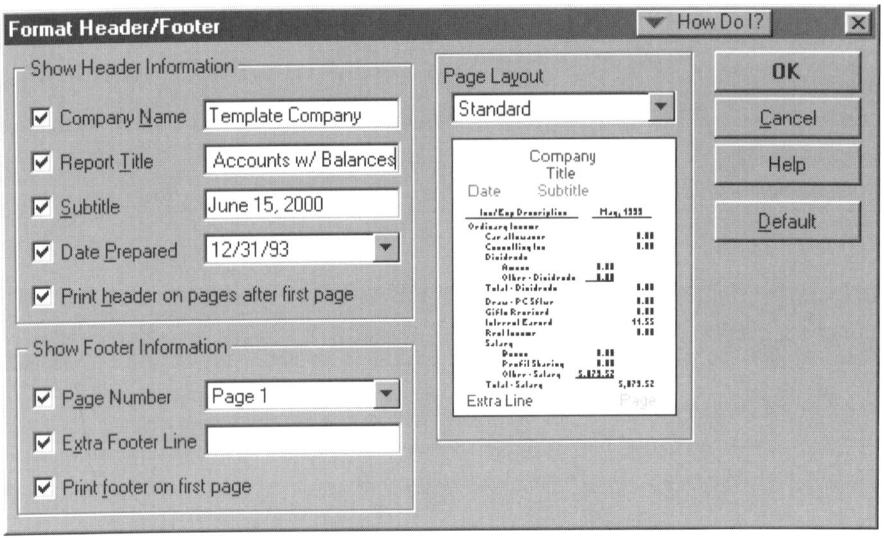

EXHIBIT 29.11 Format Header/Footer window.

5. The Columns listing will display all of the available columns containing data that can appear on the report. Use your mouse to select or de-select a column. Highlight the item you want to delete to remove the checkmark. To add a column click on the column to add a checkmark.
6. Click **OK** when you have finished making changes to the columns.
7. Set up the filters for your customized report by establishing the parameters in which the report data must fall. To change the parameters, click **Filters**.
8. The Report Filters window displays data fields that you can establish for filters. The Current Choices box shows all of the current filters for the report. For the custom report, Account Listing—Active Accounts, we do not want to see accounts with zero ($0.00) balances, so the parameter for balance totals is set for greater than one cent (>$0.01).
9. When you have finished setting the parameters click **OK**.
10. To change the title of the report click **Format** to open the Format Header/Footer window. The title will appear at the top of the report. Then click **Header/Footer**.
11. Edit the field as desired.
12. When you have finished making changes, click **OK** *two times* to see the report you have created.

13. When the report looks just the way you want it to and contains the information you need, click **Memorize** to save the report.
14. In the Memorize Report dialog box enter a new name for the report. The report name will appear in the title bar of the report window.
15. Click **OK** to save and close the custom report.
16. When you have finished customizing your report, you can:
    - Display the report on the screen if you do not want to print the report.
    - **Print** the report on the default printer.
    - **Export** the report data to an Excel spreadsheet.
17. In the future, when you want to use this report again to show current data, follow these steps.
    - Click on the **Reports** button in the taskbar.
    - Click on **Memorized Reports**.
    - You then will see a screen showing all of your memorized reports. To run a report that you have memorized previously, double-click on the report name and customize as desired.

This feature will save you from having to repeat the report design and selection process over and over again. It is a particularly useful feature when you have reports that you want to see frequently, but that take some setup and selection time to prepared. They are now just four mouse-clicks away!

CHAPTER 30

Adhering to file management procedures

We are constantly besieged with warnings regarding Internet security issues, hackers, viruses, and other technology disaster scenarios. In my real-life experience, however, these threats are insignificant compared to the amount of financial data this is lost, misdirected, or misplaced because users don't understand basic QuickBooks Pro 2000 file management and security procedures. As a result, a tremendous amount of time is spent or fees are paid to re-create or recover information.

Data loss due to lack of knowledge or slipshod procedures can be very, very costly. Now that you have committed to investing in your financial information system, you will find this chapter critical to safeguarding your newest business asset!

Creating files

You can create as many QuickBooks Pro files as you wish and open each one as you need to enter information. For example, if you have three different companies you can create three differently named QuickBooks Pro files.

> **NOTE** Be careful when you name the separate company files. For example, if you give the file for your third company the same name as the file for your first company, it will over-write and delete the first company file.

Working file names

Most QuickBooks Pro working files consist of the company name followed by the extension .QBW, such as, Dreamscape.QBW. If you create a new file for a different company, you can change the name of the old company to something like zOLD_DRSC_DBA.QBW, with the "z" preface taking the old file to the bottom of the directory listing. Then you can make the name of the new company file something like 1New_DRSC_LLC.QBW, with the 1 taking the new file to the top of the directory listing for quick access.

If you have more than one company with a similar name, however, the best naming system clearly distinguishes the companies. For example:

- DRSC_**LLC**.QBW
- DRSC_**INC**.QBW
- DRSC_**DBA**.QBW

File types

QuickBooks Pro creates different file types to store each file at various stages in its "life." You can create any name you like for any of the files, and then QuickBooks Pro assigns one of the following file extensions (the three letters following the dot) depending on the type of file you are creating as shown in Exhibit 30.1. You do not need to include the extension when you tell QuickBooks Pro the name you would like for the file. QuickBooks Pro will add the extension automatically.

Type of file	File extension
QuickBooks Pro working—current file	.QBW
QuickBooks Pro "image" file—only open when your working file is being used—displays the same pre-extension name as the file that is open	.QBI
QuickBooks Pro backup file—a condensed "save" of your working file	.QBB
QuickBooks Pro accountant's copy to give to accountant	.QBX
QuickBooks Pro accountant's copy after accountant opens	.QBA
QuickBooks Pro accountant's copy ready to be merged into original	.AIF

EXHIBIT 30.1 File extensions assigned to QuickBooks Pro files.

File locations

QuickBooks Pro may attempt to place your working file and your backup file within its own directory, typically drive C\Program Files\Intuit\QuickBooks Pro\your file. If you have only one computer, and the entire computer is backed up on a regular basis, this may work well for you. Many individuals with one computer, however, prefer to store all QuickBooks Pro data files within a folder or directory named QB, which they create in the C drive, or inside the My Documents folder. The files are quickly accessible and easily selected for backup on a tape drive or copying to a zip disk when you don't want to go through a full computer backup.

If your computers are networked and if you will be accessing the file from various computers, then your QuickBooks Pro data file must be stored on the server or on a commonly accessible computer. As above, I suggest a quickly accessible directory with a simple folder name such as QB.

NOTE If networked, each computer will need to have its own separate QuickBooks Pro software package installed.

Closing and opening files

There are two ways to close and open each user's QuickBooks Pro data file.

1. When you leave QuickBooks Pro, if you choose the File/Close company option to exit the system, when you re-open QuickBooks Pro you will need to choose the File/Open company option to get back into the file. Then locate the correct current working file within your computer or server and ask QuickBooks Pro to open the file.

TIP As a matter of course, quickly check one of the most recent transactions entered to confirm that you have opened the correct working file.

2. Alternatively, when you leave QuickBooks Pro, if you choose the File/Exit command, click on the "X" in the upper right hand corner, or click on the top left icon and select Exit, the next time you re-open QuickBooks Pro, the system will automatically re-open to the same file. If this approach stops another user from switching into the single-user mode required for certain maintenance or list activities, that user will need to re-open QuickBooks Pro and simply cancel out of the log-in process.

NOTE If the program or your computer shuts down or locks up in the middle of a transaction, when you re-open the program QuickBooks Pro gives you a message that says it is rolling back to the last recorded transaction. You can then search, using the Find feature, for the transaction you were entering when the system shut down. If it doesn't exist, go ahead and enter it. If it does exist it may be damaged, all the links may not be set, and so on. To be safe I suggest you delete it and then re-enter it.

"Lost" files

If you or one of your staff can't locate the working file, follow the first seven steps listed in the following "Rogue files—cleanup" section making sure that you search the correct location. When you believe you have located the correct file, first look at the detailed view of the file that shows its type, file size, date last modified, and so on to see if the information looks reasonable. Do this before you open the file! After the file is opened look for some recent data entries to confirm that you have the right file.

Opening versus restoring files

When you *open* a file by selecting File/Open, QuickBooks Pro looks for a working file with a .QBW file extension. You tell it *where* to locate the file, such as on your computer or on the server.

The process of *restoring* a file means that you are converting a condensed, backup file into a new file with current working file status. You may want to do this to:

- Facilitate the transfer of your file from one place to another (For example, you created a condensed backup file for work at home, and now you want to restore your updated file to its original location.)
- Resolve a problem of some sort with your current working file or hard drive and you want to restore your backup file into a new working file
- Look at an older file for some reason and to see the transactions you need to restore the backup file to a new working .QBW file status
- View a file that has been transferred to you via the Internet

Managing rogue files

QuickBooks Pro allows you to have more than one .QBW working file per company stored in various places within your computer system. This can be a very hazardous situation for the health of your financial data, so I refer to the extra files that are loitering about in your computer system as *rogue files*.

NOTE **Warning!** It is extremely important that you thoroughly understand the dangers of having more than one .QBW working file.

Rogue file creation

It is very easy to create rogue files. For example, *you* open a working file and spend the day (or week) making entries into it. At the same time *someone else* in your organization opens a second newer, older, or differently located version of your QuickBooks Pro file and makes entries to it. Now you have two different sets of information; one file has your transactions in it, and the other file has that person's transactions in it! In addition, either one of the files may include or may be missing other recently posted data.

You now have a difficult situation because there is no way to merge the two files together. All you can do is research which of the files is the most complete and then re-enter the remaining data into that file.

TIP Because QuickBooks Pro is a database, every time you open a QuickBooks Pro working file the date shown on your computer system for that file will change to the current date and time, whether or not you make changes to the file. This dating methodology is different than the way many other Windows software programs assign dates. (Usually file dates are assigned based on the last file save date.)

This is an important fact to remember if you have several working files and are attempting to determine which of them is "most current" or "most correct." As soon as you open a file to look at it the date will change! The file date therefore is *not* a good way to draw any conclusions about a file's contents. The best way to distinguish between several rogue files is to re-name each of them temporarily in a clearly identifiable manner, such as Kate.file.QBW, Fred.file.QBW so that you can make notes about the contents of each and then decide how best to proceed.

Rogue file cleanup

After you have a clear picture of exactly which file is the most correct working file, move it to the correct location in your computer system and rename the file with your standard working file name. Your next mission is to get the missing data re-entered as quickly as possible. To assist in efficient re-entry, print out any missing information that was mistakenly entered into the rogue file.

At the conclusion of the battle don't forget to go on that final "disarm or destroy" mission! Bury the bad files so that you do not repeat this costly and frustrating situation. Disable or eliminate the remaining rogue .QBW files from each computer by performing the following steps:

1. Go to the Windows **Start** button in the lower left-hand corner of your computer screen.
2. Go to the **Find** or **Search** button.
3. Select **Files** or **Folders**.
4. Go to the Named field and enter ***.QBW**.
5. Select the drive to search and check the **Include subfolders** box.
6. Click on the **Find now** or **Search** icon.
7. The screen will display any .QBW files it finds.

To disable the files so that no one can open them from QuickBooks Pro, change the filename extension to .QBR. The advantage to this process is that if you need to go back to this data at some point in the future you can change the filename extension back to .QBW and access the file again. The disadvantage is that you are using hard drive space for a file that may have no value. If you have enough hard drive space, however, this is the safest temporary alternative.

To completely delete the file when you're absolutely certain that this is a rogue file that you will no longer need to access, follow these steps.

1. Click on the file to highlight it.
2. Click with the right-side mouse button.
3. Select **Delete**.

Rogue file policing actions

It is a good idea to periodically perform this rogue file search-and-disable-or-destroy procedure, especially if your files are moved regularly from computer to computer or you restore from backup copies, even occasionally.

If employees or owners, who sometimes are the worst offenders because they are always in a hurry, do not pay *very* strict attention to how they open, backup, and restore files it is relatively easy to create rogue files. The time saved is not worth the time and expense to resolve the related problems. Be careful!

Transferring files

There are several ways to move your QuickBooks Pro files from one location to another. Each method has its merits and drawbacks. The following information is based on my own experience, as well as the experiences of my clients.

E-mailing

E-mailing a copy of your file backup is one way to convey your file to another user, such as your accountant or to yourself at home. This process normally is fairly successful, but I wouldn't rely on it to be your sole method of file transfer. To send a copy via e-mail, follow these steps.

1. Create a backup copy of your password-protected file. The backup process condenses the file so it is ready for the trip. Note exactly where the backup file is housed on your computer system.
2. Create your e-mail message.
3. Go to your Attach Files feature, and then browse to locate your new .QBB file. Highlight the file, and attach it to your e-mail.
4. Send the e-mail. If the file is large it may take a while to complete the send process and for the recipient's e-mail system to receive the file.

Some of my clients are not comfortable transmitting their financial data over the Internet. Discuss any security concerns with your computer consultant.

Manual transfer or electronic access

You can make backups of your working file at work, take the backup disks home, restore the file backup on your home computer, enter your work, create new backup disks at home, take the backup disks into work, and then restore the backup file to its original location. The drawbacks, however, are that the process:

- Is especially prone to failure with diskette backups for reasons I discuss in chapter 31
- Can be time consuming, especially if your files are large
- Is subject to "Uh oh! I left the files at home or in the back seat" goofs
- Increases the risk of rogue file proliferation as previously discussed

Although the use of a zip drive would simplify the process and make the data transfer more reliable, I prefer, if at all possible, to keep the working file locked in one central, secure location.

If you need to access your work computer from your home computer, contact your computer consultants and ask them to look into computer-to-computer communications programs such as PC Anywhere or other possible solutions that will allow you to access your network from other locations. If you really do need to work from home, or to access your financial information from remote locations, it's probably time to investigate computer-to-computer connections. If your company is technologically sophisticated, see your computer support team for other ways to transfer your QuickBooks Pro files or to access your work computer from home. Break the sneaker-netting habit and make technology work for you!

Accountant's copy

QuickBooks Pro allows you to create an accountant's copy of your file. The concept is that your accountant can have a copy of your file while you continue with your day-to-day data and transaction entry. The accountant can make some very limited entries and then send the modified file back to you where you can merge that file into your working file.

I am not fond of this process for several reasons. As a QuickBooks Pro consultant and accountant, I usually work with my clients to:

- Modify or upgrade the lists and links or other design aspects of their files
- Correct or add transactions within the QuickBooks Pro subsystems
- Update reconciliations

None of these critical activities are allowed in the accountant's copy. About all your accountant can do in the accountant's copy of the file is to post journal entries. If you are concerned about getting the correct results in your job-cost and special reports, you will soon discover that certain journal entries actually create erroneous reports! Additional areas of concern that lead me to advise my clients *against* using the Accountant's Copy feature when conveying information to and from their tax preparer follow.

Sometimes the accountant does not receive a *closed* copy of the file, meaning entries continue to be made during the timeframe covered by the accountant's copy. Both the accountant and a user may make the same or different entries during the same time period. Then, when the files are merged, there are multiple entries.

A typical example is this. The user finds and corrects transactions in the prior period without telling the accountant, who at the same time is making his or her own changes. The result is tax records and book records that do not agree. I have seen a number of situations where this issue has created time-consuming and expensive balance-resolution problems. And sometimes those problems do not become evident until the following year when the tax preparer "proofs" last year's numbers. Not a pretty picture!

Another reason I don't recommend using the accountant's copy is that all parties involved must have a thorough understanding of the three levels of the accountant's copy files, and how they should be created, opened, and merged. Without this training there is a good chance you'll have major problems transmitting files, making entries, and merging files. There are too many points along the path that provide opportunities for error, misunderstandings, and re-work.

The end result is that, unless the entire process is tightly controlled, the experience can be very frustrating, time-consuming, and expensive. I believe it is better for your accountants to work in your "live" file—either at your office or at theirs. If you can find or structure a few days when you won't be making entries into your file, give a backup file to your accountant to work in and then transfer the file back to your office.

Training all users

It is *imperative* that you train everyone who will be using the QuickBooks Pro files in the following areas:

- Basics of paths, drive names, folder names, file names, and file extensions
- Navigation through drives, paths, and folders from various views within Windows® and QuickBooks Pro
- "Detail" view and information about the files, such as file name, size, type, date last modified, and so on from various views within Windows and QuickBooks Pro (This information can be extremely valuable when you have a number of files to choose from.)
- Importance of reading the screens carefully and understanding exactly where they are located within the computer system when opening working files, creating file backups, and restoring files
- "Mapped drives—what they look like and how they work (Mapped drives are required for QuickBooks Pro multi-user network environments.)

CHAPTER 31

Backing up your files

QuickBooks Pro 2000 automatically updates the file each time you enter a transaction or change a list, so there is no need to go through the same kind of File-Save procedure required by many software programs.

A QuickBooks Pro backup, however, is a variation of the File-Save concept.

- A backup creates a copy of your file but with a .QBB file extension. You can accept the suggested name or create a new name. QuickBooks Pro automatically adds the extension .QBB.
- You tell the computer where you would like the backup to be stored, such as on your personal computer hard drive, floppy diskettes, or a zip drive.
- You can back up your file on different dates and create multiple backups with different names. These copies retain their backup dates, and multiple copies, if properly named and dated, don't have the same inherent risk of creating rogue files.
- A backup is a *condensed* version of your file, so the file size is significantly smaller.

Establishing backup policies

The backup process is so critical that I believe you should make some policy decisions early on in the life of your accounting system, document your decisions, and then periodically follow up to see that your policies are being faithfully implemented. Some of the issues to consider before you make those decisions follow.

Backup medium

Which computer medium will you use for backups? My recommendations follow but first a warning. You, or someone you consider responsible, should review your backup report after each backup is complete to confirm that everything worked properly. I know of more than one occasion when management *thought* that backups were being performed, and they *were* being performed, *but* there was also an accompanying software report that said something wasn't working quite correctly. No human being followed up, and the next time there was a major problem and they needed the backed up copies, guess what? No recent backup copies could be found! And that little report had been telling them so all along, but no one took the time to look and resolve the problem. Don't let this happen to you or you'll be feeling like the character in Exhibit 31.1!

Automatic, timed tape backups

My preference is to have timed, automatic tape drive backups of the drives on your personal computer or server where you store your QuickBooks Pro working file. Automatic, timed tape backups do not rely on a tired, stressed human to remember and to take the time to perform a backup each day, week, or month.

EXHIBIT 31.1 Sometimes you feel like. . . !

> **TIP** Periodically test your ability to restore from your backup tapes because problems have been known to occur during the tape restoration process. (Trust me, you'd rather find out now, when it can be fixed!)

Zip drive backups

My second choice for backing up your QuickBooks Pro working file would be to use a zip drive to back up your files. Zip drives are fast, each disk holds large amounts of data, and the backups are far more reliable than 3.5-inch diskettes. Although the disks are more expensive they are less subject to problems than the standard 3.5-inch diskettes. If you're going to regularly, manually transfer your files from computer to computer, and don't yet have a zip drive unit in your computer, I recommend that you add one. They are relatively inexpensive and easy to install. Talk with your computer consultant.

3.5-inch diskette backups

3.5-inch diskettes come in as only my third choice for backups and file transfers. If you have a large file it can take what seems like a very long time to complete the backup process. We also have experienced an uncomfortably high failure rate when restoring QuickBooks Pro files from diskettes. (I always ask my clients to bring *at least* two backups with them if they are bringing their files to our office!)

When you have a large data file QuickBooks Pro asks you to insert additional diskettes into your A drive until the file is backed up completely. If you have a fairly large file, your backup could require 6, 8, or even more diskettes. And of course, the more diskettes involved in a backup, the higher the likelihood that one of the disks will fail to properly read or restore the data.

> **TIP** If you do use 3.5-inch diskettes, immediately label and number them so you can insert them in the proper order when restoring the file.

Intuit also advises that your diskettes be fully **re-formatted** each time you use them for QuickBooks Pro data file backups. This, of course, also can be a time-consuming process, especially if you need to format 12 to 16 diskettes each time so that you can have two backups.

If you are going to use diskettes for backups, I suggest that you use high quality, new diskettes. Don't give in to the dangerous practice of using the same diskettes over and over again for months! After you have used the diskettes for 10 to 20 backups I'd move the diskettes on to a less-critical environment and replace them with new diskettes.

Hard drive backups

Hard drive backups are fast and handy. I never recommend them as a first choice for backing up your files, however, because one of the primary reasons that you would need to restore your QuickBooks Pro file is a hard drive failure or computer theft. If your hard drive fails or disappears with your only backup on it, you're pretty much out of luck!

For speed and convenience, however, and because, like chicken soup, it can't hurt and it might be useful, I usually suggest placing non-critical and secondary backups on your hard drive. Just don't make that hard drive backup your only safety net—the landing could be pretty brutal.

Intuit offsite backups

Another option to consider, although it doesn't seem to be an exceptionally popular option with my clients yet, are the offsite backup services offered by Intuit. Even though all sorts of security features surround the process, I believe most small businesses aren't excited about the idea of sending their financial data to a large data center. You may want to look at the program and draw your own conclusions. If you have experienced problems with backups, or feel that your technology is deficient, this may just be what you need.

Frequency of backups

I borrow the following quote from a list of famous last words that I received over the Internet.

"Backups? Backups? We don' need no stinking backups!"

The general guideline is to decide how much data you're willing to lose—1 day's worth, 1 week's worth, and so on. Then perform a backup that often. If you're a very small company and only work in your QuickBooks Pro data file once a week, a weekly backup may work well for you. If you're a larger company and make regular, daily entries to your records, I recommend that you back up your QuickBooks Pro file at the end of each day. You'd be surprised how many entries flow through the books and how much time it would take to re-create all of the details.

Backup storage

Decide which of these commonly utilized storage areas would be the best place to keep your tape, zip, or diskette backups:

- A box next to the bookkeeper's computer
- Your top desk drawer (the junk drawer) at work
- A safe at work
- Your vehicle's glove compartment, back seat, trunk
- Your accounting assistant's purse, briefcase
- Your briefcase
- Your home, 30 miles away

Although the first six are, by far, the most frequent choices that business owners, when pressed for honesty admit to, they are by far the riskiest. Storing critical backups in your office makes them subject to most of the same risks as your hard drive data—theft, fire, water damage, or vandalism. Away from the office is better, but:

- Your vehicle is subject to heat, moisture, and dust—all enemies of electronic storage
- If your accounting assistant is called out of town, becomes ill, or decides to work somewhere else the backup won't be available to you
- Your briefcase could be lost or stolen, or temporarily stashed at that risk-ridden office

"Sure", you say. "What are the chances of losing my working file or hard drive and my backup at the same time?" My response is that if you are that much of an optimist, you should sit down and study Murphy's laws every once in a while, for a periodic reality check. Better to plan for a worst-case scenario and be pleased that you didn't need the backup, than get into a situation where several related things go wrong at the same time.

Short of a safe deposit box, I recommend, of course, the final option. Although it would be inconvenient to go home to get the backup, you'd lose only a moderate amount of time even if you had to make a trip in the middle of the day. If you're going to be out of town, of course, you'll need to make alternate arrangements.

Of course, if your office is in your home where offsite backup is more difficult to arrange, or if you are concerned about the possibility that your entire community could encounter a disaster, you may want to look further into the remote backup services offered by Intuit.

My preference is a backup storage plan that blends convenience with a reasonable level of security. Be sure to clearly label and date each backup tape, disk, or diskette, and then follow these steps.

1. Monday through Friday create backups at the end of each day.
2. On Friday, hand off the Thursday backup for offsite storage.
3. In the middle of the next week, reclaim the prior week's backups for re-use the *following* week.
4. After the month-end close, create a second month-end backup at the end of the day for offsite storage. Reclaim the weekly backups from that month to re-use.
5. After quarter-end close create a second quarter-end backup at the end of the day for offsite storage. Reclaim the monthly backups from that quarter to re-use.
6. After year-end close create a second year-end backup at the end of the year for offsite storage. Reclaim the first and third quarter backups from that year to re-use.

This requires a total of 19 backup units for the first year's use—8 for 2 weeks' worth of daily onsite backups, 5 for weekly offsite backups, 2 for monthly offsite backups, 4 for the 3 quarters plus annual offsite backups less the reclaimed 1st and 3rd quarters. The second and subsequent years will require fewer new tapes, zip disks, or diskettes.

NOTE

If you are purchasing tapes this can come to a fairly hefty sum in the first year. If you use zip disks it might be a little less expensive. Diskettes, of course, are fairly inexpensive, but you'll also need more of them due to their limited capacity and life. If you're cringing at the expense, just contrast it with what it would cost to re-create your missing data. The cost of backups just became much more reasonable, didn't it?

Backing up your working file

Now that we've made all of the critical decisions regarding your backup process, please document and share your new procedures with your staff. Then you can show them how to actually create a backup of your working file.

To back up your working file, follow these steps.

1. While your QuickBooks Pro file is open, select **File/Backup** from the task bar.
2. The Back Up Company To window will appear (see Exhibit 31.2). Select the drive and then the directory where you want your backup file to be stored.
3. Name or re-name the file. QuickBooks Pro will add the .QBB extension if you don't include it in the name.
4. Click **OK** to complete the backup.

TIP

I like to code my backup file with the date I created it, or with the date I'm using as a closing date on the books. I include the last two digits of the year, two digits for the month and two digits for the date, followed by identification letters for the company.

The window in Exhibit 31.3 shows the first backup stored in this directory. I'm about to add another one as of 12/18/2001 for Westfield Homes. QuickBooks Pro automatically displayed the previous name of 011215WH.QBB in the File field, but I modified it to the new file name that I wanted.

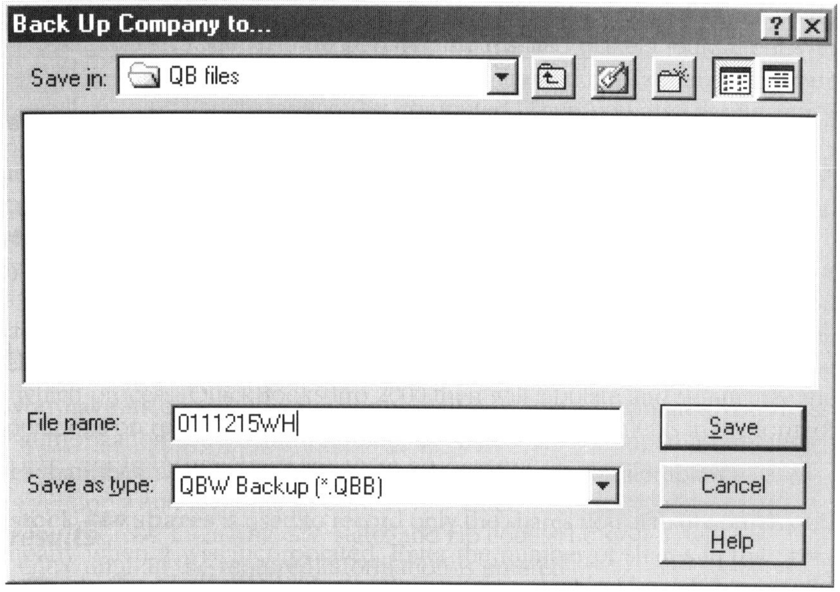

EXHIBIT 31.2 Back Up Company To window.

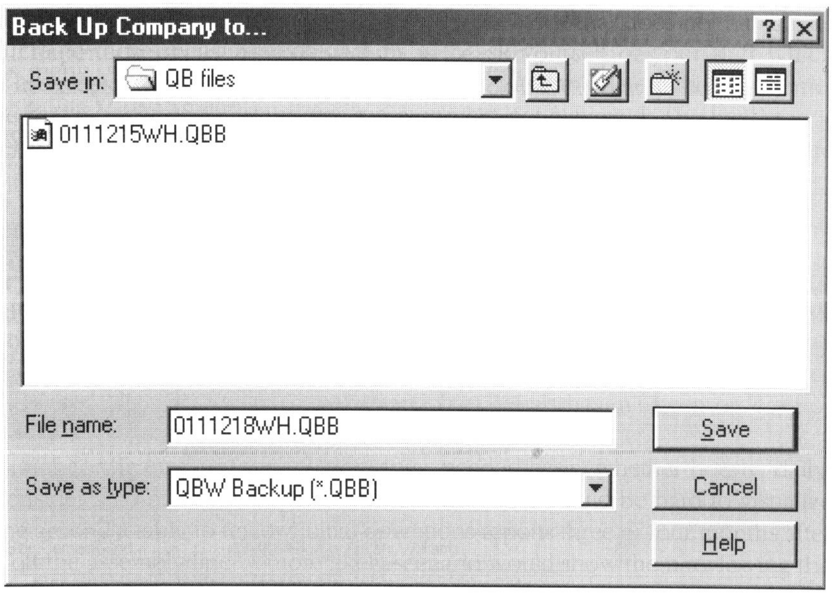

EXHIBIT 31.3 Backing up second file with unique name.

TIP: If I pressed the Details icon in the window in Exhibit 31.3 the screen also would display the file size, file type, and date last modified.

Restoring a file from a backup

Restoring a file from a backup is a two-step process.

1. After you go to File/Restore, QuickBooks Pro will ask you where you want to restore ***from***. The system will recognize only files with the .QBB file extensions. Select the correct drive, directory, and file to restore ***from*** (see Exhibit 31.4). Then click the **Open** button.

2. The next screen will ask you where you want to restore ***to***, and the name you want to give the new working file. Select the correct drive and directory to restore ***to***.

NOTE **Warning!** Be very careful to select the correct drive and directory!

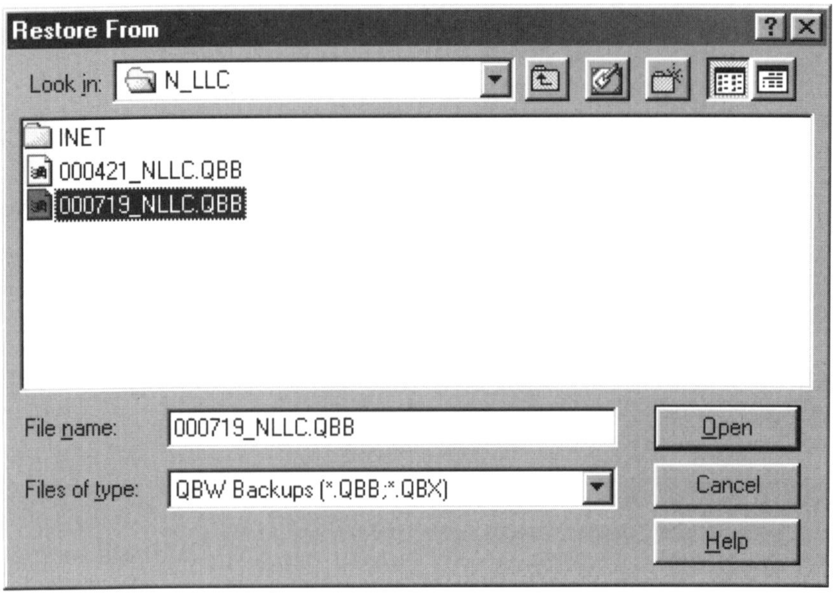

EXHIBIT 31.4 Restore From window.

The window in Exhibit 31.5 shows the correct path to the correct drive (C), and the correct folder (N_LLC).

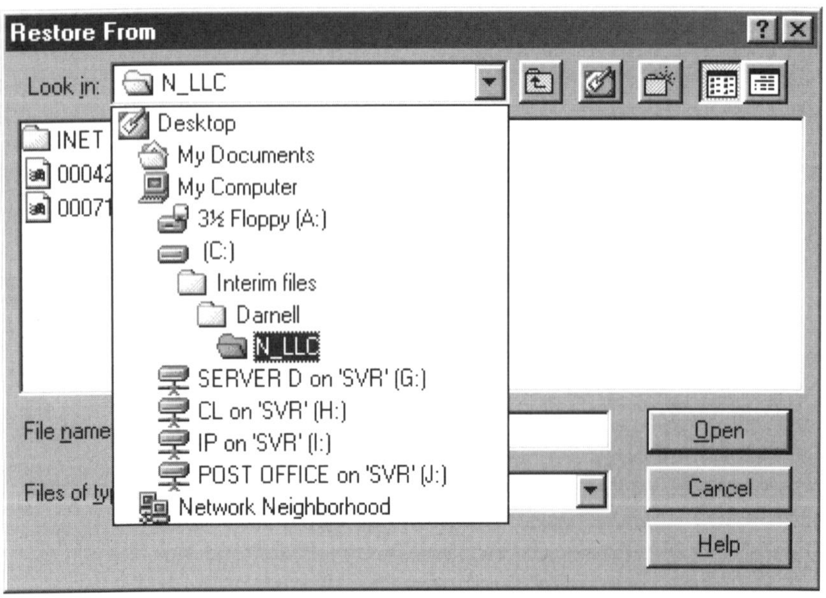

EXHIBIT 31.5 Restore From window with correct drive (C) and directory selected.

After you have clicked on the correct folder, the screen will clear and you will move to the file name field. Enter the name that you want to use for this working file. *Always create a new, clearly identifiable name.*

Heeding warnings

If you accept the name QuickBooks Pro provides, and it is your actual, current working file, the backup will over-write it and destroy it! You can lose lots of data this way! Even if you are replacing your current working file, give the new working file a new name. If you don't and the restore process fails mid-stream, you will lose *both* the backup file, and the previous working file!

Once you are certain that the newest working file has been restored successfully, disable, but do not destroy, the old working file. This is a critical step! If other individuals use the file, their user name may still be linked to the old file, and they could inadvertently create rogue files if you don't prevent them from accessing the old file by changing the file extension.

CHAPTER 32

Securing your financial information

QuickBooks Pro 2000 offers you a variety of security levels and security features that many users and their accountants may not be aware of.

Defining the Admin role

The administrator has the highest level of system access and control in your QuickBooks Pro data file. This individual can change the structure or lists of your QuickBooks Pro data file, access all of the transaction and reporting functions, set closing dates, post entries into "closed" periods, create other QuickBooks Pro users in the data file, and set the levels of access allowed to those other users. In addition, the administrator is the only user who can export or import data, change setup information about your company, or change company preferences.

> **NOTE** The Admin and other security features are embedded in your company's data file, not in your accounting software. This means that one individual could be the administrator in one company's data file, but have limited or no access in another company's data file.

Every circumstance is different, and although there are some exceptions, I normally recommend that the Admin function be accessible to:

- At least one of the company's owners
- The company's accountant
- The accounting system's primary caregiver

The reasons for including the first two individuals are fairly evident. And, of course, if the primary caregiver is also the business owner, there is no question regarding full access. The reason to usually include the primary caregiver is because he or she is the individual held most responsible for system upkeep and maintenance. In this role, this individual needs to be able to perform, sooner or later, most of the Admin functions. My experience has been that most business owners rely heavily on this individual to be intimately familiar with their financial affairs, sensitive transactions, payroll, and so on. In this situation, granting the caregiver Admin access usually is not an issue.

Exceptions exist, however, and there are issues that should be carefully considered before granting or denying full access to that individual.

- If you, as an owner, do not want your accounting assistant to have full access to your company's financial information, it may not be appropriate to grant access to the Admin function.
- If you have authorized electronic bill payment or payroll options through QuickBooks Pro you may be exposing yourself to the potential for misappropriation of funds.
- If the primary caregiver happens to be somewhat careless, is unorganized, is a slow learner, does not follow specific instructions, or is generally not "in tune" with accounting concepts, you should never allow him or her access to the Admin password. I can tell stories about the significant fees some owners have paid to clean up behind these individuals. Give new or unreliable employees only limited access until they have proven themselves!
- If you choose not to grant access to the Admin password to your primary caregiver, be prepared to be on call to perform those activities that only the QuickBooks Pro administrator can perform.

Creating the Admin password

Here are some basic ideas related to the Admin and other users' passwords.

- If there is only one Admin user, QuickBooks Pro does not require you to create a password. I would, however, highly recommend that you create one.
- Passwords are not case sensitive.
- I would suggest a fairly short name since you'll be typing it on numerous occasions.
- To make your passwords more secure:
  - Embed a number or symbol in the name, such as Saw$, #Nail, L%n
  - Think of a favorite phrase, movie, or song and then use the first letter from each word, such as BIATWC—Build it and they will come, NCW—Never Cry Wolf, LABOTW—Like a Bridge Over Troubled Water
- Be sure to change passwords periodically. They get around.

Setting closing dates

Unlike many accounting programs that make you close or lock out one accounting period before you can begin another, QuickBooks Pro allows you to post transactions to any accounting period until you set a closing date.

This flexibility can be exceptionally helpful when you find accounting errors or missing transactions because you can easily make corrections or enter transactions to current or prior periods without going through secondary, correcting transactions.

NOTE In traditional programs a second, correcting entry must be posted to the current period. This means that two accounting periods are misstated. For example in May, $550 gets entered into the Utilities account, rather than the Legal Fees account. When the error is discovered in July, a correcting entry is prepared to decrease Utilities, and increase Legal Fees. Although the year-to-date totals for each account will be correct, the Utilities account is overstated by $550 in May and the subsequent correcting entry reduces the actual July Utilities expense by $550. Now, although the year-to-date Legal Fees expense account is now correct, the expense shows up in July, rather than in May.

As you can see, the current year's accounting records would be more accurate on a month-to-month basis if the original transactions could simply be fixed. After the correction, the transaction detail also would be cleaner, with fewer error-correction entries and explanations muddying up the accounts.

I believe that QuickBooks Pro's ability to modify prior accounting periods offers many advantages over older, more highly controlled, accounting systems. On the other hand, this degree of flexibility also can create major problems. For example, problems could occur if you make additions or corrections to prior periods for which:

- Data files been turned over to your tax advisor for year-end work
- Tax returns have been prepared
- Formal financial statements have been issued to outside users
- Internal reports have been used to compute draws, commissions, or bonuses (Your changes could create discrepancies that might create confusion or contention and probably will be time-consuming to research and resolve.)

To protect against this problem, QuickBooks Pro has developed a security process to set a date *before which* only authorized individuals can post transactions.

Only the QuickBooks Pro administrator can set or change the closing date. To set the closing date, follow these steps.

1. Log on to the QuickBooks Pro data file as the administrator.
2. Go to **Company/Set Up Users/Closing Date**.
3. You will see a screen in which you will enter **The date through which books are closed.**
4. Click **OK**.

I recommend that you set a closing date at the conclusion of each month-end closing procedure. For example, your closing date would be after you are reasonably sure that the transactions you have entered are complete and correct. After the closing date has been set, only authorized individuals may enter transactions dated prior to the closing date and only after encountering a system-generated warning.

Creating users and setting access levels

If you would like other users to access all or just part of your QuickBooks Pro data file, the system allows you to create separate names and passwords for each individual. This means, for example, that you could let one employee work in the accounts receivable or accounts payable section without giving them access to more sensitive areas, such as payroll or financial reports. Also, when users are signed onto the system, each transaction is coded to show the originator.

After the QuickBooks Pro Admin function is set up, you may create other users from the User List screen (see Exhibit 32.1). Use the "wizard" provided to make a variety of selections in each screen that follows. Be sure to click on the Help button for a full description of what each "privilege" entails. Be especially careful when you get to the next-to-final screen titled Changing or Deleting Transactions because the bottom portion of this screen is where you will decide if the user can change transactions prior to the closing date (see Exhibit 32.2).

I recommend that you limit the ability to post corrections or new transactions to prior periods to one or two individuals who possess a full understanding of how those changes might influence other transactions, such as commissions, bonuses, bank draws, and so on, or their affect on reports issued to various internal or external users.

Understanding the Multi-User feature

The Multi-User feature, available since version 99, allows up to five users to access your QuickBooks Pro data file *simultaneously*. You can have more than five users set up, but only five can access the file at any given time. This is a real advantage for larger companies who may have estimators, order takers, purchasers, managers who need to review reports, or more than one accounting technician.

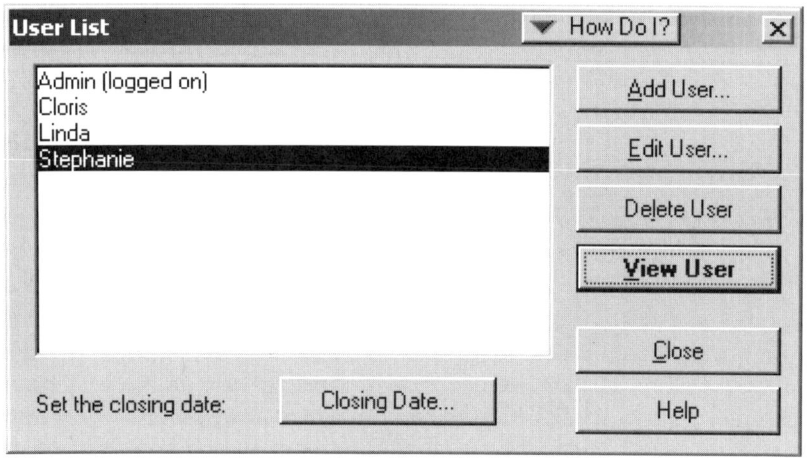

EXHIBIT 32.1 User List window.

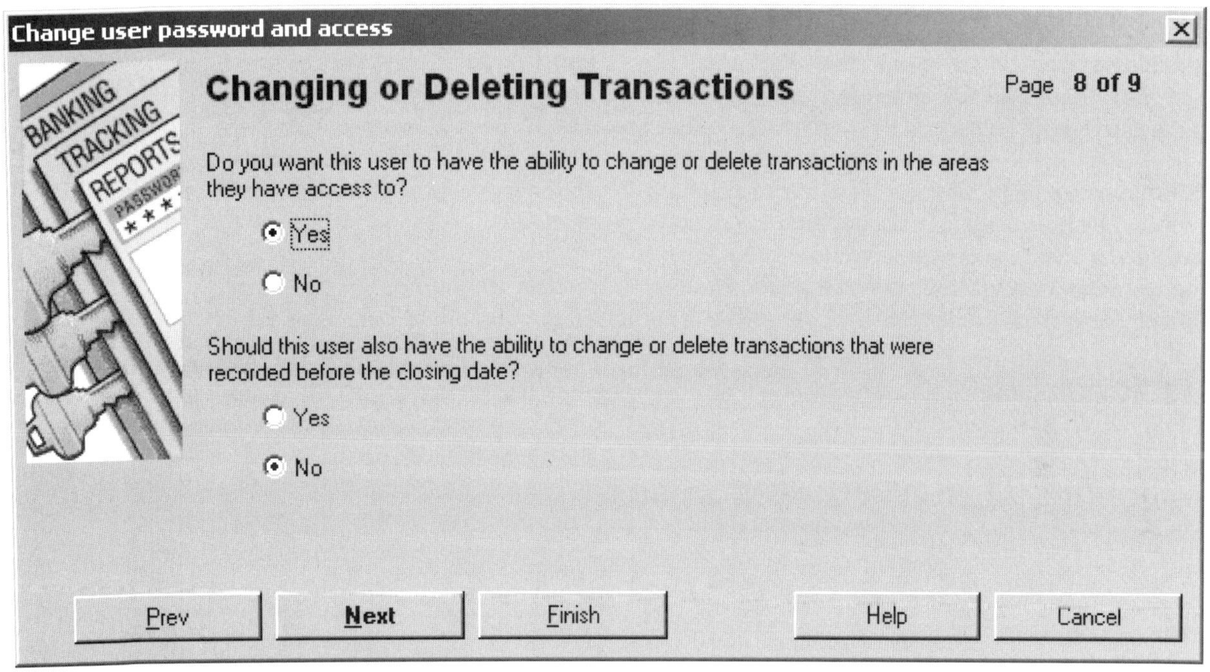

EXHIBIT 32.2 Changing or Deleting Transactions window.

I would, however, like to provide you with some insights about the various constraints attached to the multi-user environment.

- If more than one person wants to use QuickBooks Pro at a specific time, all users must select the Multi-User option.
- Many Admin file maintenance activities cannot be accessed while in Multi-User mode, such as backing up files, verifying data integrity or re-building files, re-structuring or re-ordering lists, and so on.
- System access time may be relatively slow if another user is creating a long report or is posting a large transaction.

CHAPTER 33

Troubleshooting in QuickBooks

QuickBooks Pro 2000 is, generally speaking, a fairly stable, logical, and powerful program. When something isn't working properly, however, it's very easy to lose perspective and get upset or angry with the software, and think that there has to be (check all that apply) a:

___ better ___ smarter ___ more logical, or ___ easier

program to use. In my experience, the problem only occasionally lies with the software itself. In most cases, the problem is in the system design or due to an operator error or misunderstanding.

When you experience "strange results" in QuickBooks Pro, please take the time to understand and research the problem. The following information discusses some of the more common problems my construction clients have encountered, as well as the QuickBooks Pro Help feature, to help you more fully understand the software and its design. As in martial arts, if you attempt to move *with* the flow of the system to beat the problem, rather than blindly working against it, you will find that the problem-solving process actually can be very interesting and rewarding.

Coping when report results don't agree

If, for example, your profit and loss total doesn't agree with the net income shown in your balance sheet equity section, follow these steps.

1. Check to see that both reports cover the same period. For example, if you run a profit and loss report for the month of August, the total will not agree with the amount shown as net income in the equity portion of your balance sheet because that line shows year-to-date totals.
2. Verify that both reports were prepared based on the same method of accounting. To check, from the report window, click on Customize, and look at the Report-basis buttons. If one report has been prepared using cash basis and the other report using accrual basis, they almost never will agree with each other.

TIP This happens most frequently when you access a memorized report that is set to prepare reports in one mode, while your file preferences are set to prepare reports in the other mode. As previously mentioned, I recommend that you set all reports to run as accrual-basis due to the unpredictability of the cash-basis report results. If you have conflicting settings, check both your company preferences in the Reports and Graphs section, and the settings on memorized reports. Modify your settings, as warranted.

There is also a possibility that results don't agree because there is a data integrity problem in your file. The following section addresses this problem area.

Resolving problems due to broken database links and pointers

In non-technical terms, QuickBooks Pro is a database comprised of many fields and records that are tied together with various pointers and links. On occasion, some of these links become corrupted or broken. (I like to point the finger at "electrical" hiccups.) When this happens you may get some unusual numerical results, such as a balance sheet that won't balance, or the net income shown on the balance sheet doesn't agree with the net income shown on the profit and loss report. If you should encounter these kinds of odd balances or results try the following.

Re-sort your lists

Although it's exceptionally convenient to manually move accounts, items, customers, or jobs to different positions within your lists using the mouse, if you've done a lot of that type of reorganization, some of the relationship pointers may have become broken or confused. Re-sorting your lists may resolve the problem.

TIP Before you run the re-sort, if the current order of the list is important to you, print out a list report, which usually is shorter than the actual list. If the re-sort solves the problem, but you want your list back in the previous order, you will have a copy to refer to. If the natural re-sort order doesn't meet your needs, devise a numbering or lettering system for the names within the list that will force the list, when re-sorted, to sort into the order you desire.

Verify data in your file

QuickBooks Pro's File Verification feature is designed to detect problems with the data and related links. To use this data verification, follow these steps.

1. Close all open reports, windows, and Navigators.
2. Click on **File** on the task bar.
3. Select **Utilities**.
4. Select **Verify data**. Let the QuickBooks Pro process finish. Don't interrupt the data verification process.
5. You will see one of the following messages:
   - *QuickBooks Pro detected no problems with your data*
   - *Warning: Data has lost integrity. Close all windows and choose* ***Utilities*** *and* ***Rebuild Data*** *from the* ***File*** *menu. If you see this message again, call Intuit Technical Support.*
6. If you see the second message, you will definitely want to use the Rebuild command as indicated above.

NOTE I sometimes use the Rebuild feature even when I get the first message, as experience has shown me that the rebuild process has fixed one or more transactions that weren't detected as problems by the Verify data process.

When you choose the Rebuild Data command, you will be forced to back up your data file. You may want to name this backup BAD or REBUILD so that it's readily identifiable.

QuickBooks Pro will rebuild your data file in two passes. If your file is quite large, this may take a while. I've noticed that the file typically is smaller after a rebuild, so I believe that the process also re-organizes the data in a more efficient manner.

The rebuild process should, but isn't guaranteed to, re-establish all of the links and pointers. When the rebuild is complete, QuickBooks Pro normally will tell you that the rebuild was successful. The system *does not*, however, tell you that your data file has been verified and that no problems exist, so you should run the Verify data utility again.

Locate and correct damaged transactions

If you get the *Data has lost integrity* message again, look for remaining damaged transactions by performing the following steps.

1. Press **Ctrl+1**, then **Ctrl+2**. This will take you to a window called Tech Help.
2. Click on the **Open File** tab.
3. In the middle of the window you will see a list of files. Highlight the line that ends with **QBWIN.LOG**. You may need to move the bottom arrow to the right to be able to see the file names at the end of the paths.
4. Double click on that file or click the **Open File** button.
5. The document you will see is in Wordpad. This may be a long document because it records and adds a new section each time your QuickBooks Pro program crashes, each time you verify your file, and each time you rebuild your file.
6. To go to the end of the file to see the results of your file verification and the results of your rebuild, drag the button on the right side of the window to the bottom of the window. Review the Verify log and the Rebuild log sections.
7. Damaged transactions in the Verify log section will be marked as Error transactions. If you find damaged transactions that were not fixed in the Rebuild process, delete the original transactions and re-enter them. To do this:
   - View the log and note all of the pertinent data about the transaction, such as transaction type, transaction number, date, document number, the amount, and the name
   - Use the Find feature to locate and view the original transaction
   - If it affected a balance sheet account that may carry a "cleared" designation, such as a checking account or a credit card account, also look at the transaction in the Register to see whether it has the "cleared" checkmark. If it has cleared, make a note to yourself to also clear the new transaction after you have entered it
   - Re-enter the transaction using all the original information (be especially careful to use the original date!)
   - If the transaction needs to be cleared, go to the Register and manually clear it by clicking in the column with the checkmark at the top; then click the **Record** button
   - Go back to the original transaction and delete it
   - Repeat these steps for all remaining error transactions
8. Re-verify the file, and re-check the QBWIN.LOG Verify log section to confirm that it no longer contains error transactions.

Clearing balances in the Opening Balance Equity account

The Opening Balance Equity account is used only to enter opening balances into your accounts. After the process is complete, the balance should be $0. Because no one should make any further entries into this account, inactivate this account immediately following the opening process, so that it is no longer visible in your Chart of Accounts. Thereafter, it acts as a control account that automatically receives postings resulting from certain "problem-indicator" entries.

NOTE **Important!** When a balance sheet report shows a balance in this account, investigate the cause!

One possibility is that someone used the account in a transaction or a journal entry. Although your inactivation of the account makes this much less likely, if this does happen, determine which account should have been used and correct the transaction.

Another reason for a balance in this account is that someone, after opening balances are entered:

- Adds a new customer to QuickBooks Pro and also enters an outstanding balance in the Set Up window
- Adds a new vendor to QuickBooks Pro and also enters an outstanding balance in the Set Up window

In either of these cases, the amount for the unbalanced entries will post to the Opening Balance Equity account where they will be visible as errors that need to be resolved. To fix the problem, delete the entries and enter the new customer's "beginning balance" as an invoice, or the new vendor's "beginning balance" as a bill.

Another possible cause may be due to the reconciliation process. If an account reconciliation doesn't balance and the accountant instructs the program to finalize the reconciliation anyway, the difference posts to the "Opening Balance Equity" account as a check or deposit. If you look at the posted transaction, you will be able to see which account was not properly reconciled and that QuickBooks Pro generated a memo that says "Balance Adjustment."

After you have determined the reason for the difference, either:

- Enter the correct detailed transactions that equal the adjustment, manually clear them in the correct register, and then delete the QuickBooks Pro-generated adjustment
- Go back into the check or deposit and select the correct account or item

Either of these processes should clear the transactions out of your Opening Balance Equity account so the balance goes back to zero.

NOTE Should balance adjustments arising from reconciliation problems cause you concern? In my experience and opinion, *accounts should reconcile—right down to the penny—almost all of the time!* The reconciliation process is just a large matching game where you match your accounting entries to the bank's entries, and then you list and review the non-matching items, such as the outstanding checks and deposits that are left. Once or twice a year, if you're off a small amount, such as $.01 to $5, and you have 100 or more checks clearing, in the interest of using your time effectively, it may be acceptable to post an entry to Miscellaneous Office Expense for the minimal unreconciled difference. If you find that your accounting assistant regularly has small, unresolved differences, be vigilant and let him or her know that you are reviewing the reconciliations closely. If you see large unresolved differences, follow up immediately!

Determining why your bank reconciliation opening balance changed

This is absolutely the most difficult problem-solving scenario that I run into with clients because, short of an extensive review of the audit trail, changes to cleared transactions are extremely difficult to track. The opening balance you see in your reconciliation is the balance of *cleared transactions* in the account, so changes to cleared balances between reconciliations arise from one or more of these user actions:

- A cleared transaction's checkmark is removed
- An uncleared transaction is marked as cleared
- A cleared transaction is deleted or voided

- A cleared transaction's balance is altered
- A cleared transaction is moved from one reconciled account to another

Prevention

Prevention is definitely the best way to go, so here are some front-end tactics to employ to avoid this problem, or to make it easier to solve if you should encounter it.

- First and most importantly, you need to understand exactly what those little checkmarks in the Register mean, how important they are, and how much work is required if someone is careless. Some QuickBooks Pro users have a tendency to click and flip their mouse most everywhere. *Keep that mouse a long way away from those checkmarks when you are cruising around in the Register!* A simple click over the top of a checkmark will remove it, or a simple click over a blank cell will add the mark. Granted, QuickBooks Pro will ask if you want to change the transaction, but if you went into the transaction to make a different change, it's easy to innocently go ahead and save the transaction with the altered clearing indicator. It's also easy to get confused about exactly "where" you are in the Register and just click on Yes when QuickBooks Pro asks the question. And, of course, if the marker is modified, so is your opening balance.
- The second major preventive technique to employ is to be *extremely careful* when you void, delete, or alter transactions. If you are going to change a transaction, be sure to research it first to find out whether it has been marked as cleared. If it has, you'll need to find another way to make the correction you are attempting to make.
- Another effective control is to set your closing date every month after you complete the reconciliation. This will help stop the "clickers and flippers" from inadvertently turning the reconciliation markers on and off, at least for prior periods. It also hopefully will limit prior-period changes to users who are more experienced and know to watch out for reconciled transactions.

Detection and correction

If you haven't had a chance to implement these measures, and you are faced with an altered opening balance, here are some of the problem-solving steps that I follow to find the easier problems. If one doesn't solve the problem, then I go to the next step.

Step #1

1. Run a file integrity check to confirm that all of the reconciliation markers and links are properly set.
2. Review the current month's checks and deposits to see that none of them have been marked as cleared. $0 amounts marked as cleared are not a problem.
3. If this is a bank reconciliation, locate last month's bank statement and full reconciliation or print it out from the reconciliation window. Use it to:
   - Review the register to confirm that last month's outstanding items are still shown as "not cleared" in the Register
   - Review the register to confirm that all of last month's cleared items are:
     - Still shown in the Register
     - For the same amount as shown on the reconciliation
     - Still shown as "cleared" in the Register

Step #2

Use the Find feature to look for transactions that equal the amount of the difference between last month's cleared balance and this month's cleared balance. If you locate any transactions equal to the difference, investigate them to see if they are marked as cleared or not. If the transaction is *not* cleared, check it against last month's outstanding list. If it wasn't on last month's outstanding list and was issued prior to last month's reconciliation date, it should be cleared. Manually clear it. This should take care of the problem.

At this point, if you still have a difference, and the amount is fairly small, such as under $50 or $100, you may choose to enter an adjusting check or create an adjusting deposit, and then clear it manually. Confirm that this step brings the opening cleared balance in line.

Step #3

If the amount is larger, and if you want to continue the problem-solving process, pull out the second preceding month's full reconciliation and match the cleared items against your Register. You can do this a month at a time until you find the problem.

> **NOTE** Always print out full reconciliations rather than summary reconciliations. Keep them in a safe place because you may need them for this type of problem-solving situation, or for other critical reconstruction exercises.

Step #4

Peruse the audit trail for the time period following the date on which you last reconciled.

Step #5

As a last resort, if you have less than a year's worth of transactions and depending on the number of transactions you have entered, you may choose to clear the "reconciled" mark from each transaction, and re-do your reconciliations from day one. This is a lot of work, but can be very effective. (It's also the equivalent of writing "I will be exceptionally careful of cleared transactions" a thousand times on the blackboard!)

Protecting your files by avoiding a variety of other problems

Here are some snippets of advice gleaned from a number of years of working with clients and seeing some of the problems that they innocently create and encounter.

So-called computer experts

Yes, I know you don't want to insult family members or friends by hiring someone else, and the price is certainly right, but ask the following questions of anyone who is going to work on your computer hardware or networking system.

- What are your *real-life* computer experiences and qualifications?
- How do you stay current with changing technologies?
- Can you help me strategize about how to upgrade my computer system when the need arises?
- Do you have experience in my specific network environment?
- Can you explain, in language that I can follow, what you are doing and why?
- Will you be available when I need you? What is your response time for critical problems? For less critical problems?
- Do you use standardized checklists?
- When you make changes to my system, do you test my programs and printers to verify that they still work properly?

- If something isn't working properly, will you creatively persist in the problem-solving effort until the problem is resolved?
- Will you show up when you say you will?

Search long and carefully for your computer consultants. If you aren't happy with your first choice, keep searching. It's worth the effort to find the right computer support.

Frequent error messages and system crashes

When these happen, something is wrong. Talk to your computer consultant to get the problem fixed before it causes your employees any more lost time and frustration, gets progressively worse, or actually disrupts your operations at a critical time.

Computers assembled from pieces and parts

There's an old country and western song about the fellow who worked in the auto factory. Over the years he brought home enough pieces from various car models to build his own—and I'll bet you can imagine just what he ended up with!

It's so tempting to build that inexpensive computer from the pieces of other old, unused computers, or from the remnants of those smoke-damaged computers that someone's recycling. But if you value your time, your money, or your data—don't do it! Computer components, under the best of circumstances, aren't always compatible with each other. Don't add to the fray with a homegrown attempt to build a mix-and-match computer system.

Non-functioning computer components

If your floppy drive doesn't work, please get it fixed. If you think your modem fried during the most recent thunderstorm, remember that you need it to connect to the Internet and payday is only three days away. If your CD-ROM drive keeps running when it shouldn't, you might have a problem soon. If your hard drive or zip drive even begins to make clicking noises (often referred to as the "click of death") contact your computer consultant *now*! Don't drive a broken computer. The results could be hazardous to your financial health.

Bargain diskettes

200 diskettes for $10. I'm not quite sure where these are manufactured, but you'll live to regret the day you bought them. Believe me, I have. If you use diskettes, invest in name brand, preformatted, high-quality diskettes.

Power outages and power fluctuations

Power outages create computer damage from at least two directions. When your computer suddenly loses power, it can't shut down a number of operating files properly, which can result in various problems. Secondly, when the power resumes, you can get some fairly powerful electrical surges that can cause damage to sensitive equipment. For your major computer components, including servers and primary, critical computers, I recommend you invest in an uninterruptible power supply (UPS) unit. These provide enough power during power outages to let your computer shut down without doing damage, and act as surge protectors for when the power resumes.

For less-critical workstations, I prefer to use the type of surge protector that comes with a lifetime guarantee and coordinates the power to all of the workstation's components, including CPU, printer, monitor, extra peripheral, and phone line. These are generally flat units, about 15×15×1 inches in dimension. At $70 to $90 each, they are actually very inexpensive insurance.

> **TIP** If you have several computers, buy an extra one of these units. When one sacrifices its life to save your computer (you'll know that it's happened because the unit will be dead), you can send it in for replacement and use the extra as a substitute in the interim.

Talk to your computer consultants about the various types of UPS units and surge protectors that are available and how the units should be used for your various computer components.

Static electricity

If you touch your computer or keyboard when you are charged with static electricity, you *can*, believe it or not, cause damage to your computer or files. If your office is static-prone during the winter months, in addition to the traditional anti-static devices such as special floor mats and humidifiers, look in the laundry section of your grocery store for the anti-static spray. Spray down the area *around* your workstation as needed. It really works!

Turning off the computer in mid-stream

"Oops! I made a mistake. Better turn off the computer before anyone finds out." Doing this midstream during a QuickBooks Pro backup is a guarantee for data file disaster. Make certain that anyone who might touch a computer knows that they *must not* turn the computer off during any file opening, closing, or saving activity. A mistake can be corrected. A destroyed file poses far greater hazards.

If, in the normal course of working on your computer, your computer fails to respond, there are special shut down processes to follow—see your computer consultant for further advice.

Children and your business computer

If humanly possible, I suggest the furthest possible separation between children under 18 and your business computer. If they should inadvertently delete any of your critical files, or damage your computer system, it could be fairly costly to recover. If money for a second computer is an issue, try buying a used system or get one of the low-priced current technology machines.

If it is absolutely necessary for them to use your business computer (or for you to use their computer?), create a separate directory for your financial and business data. Maybe you could label it TIMEOUT, GROUNDED, or NO CAR as a reminder (just kidding!).

Liquids, food, dirt, and your computer

Liquids, food, and dirt (LF&D) can be closely related—but not confined to—the topic just discussed. Although it should go without saying that sensitive electronic equipment and LF&D do not mix well, I have seen circumstances where it seems that they have cohabited for far too long. In the interest of a cleaner, more pleasant, and more professional work environment, do your best to segregate your computer from these "bad elements."

> **TIP** Rubbing alcohol, available at the grocery or drugstore, is a very inexpensive and effective cleaner for keyboards, mice, telephones, and other plastic surfaces that pick up body oils and dirt.

Viruses

We hear so much about computer viruses. How can you best protect yourself?

- Get educated. Learn about computer viruses, how they are commonly transmitted, how you can be alerted to particularly dangerous ones, and how virus protection software works.
- Invest in virus protection (VP) software. Talk to your computer consultants and ask them which virus-protection software they recommend for your particular situation. My experience has been that these programs can cause some other, sometimes undesirable, effects on other parts of your computer system. The virus protection software also often comes bundled with a number of other utilities that can cause system conflicts, so coordinate this protection with your computer expert, rather than attempting an off-the-shelf solution on your own. Obtain a commitment from your computer support team that they will troubleshoot you through the process until you are a satisfied with the results.
- Use and update the virus protection software. The software is only good if you use it periodically to scan your files. On the other hand, it can really interfere with your computer's processing speed if it's running constantly, so you'll want to coordinate the optimal solution with your technicians. Also, because new viruses hit the computer world every day, the software needs to be continually updated. Make sure you get those updates on a recurring basis. Some of the software will update itself automatically each time you go to the Internet.
- Avoid situations where infection is likely. As always, prevention beats detection and correction, so here is a list of situations to avoid. Feel free to add your own to this list:
  - Bringing in files from non-virus-checked home computers
  - Opening e-mail from unknown senders
  - Downloading files from questionable web sites—notorious as virus-carriers (Impress this restriction on employees by establishing clear company guidelines for acceptable Internet use.)

Maximizing response time

A QuickBooks Pro data file can handle up to two billion transactions, so you are less limited by the software than you are by the constraints of your computer memory, processing speed, and other considerations. If you are experiencing a slow response when you post transactions, or reports seem to take a long time to run, look through this list of enhancement possibilities.

Hardware issues and potential fixes

If your hard drive is too full, your system speed will significantly degrade. At least 10 percent of your hard drive should be open. If at least 10 percent isn't accessible, move or delete unnecessary files. If this is an ongoing problem, you may need a larger hard drive.

An individual computer's RAM (computer memory) may be inadequate. You need at least 64MB for good performance. Computer memory is relatively inexpensive, so if this is the problem the solution shouldn't cost a lot. Also, an individual computer might be running too many other programs and utilities that are consuming RAM and processor time. Check with your computer consultant for cleanup options.

If you have a network, ask your computer support team to check the speed of your network cards, cabling, and network configuration.

An individual computer's processing speed may be too slow. As technology changes, the recommended minimums increase. It may be time for a faster, more powerful computer.

The speed of the hard drive where your data file is located also may not be optimal. Since hard drive speeds are dependent upon a variety of factors, consult with your computer advisor for insights.

User and file issues

It's best to set up office policies about how the computer is used and the Internet accessed.

- Internet browser usage. Have you been on the Internet since you last powered up your computer? If so, Internet browsers may have left your RAM in less than pristine condition. Power your computer down and count to 10. This allows your RAM to clear. Then power up again.
- Lots of open reports and transaction entries. If you keep the Auto-Refresh feature on, which I highly recommend, and you have a variety of reports open, QuickBooks Pro re-runs *all* reports each time you enter a transaction. Try closing all but the vital reports until your transactions are completed. Then re-run the other reports.
- Recall Last Transaction feature. When the Recall Last Transaction feature is engaged, QuickBooks Pro searches for the last similar transaction and fills in the detail. This is a very useful feature. I wouldn't disengage it unless absolutely necessary.
- Very large working file. Do you have an extremely large file? Whenever you run reports or perform various "find" activities, QuickBooks Pro reviews your data file within the parameters you set, so the larger your file gets, the longer it takes. Files larger than 50MB are not that uncommon, but are starting to get fairly large. Alternatively a file that consumes more than 10 percent of your hard drive is considered a big file. If you have ruled out hardware considerations, and if a large file is the primary cause of your slow speed, you ultimately have two options available: condense your file or create a new file.

Condensing a file

In the Condense Data procedure, QuickBooks Pro consolidates all of the individual transactions that were posted to an account for each month and then posts the total amount as a single journal entry. The original detailed transactions, with certain exceptions, are deleted. For example, if there were 68 transactions that posted into the Trade contractors account in July of 1999, and the net total of the transactions was $34,566.89, after the condensing process you would see only one journal entry for the month in the amount of $34,566.89. The system will, however, leave any "unlinked" or "outstanding" transactions open. For example, outstanding checks, invoices, and payables—and any transactions linked to them—will still be shown at the detailed level. As you can imagine, this process will truly make your file much smaller!

NOTE

Warning! In a Condense Data procedure, you set the timeframe covered by the condensing procedure. Within this timeframe *you will lose all of your job-cost detail!* Your file will become a one-dimensional general ledger system *for the period that you have condensed.* You will no longer be able to prepare job-cost reports for the period condensed, or access the detailed transactions from the newly condensed file.

Coping with condensing

There are two significant ways to work around this potential loss of detail that you've worked so hard to create. The first option is to keep or create a copy of the original working file so that you can always access it at a later time.

Before you are allowed to condense the file you will be *required* to create a backup. Immediately use the backup to create a separate, uniquely-named, previous-period working file in which you can access all of the detail from, and create reports for, older jobs. As long as you properly protect that file, you will always have access to the original transactions.

TIP Make a couple of additional backups, and name and store them carefully!

Specify the ending date for the period that you want to condense. Limit the period to what is useful and convenient for you. For example, let's say that it is April of 2001. You started your QuickBooks Pro file in January of 1998, and you believe that it should now be condensed. Generally, your jobs are completed in a six- to eight-month timeframe. In this circumstance, I would recommend that you condense only 1998 and 1999. This would leave the full prior year's detail, as well as the current year's detail accessible from your current working file.

If you need to research a job that started in 1998 and finished in 1999, or started in 1999 and finished in 2000, you can go back to the previous detailed file. On the other hand, if you need to print a report on a job that started and finished in 2000, or started in 2000, but is still in process, you can use your current file.

If your jobs stretch out over a longer period of time, you will need to study the condensed dates more carefully. If you really need to condense, and you have jobs that extend back into the period you want to condense, the worst case scenario is that you would need to lay two reports side by side—one from the old detail and one from the new. That isn't a pretty solution, but it could be workable.

Proceed with caution

If you choose to condense, read all of the Help information provided by QuickBooks Pro and think it through very carefully before proceeding. You also will have some options regarding items to remove, so if you are no longer using them in the current portion of your file, this is an opportunity to clean house.

Exhibit 33.1 shows the window at the beginning of the Condense Data cycle.

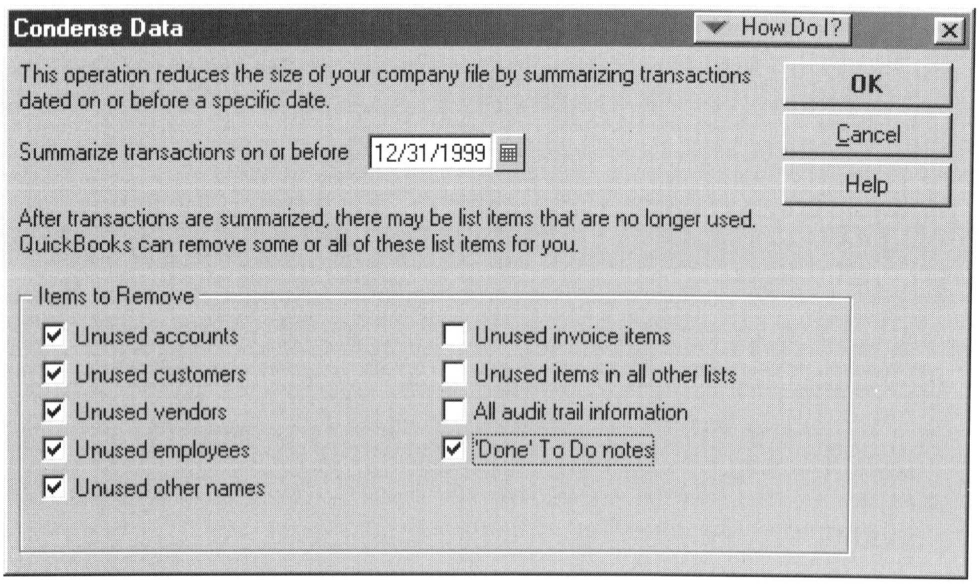

EXHIBIT 33.1 Condense Data window.

To check the results after the Condense Data procedure, immediately go into the new file, run various reports, and look at the results in the condensed and the current period to ensure that you now have a file that looks and acts the way you want it to.

The process can be somewhat challenging, but the good news is that if, after you carefully check the condensed working file, you find that you've made an error in judgment, you can go back immediately to your pre-condensed file backup, restore it, and re-do the process.

Creating a new file from a large file

Eventually, even if you utilize the Condense Data feature, you probably will need to create a new QuickBooks Pro file. If you track many detailed transactions, you may need to create the new file in three to four years. If you have a smaller company, you may not need to create a new file for six, seven, or more years. Developers, who have a tendency to create a number of separate limited-timeframe companies, may never need to create a new file for an old company.

If and when you're ready for a new file, QuickBooks Pro helps in its creation by providing a very simple process that lets you export lists from your old file so that you can import them into your new file. You can export, then import, the following lists shown in Exhibit 33.2.

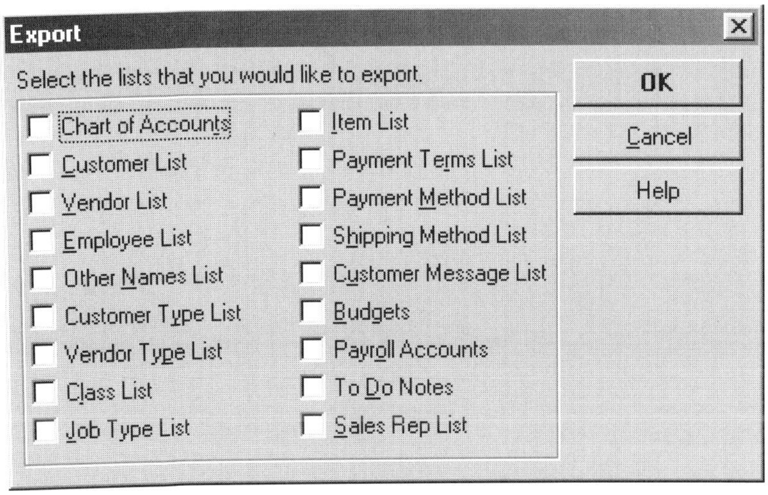

EXHIBIT 33.2 Export window.

After you have imported the lists into your new file, you probably will want to clean up your lists by deleting accounts, vendors, customers, employees, and so on that you no longer need to keep in your current file. Also, if you notice duplicate names, this is a good time to merge them.

Creating a new file can be a planned event, so I encourage you to create your new file as near to the beginning of the year as is reasonably possible. Refer to chapter 14 for help entering opening balances.

Contacting troubleshooting resources

Although I have attempted to provide insights into some of the most common problems that I find clients encountering on a day-to-day basis, every company is unique and encounters software challenges. Additional troubleshooting resources to contact include:

- Intuit for technical support such as lost passwords, window messages advising of program or printer errors, questions related to the software's capabilities, and so on
- A certified QuickBooks Pro advisor for accounting issues, tax questions, or QuickBooks Pro software utilization, design, or reporting problems
- *www.QuickBooks.com* will help to guide you to either resource

CHAPTER 34

Reassigning and allocating developer costs

I have developed a special approach that lets developers allocate a number of other job costs by using certain features within QuickBooks Pro 2000 in a creative way. You may need to follow up with a simple journal entry or two, but this creative approach certainly beats computing, preparing, posting, and proofing a large number of manual entries!

Because I dislike tedious work that is especially vulnerable to errors, I decided to use the power inherent in QuickBooks Pro to create a better way to have the system allocate the costs. So, while a few special costs, such as anticipated costs, still must be allocated manually using standard transaction documents such as checks or, occasionally, journal entries, I will show you tips that use memorized transactions to make even this process semi-painless.

Before we discuss the particulars of how you reassign and allocate costs, let's look at a few relevant definitions.

- **Assigning costs.** Costs are assigned to specific accounts, items, jobs, or classes as part of a standard accounting transaction.
- **Reassigning costs.** Sometimes it is necessary to reassign costs because:
  - The original transaction was incorrect, and needs to be corrected
  - The original transaction was posted in a way that needs to be changed
  - The transaction then is re-posted so the ending result is correct
- **Allocating costs.** An allocation is a variation on the reassignment of costs in that it also includes:
  - A logical assumption or estimate about costs
  - A logical assumption or estimate about cost drivers
  - A calculation based on the assumptions or estimates

NOTE Another way to say this would be that additional information has been introduced and calculations have been performed. Based on that information and these calculations, we want to assign costs differently than we currently see them on the books.

- **Allocation methodology or cost-drivers.** A logical or reasonable approach used to mathematically relate certain costs to their source or to their use.
- **Posting method.** The results of the calculations are posted to your accounting records using various automated or manual posting methods.

Allocating general and administrative costs

Although general and administrative (G&A) costs should be computed and included for estimating and pricing decisions, I normally recommend that construction service companies do not allocate *actual* G&A costs back to individual jobs because:

- There often isn't a clear relationship between G&A costs and job costs
- If you're going to do it properly, allocations can be complex with many cost drivers
- The G&A allocation process can be costly and time consuming
- The results typically are not all that useful to owners or management
- G&A allocations are not required for tax or other financial reporting purposes

On the other hand, when a company has been created expressly to accomplish certain development or building tasks, such as those created by developers and larger spec home builders, it is entirely appropriate to allocate G&A costs back to specific units.

Allocating tips for developers and multi-site home builders

The following information is designed for company owners who create a company for the purpose of developing a specific section of land into individual home sites. The company may then sell the lots, or may build some or all of the homes on the lots.

Due to legal considerations, and the ever-changing flow of investors, these companies are formed on a specific-project basis. When the project is complete, and the useful life of the company has expired, the company is terminated. Developers then form new companies for subsequent projects.

These kinds of companies are unique and, accordingly, I like to employ some special allocation techniques to accommodate their unusual circumstances. I recommend that unit job costing for regular construction costs be handled in the same manner as that described for other allocations. Certain costs incurred in the process of creating and holding a development, however, are unique in nature and thus ideal candidates for allocations. These types of special costs and situations are described below.

Large costs, unclear relationship

In a development situation, not all costs can be assigned directly to units. For example, interest from large development and construction loans can't be traced to individual units because the loans are for the development and construction project as a whole, not for individual lots, homes, or condos. Also, the life of the loans and the life of the individual units bear no direct relationship to each other. Loan repayment due dates and amounts don't track cleanly or proportionally with cost accumulations or income from the sale of units. These types of costs, therefore, should simply be allocated to each unit in a reasonable fashion.

Large costs, unpredictable timing

A significant portion of development work typically continues after a number of units are sold. The costs benefit all units in the development, so total development costs, therefore, should be allocated or attached to individual units.

Clear relationship, unpredictable timing

Some costs, such as warranty work, can be traced easily to an individual unit, but the cost may occur long after the fact, so it's too late for the information to be of any real value. It makes more sense to estimate total warranty work for the project and allocate the average warranty cost to each unit.

Clear relationship, small uncontrollable costs

Although some specific costs incurred during the life of the development and construction process can be traced to individual lots or units, assigning those costs to individual units usually doesn't provide useful information for management or decision-making purposes. For example, the last unit sold will have incurred more actual taxes and electricity costs than the first unit sold, but is that really useful information to anyone? When there are a large number of units, it often doesn't make sense to go to the effort and expense of posting these types of holding costs to each unit. These costs, although assignable, are excellent candidates for allocation.

Management needs

Although these costs are allocated to individual units, monitor and manage them closely for the development as a whole. For example, you might want to review estimated vs. actual cost reports for the special allocation-related jobs that you create, or you may want to closely monitor the budget vs. actual reports that track your actual costs vs. budgeted costs over your anticipated timeframe for the development.

The detailed method that follows is based on the following premises.

- Certain costs are incurred for the development as a part of the total project.
- These costs should be allocated out to individual units at the beginning of the project.
- Allocated costs should be based on detailed beginning estimates.
- If, part way through the project, it is determined that the beginning estimates were erroneous, the allocations attached to the remaining units should be adjusted to reflect the corrected estimates.

This approach allows owners to:

- Review estimated total costs, or budgeted costs spread over anticipated timeframes, versus actual results for the "big picture" costs associated with the development and certain construction costs (This gives owners information about the project as a whole and provides critical time/cost information to help manage the project.)
- See and account for each unit's estimated portion of the development's total cost, regardless of actual sale date
- Modify pricing for units in the current development
- Monitor each unit's actual construction costs
- See (relatively) final profit and loss reports on individual units at a date near closing
- Anticipate and include in the cost at the time of sale an average "allowance" for future warranty work for each unit
- Better anticipate and estimate costs for future developments

Before beginning, you will need to determine the cost driver you will use to allocate the costs. Options to choose from include:

- Assigning costs *equally* to units—an appropriate cost driver when lots are approximately equal in size, price, and development cost
- Square footage—a cost driver that may be applicable if specific development costs vary based on the size of the lot
- Anticipated sales price—which may apply if lots or units will vary widely in sales price
- Another logical method (see the following note)

NOTE

Caution: Sometimes allocation calculations are dictated by generally accepted accounting principles, or by tax laws. A logical, reasonable, consistent approach, however, typically is acceptable or can be modified easily "on the side" by your tax advisor. If you have questions about a particular allocation methodology that you want to use, see your tax advisor.

Creating special jobs and items

To allocate these costs, you create special jobs, items, and accounts. The special *jobs* initially receive the allocation offsets for certain anticipated costs to be incurred during the life of the development, and then are used to hold the actual costs for each of the cost centers. The special *items* are used to assist in the allocation process and link to the special cost of goods sold (COGS) allocation accounts (see chapter 8).

Special jobs

- **1.NJS-cnstr.** Assign NJS *construction* costs to this job. Use the items from the NJS portion of the Item List to assign these direct costs to this job. If you have employees, allocate these costs out to jobs using the manner described for payroll allocations. If you do not have employees, follow the same process described in the following sections.
- **2.Int/loan fees.** *Process:* At the beginning of the development, create an entry to *allocate* estimated total interest and loan fees to each unit. Use items from the "Allocations" section of the Item List in a $0 check.

> **NOTE** The $0 check records the allocations at the item level, which in turn also posts into accounts in the Chart of Accounts.

The allocation shows up as an allocated cost in each unit, and as an allocated cost offset in the 2.Int/loan fees job, with the net impact of $0 for the company in total. An abbreviated example for a 10-unit development is shown in Exhibit 34.1.

Then, assign all *actual* interest and loan fees for land and land development loans to this job using the interest/fees items listed in the ".LND/DEV" section of the Item List. Also assign all actual fees and interest for construction loans to this job using the 944d.Interest/fees item listed in the ".Othr Cnstr" section of the Item List.

An abbreviated spreadsheet version of a profit and loss by job report showing the results of this posting, and the posting of actual costs, is shown Exhibit 34.2. (Not all 10 units are shown, and the accounts have been filtered to show only the interest-related accounts.)

> **NOTE** In Exhibit 34.2, allocated costs are posted to individual jobs, and actual costs are posted only to the 3.Int/loan fees job, rather than to individual jobs. At this point, there has been $15,710 more interest allocated to jobs than has actually been incurred to date. You also can prepare an actual-to-date vs. budget-to-date report to monitor interest and fees costs against the planned totals at any point in time for the whole development.

- **3.Land and dev costs.** Assign all costs associated with land, land development, engineering, legal fees, general permits and paperwork related to the development, infrastructure costs, general operations, association fees, taxes through closing, utilities through closing, general maintenance, and repair costs to common spaces to this job using items listed in the ".LND/DEV" section of the Item List.

  Process: In a manner nearly identical to the previous process, create one entry to *allocate* land cost and another to *allocate* estimated total development costs, excluding interest and loan fees, to each unit in $0 checks at the beginning of the project. Use items from the "Allocations" section of the Item List. The allocations will show up as an allocated cost in each unit, and as an allocated cost offset in the 3.Land and dev costs job, with the net impact of $0 for the company in total.

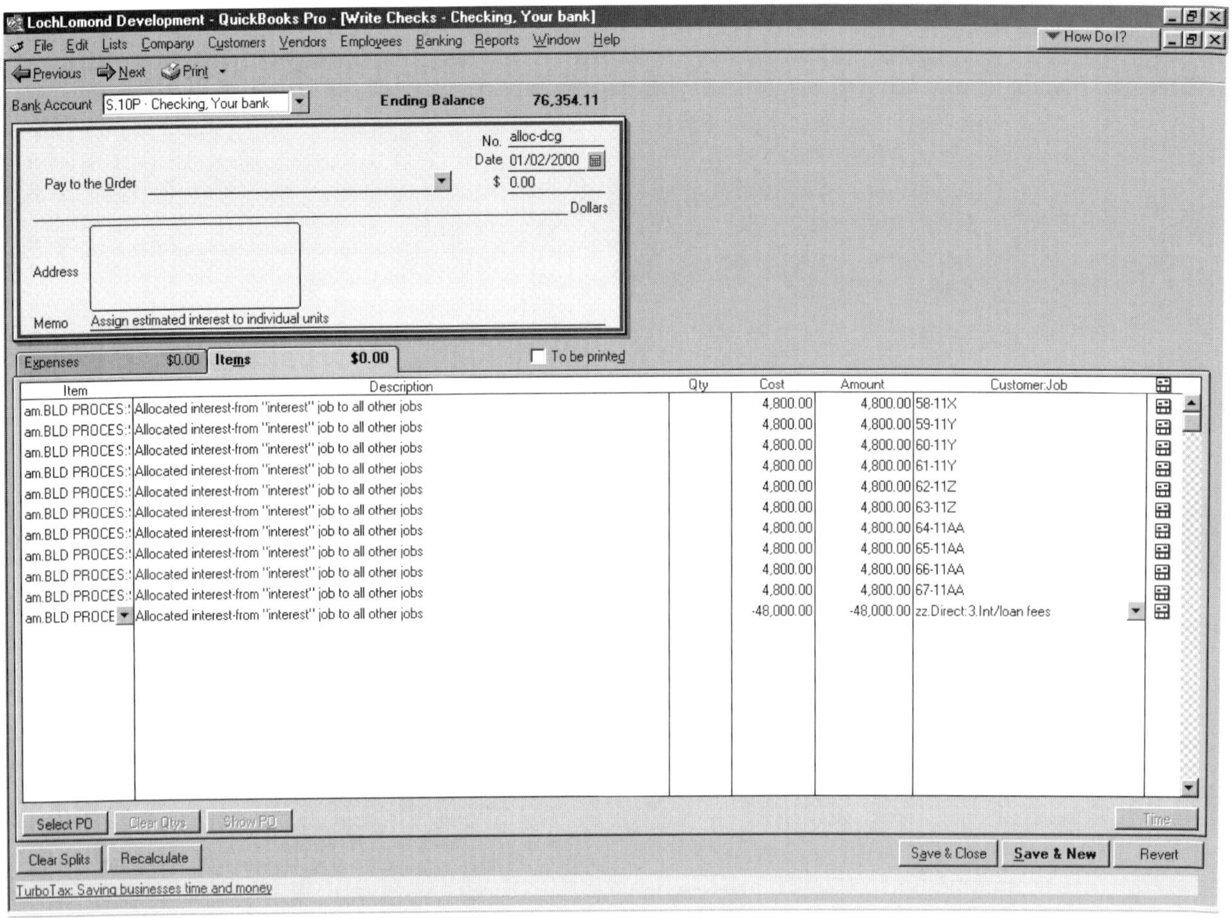

EXHIBIT 34.1 Abbreviated 10-unit example of allocations to special job 2.Int/loan fees.

> **NOTE** Prepare a profit and loss by job report to show costs to date, as well as the allocated total. Prepare an actual-to-date vs. budget-to-date costs report to help monitor land and development costs against the planned costs for the development.

- **8.G&A and warranty.** G&A costs include office supplies, administrative salaries, occupancy costs, general professional fees, and other costs associated with keeping the company operational.

  Process: At the beginning of the development, create an entry to *allocate* estimated total G&A costs to each unit in a $0 check. Use items from the "Allocations" section of the Item List. The allocation will show up as an allocated cost in each unit, and as an allocated cost offset in the G&A job, with the net impact of $0 for the company in total.

> **NOTE** As various G&A costs are incurred over the life of the development post them directly to the QuickBooks Pro overhead or expense accounts. *It is not necessary to assign these costs to any job.* At the end of the accounting period, create a $0 adjustment check to offset the period's G&A expense and move it into the COGS section as follows:

	64-11AA	65-11AA	66-11AA	67-11AA	3.Int/loan fees (zz.Direct)	Total zz.Direct	TOTAL
Ordinary Income/Expense							
Cost of Goods Sold							
S.600 · Direct interest							
6002 · Interest & fees-development In	0.00	0.00	0.00	0.00	9,270.72	9,270.72	9,270.72
6004 · Interest & fees-construction In	0.00	0.00	0.00	0.00	23,018.49	23,018.49	23,018.49
6006 · Interest allocation to units	4,800.00	4,800.00	4,800.00	4,800.00	(48,000.00)	(48,000.00)	0.00
Total S.600 · Direct interest	4,800.00	4,800.00	4,800.00	4,800.00	(15,710.79)	(15,710.79)	32,289.21
Total COGS	4,800.00	4,800.00	4,800.00	4,800.00	(15,710.79)	(15,710.79)	32,289.21
Gross Profit	(4,800.00)	(4,800.00)	(4,800.00)	(4,800.00)	15,710.79	15,710.79	(32,289.21)
Net Ordinary Income	(4,800.00)	(4,800.00)	(4,800.00)	(4,800.00)	15,710.79	15,710.79	(32,289.21)
Net Income	**(4,800.00)**	**(4,800.00)**	**(4,800.00)**	**(4,800.00)**	**15,710.79**	**15,710.79**	**(32,289.21)**

EXHIBIT 34.2 Profit and loss by job report.

NOTE
Continued

1. In the Expenses tab, enter account S.89P-(Less) G&A allocation to COGS and a *negative* amount equaling the period's total G&A expense.
2. On the next line, select account 6800.G&A costs from expense and enter the same amount—as a positive number.
3. This particular entry could be created with a journal entry as well. For the sake of consistency in this process, however, a $0 check serves the same purpose.

Warranty and repair

This approach to preassigning warranty and repair costs to individual units takes into account the unpredictability of individual repair work and measures the results for the development in total versus the amount estimated for the development in total.

Process: At the beginning of the development, create an entry to *allocate* estimated total construction warranty and repair costs to each unit in a $0 check. Use items from the "Allocations" section of the Item List. The allocation shows up as an allocated cost in each unit, and as an allocated cost offset in the Interest/fees job, with the net impact of $0 for the company in total.

Assign actual warranty and repair work on specific units here using items in the "96.Warranty" section with a memo regarding the unit for which services were performed. Then, to help you compare actual G&A and warranty and repair against the planned costs for the development, prepare:

- Profit and loss by job report that shows costs to date as well as the allocated total
- An actual-to-date vs. budget-to-date costs report

Creating front-end allocations

Although the front-end allocations initially may seem to be a large job, if you do these allocations up front they actually are less work in the long run than trying to assign many different costs out to individual units on an ongoing basis.

First allocation

Your first $0 allocation check might seem a rather tedious document to prepare, especially if you have 75 to several hundred units in a project! Here is a way to really speed up the input process for your first allocation:

1. Set up all of your units as jobs. (See chapter 11 for tips about making job numbering work most efficiently for you.)
2. In Banking, select **Write Checks**, or in the **Banking Navigator**, click on the **Write Checks** icon.
3. Enter all of your units in the Customer:Job column. Use the down arrow and enter them one after the other.
4. In the first line in the Item column, enter the correct item. After you enter it, press **CTRL/C** to copy it to the Windows clipboard.
5. Arrow down to the next line and press **CTRL/V** to paste the same item to that line.
6. Repeat the arrow down and CTRL/V for all units.
7. Go to the first line in the Amount column and enter the amount of the allocation. Repeat the **CTRL/C** on the first line and the **CTRL/V** for all remaining lines to enter the amount in all lines.
8. Look at the top of the Items tab to see the total amount entered so far for all of the units.
9. Confirm that it agrees with the total that you want to allocate.
10. If it doesn't agree, find the error and correct it
11. On the last line, enter the same item. Go to the Amount column and enter the total allocation amount with a *negative sign* in front of it.
12. Enter the appropriate allocation job, such as 2.Int/loan fees.
13. Your total now should be $0 on the check.
14. Be sure the To Be Printed box is not checked.
15. Be sure there is no name in the Pay to the Order Of field.
16. In the number field, enter **Alloc-** followed by your initials.
17. Before you leave the transaction, memorize it.
18. Name it Allocations.
19. Choose **Don't remind me**.

Second allocation

For your second allocation, double-click on the memorized transaction. When it appears on your window, you will need to:

- Modify the items
- Modify the amounts
- Change the job on the last line

If you use the CRTL/C and CTRL/V approach described previously it should take only a few minutes to create your second allocation. The remaining allocations are as simple, and (voila!) you hopefully are done with allocations for the life of the development.

> **NOTE** The exception to the "done" scenario is that at the mid-development stage the allocations might need to be modified for remaining units. If so, pull up the memorized transaction, use CTRL+ Delete to remove the lines for the units that were finalized during prior financial periods, re-memorize the transaction, and then modify it as needed.

Using results to affect costs and pricing

The budgeted-to-date vs. actual-to-date report provides critical information on time-sensitive costs such as interest. Detailed estimate vs. actual reports on individual jobs give you the opportunity to keep an eye on detailed costs that, if you didn't see them until later, could add up to significant decreases in your bottom line. So, when you can see individual line items relating to your development costs as they were estimated and as they actually shape up, you can stay on top of them before they spiral out of control. You also can make immediate pricing decisions on units that have not yet been sold with this data.

A third view of information that is exceptionally useful is a profit and loss by job report with gross profit margin columns next to each job. Individual profit margins accompanied by the total profit margin on all units shown can be an eye-opening experience.

When owners see all of their actual construction costs, as well as all of their allocated development and construction costs, they often find their profit margins are not as high as they thought. A typical response to this information is to make immediate adjustments in pricing. This may or may not be the case for you, but wouldn't it be nice to see all of those costs arranged in nice rows and columns so you can immediately see where you stand?

Just think what might happen if you find you need to change your pricing by $5,000 per unit on the remaining 150 units to achieve your target profit margin. That's $750,000 you might otherwise have lost. How much are you willing to invest in a good measurement system that will provide you with information that can earn or save you that much money and stress?

CHAPTER 35

Creating an efficient work environment

A good subtitle for this chapter would be: E^4 or the efficient and effective entry environment, a.k.a. No pain = big gain.

What I'm talking about are ways to relieve the pain that thousands of keystrokes a day can cause. As a builder or remodeler you know that repetitive, unrelieved construction tasks can cause tissue injury. But did you also know that the repetitive motions involved in computer work also can cause tissue strain and pain? As an employer you need your accounting assistant to be on site, healthy, and able to work long hours on the computer without developing debilitating headaches or experiencing other major discomforts or inefficiencies. In this chapter I discuss some easy-to-achieve things you can do to help everyone in your office who sits in front of a computer for most of their day.

You can help protect yourself, especially in today's litigious environment, and your investment in your staff by providing a work environment that encourages, and even insists on, some basic computer-related, personal-health practices.

Taking regular breaks

Make it a *written* policy in your office that computer operators must move away from the computer, even if only to do other office tasks, for at least 5 to 10 minutes per hour. This also (reputedly) will help alleviate the dry eyes and eyestrain that can accompany long sessions at the computer.

Stretching out the kinks

You might go to your local or Internet bookstore to locate some information on quick, helpful, computer-relief stretches that you can post in your office. This might sound a bit extreme, but it *is* in your best interest to keep your office support team healthy. We actually have found inexpensive specialty software that periodically reminds you to stretch those tense body parts, and illustrates some quick arm, neck, and shoulder stretches that can be performed right at your desk.

Taking control of your environment

Do whatever it takes to manipulate the various elements of your work environment so that you and your staff can stay healthy and comfortable while you work. You need to be in control of your environment, not vice versa. Here are some suggestions.

Seating

Working from the wrong chair, or from a good chair that is poorly adjusted, can cause neck pain, shoulder pain, arm pain, elbow pain, wrist pain, back pain, hip pain, leg pain, eye strain, and headaches (for a start!). All the team members in my office work on computers constantly. We were experiencing, at various times, most of these types of symptoms. I consulted with an ergonomics specialist who helped us immensely by studying our workstation layouts. She recommended extensive changes so that our surroundings would properly support us, and work *for* us rather than *against* us.

She especially recommended that we adjust our seat height to a level that generally was higher than we were used to. She also suggested seat angle adjustments. You wouldn't believe the difference that a simple chair wedge can make. She helped us to understand which seat adjustments are most critical and how we could modify our workstations in subtle, and usually inexpensive ways, to enhance our short and long-term health and comfort.

Monitor placement

Your monitor should be directly in front of your keyboard. The screen should be clearly visible with the middle area of the screen about 15 to 30 degrees below horizontal from the user's eyes.

Monitor size

The prices on 17-inch monitors are now extremely reasonable. If you can afford to have even one 17-inch monitor in the office, please give it to the person who's working the most on the computer.

Keyboard placement

Place your keyboard just above lap level so that your elbows and wrists can drop naturally from your shoulders.

Keyboard style

Some people like the curved ergonomic keyboards, but they require adaptation time and, if you have to switch between computers on a regular basis, they can be downright disconcerting. Most people have very strong feelings one way or the other about the kind of keyboard they like to work on, so it's probably worth checking with the primary user.

Grimy mice

Dirty mice and grimy mouse pads can really slow down data entry, and are frustrating to users as well. As with the live, furry version, fighting with a computer mouse to get it to go where you want it to go is probably not how you want to spend your, or your staff's, time! Here are some suggestions.

- Clean your mouse's innards regularly. If you do it yourself, just be sure you know how to re-assemble the mouse properly.
- If your mouse is just old and tired, send it to the retirement drawer with good cheer; strong, young, able mice are available at bargain rates at your nearest office supply or computer supply store.
- Consider purchasing one of the new varieties of mice that use light beams instead of internal track balls. This breed of mouse doesn't even need a pad. Laps and table tops are equally acceptable tracking surfaces.

Mouse placement

Your mouse should be at the same height, or slightly below your keyboard, and to the right (or left) side depending on your dominant hand. It *should not* be situated higher and/or in front of your keyboard. Hold your hand out at that angle for a minute or so and you'll see what that placement requires of your neck, shoulder, and arm!

Mice versus track balls

Most of the people I speak with believe, and I agree, that mice generally are easier to use over long periods of time.

Wrist rests

Resting your wrist on a hard surface over a long period of time while performing repetitive motions can cause discomfort, pain, and nerve damage. Find wrist rests for your keyboard and your mouse pad that you like. I'm a big fan of the cushy gel pads, but everyone has their own preference.

Document placement

I've seen clients try to work with documents on their lap, spread across their keyboard, sitting on a table on their far right or left, or held in one hand while they type with the other. I've never been a fan of the mass-marketed copy-holders that hang, or stand, off to the left or right side of your monitor, because they:

- Require you to hold your neck at an awkward angle for long periods of time
- Don't lend themselves to holding easily the kinds of documents that you normally use and sort through for accounting purposes

I believe that the ideal situation is to have your input documents directly in front of you between the keyboard and screen, with neither blocked from access or view by the documents themselves. If the documents can just sit in your copyholder with no clips to hold them in place, you can still handle or sort through them easily, or even leave them where they are while you take care of something else that demands your attention. I have found that this arrangement relieves neck strain considerably.

I often help clients create a makeshift copyholder that meets those requirements by assembling the following.

- Locate a piece of corrugated cardboard or a piece of foamboard (available at your local office supply) and cut it to approximately 13 × 15 inches.
- On the longer edge, I make a *partial* slice in the backside about 1.5 inches from the bottom so that I can bend the bottom edge toward the front side.
- After bending the bottom edge toward the front side at about 45 degrees, I fasten it into that position with tape on either end. This creates a ditch that will hold documents in place while you are working on them.
- Then I find a way to position this copy holder between the keyboard and the monitor, and to hold it in place. The keyboard itself often acts as a good "brake" to the bottom edge.
- Seen from the side, the setup would look something like the diagram in Exhibit 35.1.

The advantages of this arrangement are that:

- Both your hands are free to enter data
- Documents are directly in front of you, easily visible, and visually in line with your screen and keyboard, which is an efficient viewing format
- Paperwork doesn't disappear between other piles of papers when you're interrupted (When you come back to the task, there it is!)
- Your can focus on what you're doing rather than on how to juggle multiple pieces of paper

Amazingly enough, this simple arrangement can help you get your accounting work done more quickly and more accurately, and with less unnecessary stress.

Computer placement

Your computer should be in a spot where you can quickly and easily insert and remove diskettes, zip drives, CD ROMs, or tapes for backup purposes. Additionally, you should be able to view and access the back of the computer periodically without having to dismantle, and then reconnect, crit-

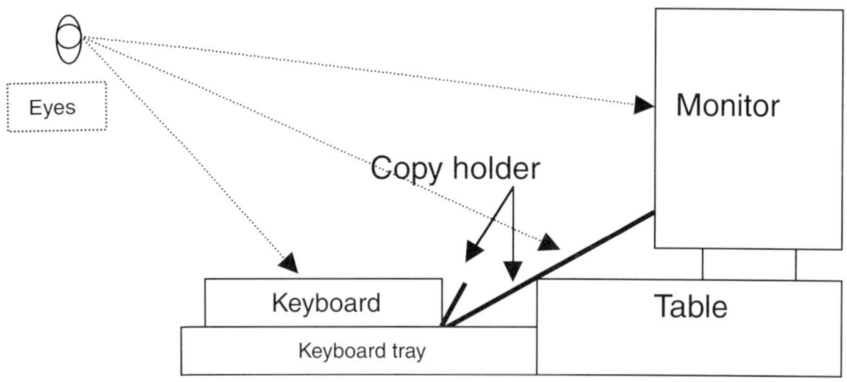

EXHIBIT 35.1 Copy holder eases eye strain.

ical components, and without having to call "Two Men and a Truck" to move heavy furniture. Make sure your initial cables are long enough to be able to pull your computer out far enough to make changes back there, because you or your computer technician will, sooner or later, need to do so.

Telephone placement

Place your phone close enough to your computer to be able to actually *use* your computer while you are on the phone. For instance, any software or hardware telephone support will require you to be able to talk on the phone and enter information on the computer at the same time. If you need to access information from your computer when clients or vendors call, you also will want your phone near the computer. If your jack isn't close enough, obtain and install additional phone extension wires.

If you'll be doing a lot of phone and computer work, consider a headset. Mobile phones or speakerphones can be helpful as well.

Calculator placement

Side calculations are an ongoing fact of life for anyone constantly working with numbers. Although you can utilize a variety of calculation functions directly within QuickBooks Pro 2000 or from other onscreen software programs, you definitely will want access to a printing calculator as well. Be sure that it is close enough to your computer to be able to move quickly and easily between the two

Lighting

Your computer screen should be free from glare and reflections from background windows or lights. If you're having glare or reflection problems do your best to resolve them. It just isn't necessary to put up with this persistent, low-level annoyance. Ideas to try include:

- Turning off an offending light that isn't critical
- Adjusting the angle of the monitor
- Adding a blind to, or closing a blind on, a window that is creating a reflection on your screen
- Adding a screen hood (think of it as horse blinders for your computer monitor)

You may want to try a screen-glare protection device, which basically is a screen that sits in front of your monitor's screen to reduce glare. Confirm that you can return the device if it doesn't take care of your problem.

If the glare is primarily from overhead ultraviolet lights, there are several types of glare-reducing light grates that you might try. Sometimes they also cut down on the overall light in the room, so again you may want to confirm that you can return them if the ultimate result is not satisfactory.

CHAPTER 36

Staying current with QuickBooks Pro

Intuit issues a new version of QuickBooks and QuickBooks Pro approximately once each year. The new version represents a major upgrade to the software. You can purchase it off the shelf in retail stores and then typically receive a rebate from Intuit as a current user of a prior version of QuickBooks Pro. Or you can respond to Intuit's advertising and order forms, which the company sends out in advance and shortly after the release of the new version. The price on the order form usually reflects at least the amount of the rebate you would receive if you bought the software at retail.

I recommend that you purchase each new version as it is released, for two reasons.

1. There usually are enough upgrades and new features to make it worth the modest cost of upgrading.
2. Each version, although significantly upgraded, usually takes very little time for you and your staff to incorporate into your system. The learning curve as it relates to the changes also isn't exceptionally high. If you wait for another version before you buy, the learning curve and adaptation time investment becomes much higher. Small changes are easier to assimilate than large ones!

If you ever decide to pass over a new version, please don't ever skip more than one. It just isn't worth the perceived savings. As a Certified QuickBooks Pro Advisor, I support only two versions—the most recent version and the immediately preceding version—because users lose out on too many features if their software is older than that.

Incorporating interim updates

Intuit issues corrections and updates, and adds new features to the current version of QuickBooks Pro throughout the life of the software, or as long as Intuit supports that version of the software, free of charge to registered software purchasers. Be sure to register your software so that you can utilize these free updates and upgrades, referred to as releases, on a periodic basis.

If you do *not* have Internet access, call the Intuit Direct Sales and Customer Service hotline to request the most current release on CD-ROM. Select Phone Directory from the How Do I? Help menu to obtain the correct phone number.

If you have Internet access, you can access and receive these updates from within QuickBooks Pro. I recommend that you set your update options as soon as your software is installed. To set your ongoing update options, choose from two settings by selecting File, then Update, and then Update Options (see Exhibit 36.1).

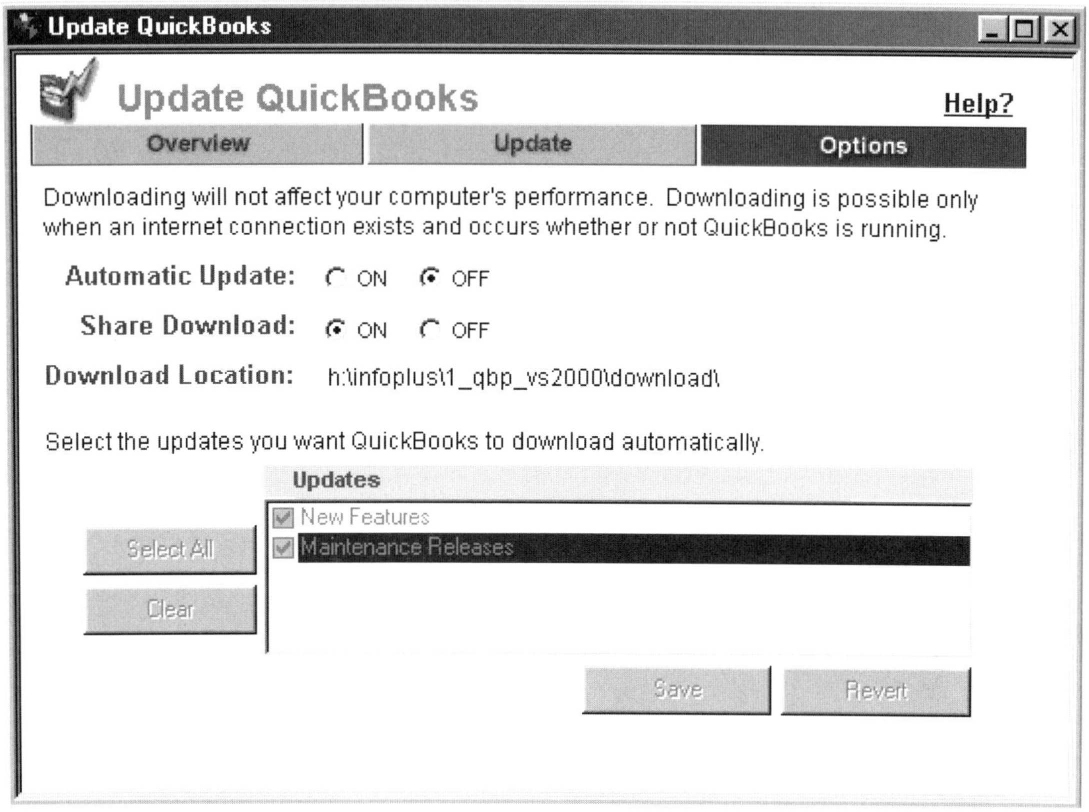

EXHIBIT 36.1 Update QuickBooks window.

Automatic updates

After selecting Update Options you will see a line that says Automatic Update. You can then click on either On or Off.

Click On if you:

- Do *not* have a network with a security setting that prevents software updates performed by anyone other than the system administrator
- Maintain a constant connection to the Internet via a dedicated line that is constantly connected to your Internet service provider
- Want to receive your updates gradually, in the background, a bit at a time, with installation when the download is complete

NOTE One of our clients discovered that this feature automatically dials your Internet connection once every hour looking for updates. This could create up to 24 extra phone calls per day, and/or tie up your modem line more frequently than you might wish.

I recommend clicking Off for most of my clients and most computers. If it's been a month and you haven't updated to the latest release, QuickBooks Pro will remind you that you haven't updated recently, and will ask you if you'd like to update now. Other benefits include the following.

- You'll be making one single connection to the Internet to obtain your update, but you'll still stay in reasonably close synchronization with the latest release. The download process may take a bit longer this way because the entire update will be transferred at one time, rather than in segments.

- If you have a network security system that allows only the network administrator to accept software updates, you can control the timing of the update process to guarantee the network administrator is available to properly administer, test, and troubleshoot the update process for all users at a non-critical time. For example, you probably wouldn't want to have the update load the morning before payday. You also may want to discuss with your computer consultants how the level of security in your network system interacts with the QuickBooks Pro update process, as well as the pros and cons of changing that particular security feature.

Shared downloads

The next line on the Update Options window will say Share Download? Again, you can click either the On or Off button. I normally click On because the Share Download feature can be set up so that each user does not need to access the Internet to update their version of QuickBooks Pro. This is especially critical in a multi-user environment where every user may not have access to the Internet. Share download means that when one user updates his or her QuickBooks Pro software, all other users will get the updated version the next time they access the file. For certain kinds of updates, QuickBooks Pro also will display a screen as you exit QuickBooks Pro that says you have received an updated release, and asks if you would like to install the update. For detailed instructions on setting this feature to work properly follow these steps.

1. Go to the **Help Index.**
2. Select **Updating**.
3. Select **QuickBooks**.
4. Select **Sharing updates in a multi-user environment**.

Updating your payroll tax table

Even if you constantly or periodically update your QuickBooks Pro software, you will need to perform a separate periodic update for your payroll tax tables. QuickBooks Pro 2000 has changed its terminology regarding tax table updates. They now refer to the background payroll tax calculations as the *Basic Payroll Service*, and to their actual payroll preparation service as *Deluxe Payroll service*. Either system requires you to use the Internet to link to Intuit to obtain the information.

Heeding update warnings

Unless you chose not to use any of Intuit's tax-calculation services, QuickBooks Pro may prevent you from entering any information in the payroll system if you have not updated your *Basic Payroll Service* recently. This is one of those cases where you'll probably want to save yourself some stress by handling the update well in advance of the cutoff! If you access the employee payroll features on a regular basis, you normally will see warning screens several weeks prior to the date when you will be prevented from entering information into the payroll system without installing the update. Be sure you update prior to the cutoff date. If the cutoff date passes, you will be required to update before you can process the payroll information.

I also have discovered that some Internet service providers (ISP) are not able to process the specific type of update files that Intuit uses for the Basic Payroll Service update. If you have an established, proven ISP for all other aspects of Internet usage, and still experience difficulties when

> **TIP**
> When you attempt to process payroll after a successful download you may get a message that says state or local tax tables have changed, and that you need to review the employee information before you proceed. You then will be shown the payroll information for each employee as you proceed through the payroll entry process. If you go to the Federal tab and click OK, you will get the same message again, so be sure to go to the State or Local taxes tab, review the information for potential changes, and click OK before you proceed.

receiving the update, contact your computer consultant to help you resolve the problem. They may need to contact Intuit for technical assistance and/or add another ISP to your system.

For specific instructions regarding the update process, follow the instructions on the screen when you attempt to access your payroll features, or select tax tables in the onscreen Help index.

APPENDIX A

Chart of Accounts

Balance Sheet Accounts

ASSETS

Current

Checking/Saving

- S.09P · Checking, your bank name
- S.100 · Checking
  - 1002 · Operating funds
  - 1004 · Payroll cks/deposits/svc chgs
  - 1006 · Job funds
    - 001 · Job 001
    - 002 · Job 002
    - 003 · Job 003
  - 1016 · Charge card deposits/svc chgs
  - 1018 · Escrow funds
  - 1020 · Clearings (bill/closings, etc.)
- S.10P · Savings, your bank name
- S.110 · Savings
  - 1102 · Regular cks/deposits/svc chgs
  - 1104 · Escrow funds
- S.119 · Cash on hand
  - 1190 · Petty cash
  - 1195 · Cash boxes/registers

Accounts

- S.12P · Accounts receivable

Other Current

- S.12S · (Less) Allowance for bad debt
- S.131 · Other receivables
  - 1310 · NSFs in transit
  - 1312 · Receivables clearing (temp)
  - 1320 · Income earned- not yet invoiced
  - 1328 · Retainage receivable
  - 1330 · Accounts receivable - other
- S.134 · Other current assets
  - 1340 · Short term investments
  - 1345 · Refundable bond payments
  - 1350 · Due from employees
  - 1360 · Due from related party
  - 1370 · Prepaid expenses
  - 1380 · S/term portion of L/term assets
  - 1390 · Refundable deposits
- S.13P · Undeposited funds
- S.14P · Construction mat'ls inventory
- S.15P · Work in process
- S.16P · Lots/homes ready for sale

Fixed Assets

- S.170 · Land & improvements
  - 1700 · Land, at cost
  - 1705 · Land improvements, at cost
  - 1805 · (Less) Accum depr-land imprvmts
- S.171 · Buildings, net book value
  - 1710 · Buildings, at cost
  - 1810 · (Less) Accum depr-buildings
- S.172 · Leasehold improvements, NBV
  - 1720 · Leasehold improvements, at cost
  - 1820 · (Less) Accum amort-L/H imprvmts
- S.173 · Office furn/fixt/equip, NBV
  - 1730 · Furn/fixt/equip, at cost
  - 1830 · (Less) Accum depr-FF&E
- S.174 · Computers/sftware, net book val
  - 1740 · Computers/software, at cost
  - 1840 · (Less) Acc deprec & amort-C&S
- S.175 · Vehicles, net book value
  - 1750 · Vehicles, at cost
  - 1755 · Construction vehicles, at cost
  - 1850 · (Less) Accum deprec-vehicles
  - 1855 · (Less) Acc deprec-cnstr equip't
- S.176 · Construction equipment, NBV
  - 1760 · Construction equipment, at cost
  - 1860 · (Less) Accum depr-cnstr equip't
- S.177 · Construction tools, NBV
  - 1770 · Construction tools, at cost
  - 1870 · (Less) Accum depr-constrn tools
- S.178 · Organizational costs, NBV
  - 1780 · Organizational costs
  - 1880 · (Less) Accum amort-orgz'l costs
- S.179 · Goodwill & NC covenants, NBV
  - 1790 · G'will/non-compete cov, at cost
  - 1890 · (Less) Accum amort-GW/NCC costs

Other Assets

- S.190 · Capitalized loan costs
  - 1900 · Capitalized loan costs
  - 1905 · (Less) Accum amort-loan costs
- S.192 · Long-term notes/interest recble
  - 1920 · Note receivable, borrower/descr
  - 1945 · Accrued interest receivable
  - 1949 · (Less) Short-term portion
- S.195 · Long-term investments
  - 1950 · Investments, at cost
  - 1960 · Cash value, co. life insurance
  - 1970 · Invstmnt in property to remodel

Accounts Payable

- ☐ S.20P · Accounts payable

Credit Cards

- ☐ S.210 · Company credit cards
  - ☐ 2100 · Co credit card name #1
  - ☐ 2101 · Co credit card name #2
- ☐ S.215 · Propr/prtnr/sh'holder cr cards
  - ☐ 2150 · CC name, owner of cc #1
  - ☐ 2151 · CC name, owner of cc #2
- ☐ S.217 · Business exp reimbursements due
  - ☐ 2170 · Reimbursements due to (owner)
  - ☐ 2171 · Reimbursements due (employee1)
  - ☐ 2172 · Reimbursements due (employee2)
- ☐ S.21P · Reimbursements due petty cash

Other Current Liabilities

- ☐ S.221 · Payroll taxes/withholdings due
  - ☐ 2210 · FIT withheld
  - ☐ 2211 · (Less) Adv earned income credit
  - ☐ 2212 · FICA/MC withheld & matched
  - ☐ 2214 · SIT withheld
  - ☐ 2216 · State unemployment taxes due
  - ☐ 2218 · State disability insurance due
  - ☐ 2220 · FUTA due
  - ☐ 2222 · Local taxes withheld
  - ☐ 2230 · 401K w/held & co. contribution
  - ☐ 2232 · SIMPLE IRA w/held & co contrib
  - ☐ 2240 · Friend of the Court withholding
  - ☐ 2242 · Garnishments payable
  - ☐ 2245 · Other withholdings payable
  - ☐ 2250 · Other contributions payable
  - ☐ S.221 · Other p'roll taxes/withholdings due
- ☐ S.230 · Other taxes payable
  - ☐ 2320 · Real estate taxes payable
  - ☐ 2375 · Personal property taxes payable
  - ☐ 2380 · State taxes on earnings payable
  - ☐ 2385 · Fed corporate inc taxes payable
  - ☐ 2390 · Other taxes due
  - ☐ 2395 · Tax-related interest/penalties
  - ☐ 2410 · Compensation payable
  - ☐ 2420 · Client deposits (unearned inc)
  - ☐ 2422 · Work billed - not yet performed
  - ☐ 2430 · Retainage payable
  - ☐ 2450 · Accounts payable, other
- ☐ S.250 · Line(s) of credit
  - ☐ 2500 · Bank #1-line of credit
  - ☐ 2502 · Bank #2-line of credit

Long Term Liabilities

- ☐ S.280 · Notes/leases/interest payable
  - ☐ 2800 · Land contract payable-prin
  - ☐ 2810 · Note payable, land-principal
  - ☐ 2815 · Note payable, development loan
  - ☐ 2820 · Note payable, construction-prin
  - ☐ 2840 · Note payable, secured-principal
  - ☐ 2850 · Note payable, unsecured-prin
  - ☐ 2860 · Mortgage payable-principal
  - ☐ 2880 · Capitalized lease-principal
  - ☐ 2890 · Accrued int due-notes/leases
- ☐ S.29^ · (Less) S/Term portion-L/T liab
- ☐ S.29P · Other long term liaibilities

Equity

- ☐ S.300 · Capital-proprietor/partners
  - ☐ 3000 · Capital-prior years
  - ☐ 3010 · Plus current-year contributions
  - ☐ 3020 · (Less) Current-year draws
- ☐ S.310 · Capital & outstanding shares
  - ☐ 3100 · Capital stock-### shares
  - ☐ 3110 · Capital in excess of par value
  - ☐ 3120 · Treasury stock
- ☐ S.35P · Opening bal equity (temp)
- ☐ S.36P · Retained earnings
- ☐ S.37P · (Less) Curr yr S Corp distrib
- ☐ S.38P · (Less) Curr yr C Corp dividends

Income Statement Accounts

Income

- ☐ S.40P · Unclassified income
- ☐ S.41P · Design & consulting services
- ☐ S.42P · Construction services
- ☐ S.440 · Income from sold units
  - ☐ 4400 · Lot sales
  - ☐ 4410 · Home sales
  - ☐ 4490 · Add-ons
- ☐ S.45P · Misc. reimbursed expense
- ☐ S.46P · Other income
- ☐ S.47P · (Less) Discounts to clients
- ☐ S.48P · (Less) Refunds to clients
- ☐ S.49P · Income from/(to) unearned inc

Continued

Cost of Goods Sold

- S.50P · Unclassified COGS
- S.510 · Construction costs
- S.520 · Direct labor comp/taxes/burden
  - 5200 · D-Regular compensation
  - 5202 · D-Overtime compensation
  - 5204 · D-Payroll taxes/specif benefits
  - 5206 · D-Allocated burden & NJS cost
- S.522 · Employee labor burden
  - 5220 · E-Workers' comp insurance
  - 5222 · E-Health insurance
  - 5224 · E-(Less) Employee-paid hlth ins
  - 5226 · E-Private disability insurance
  - 5228 · E-Life insurance
  - 5230 · E-Retirement benefits
  - 5232 · E-Employee assistance programs
  - 5234 · E-Gifts/promotion
  - 5236 · E-100% deductible meals & ent
  - 5238 · E-Recruiting and hiring costs
  - 5248 · E-(Less) burden allocation
- S.53P · Trade Contractors
- S.54P · Building materials & supplies
- S.55P · Other job-specific costs
- S.560 · NJS (Non-job specific) dir cost
  - 5600 · General liability insurance
  - 5602 · Small tools & misc job supplies
  - 5604 · Cnstr equipt/vehicles-gas/oil
  - 5606 · Cnstr equipt/vehicles-lic/ins
  - 5608 · Cnstr eqpt/vehicles-repair/mntc
  - 5650 · Uniforms & laundry
  - 5660 · Cell phones-field
  - 5670 · Depr, cnstrn equipt/tools/vehic
  - 5690 · (Less) D-NJS burden alloc
- S.57P · Cnstr costs from/(to) WIP
- S.620 · Direct interest costs
  - 6200 · Interest & fees-land prch loans
  - 6202 · Interest & fees-developmt loan
  - 6204 · Interest & fees-construction ln
  - 6206 · Interest allocation to units
  - 6208 · Interest expense from/(to) WIP
- S.650 · Developmt/infrastr costs
  - 6506 · Land development costs
  - 6508 · RE taxes/utilities/maint costs
  - 6560 · Dev cost allocation to units
  - 6590 · Developmt costs from/(to) WIP
  - 6700 · Land/lot purchases
  - 6710 · Land cost allocation to units
  - 6790 · Land costs from/(to) asset acct
- S.680 · Gen/admin/warranty costs
  - 6800 · G&A costs from expense
  - 6810 · Gen & admin allocation to units
  - 6860 · Warranty/repair alloc to units
  - 6890 · G&A & warranty from/(to) WIP

Expense

- S.70P · Unclassified expense
- S.710 · Marketing & sales expenses
- S.711 · Marketing & business developmt
  - 7100 · Events (shows, tours, etc.)
  - 7110 · Newsletters
  - 7120 · Donations
  - 7130 · Sponsorships
  - 7140 · Giveaways
  - 7150 · Referral expense
  - 7160 · Business promotion
  - 7170 · Gifts - ded'bl @ $25/person
- S.720 · Advertising expenses
  - 7200 · Business cards/brochures
  - 7210 · Yellow pages
  - 7220 · Print-nwsppr/mag/guides, etc.
  - 7230 · Specialty publications
  - 7240 · Radio/television
  - 7250 · Site signs
  - 7260 · Website
  - 7290 · Model home expense
- S.730 · Sales-comp/taxes/burden
  - 7300 · S-Regular compensation
  - 7302 · S-Overtime compensation
  - 7304 · S-Payroll taxes/specif benefits
  - 7306 · S-Allocated burden
- S.800 · Admin employment costs
- S.802 · Admin comp/taxes/burden
  - 8020 · A-Regular compensation
  - 8022 · A-Overtime compensation
  - 8024 · A-Payroll taxes/specif benefits
  - 8026 · A-Allocated burden
- S.804 · Owner's comp & benefits
  - 8040 · O-Regular compensation
  - 8044 · O-Payroll taxes/specif benefits
  - 8046 · O-Allocated burden
  - 8050 · O-Health insurance
  - 8051 · O-(Less) Owner-paid hlth ins
  - 8052 · O-Life insurance
  - 8054 · O-Private disability insurance
  - 8056 · O-Retirement benefits

- ☐ S.81P · Contracted admin labor
- ☐ S.820 · Occupancy expense
  - ☐ 8210 · Rent/lease on office space
  - ☐ 8220 · Property/liability insurance
  - ☐ 8230 · Repairs & maint, office space
  - ☐ 8240 · Janitorial/cleaning services
  - ☐ 8250 · Personal property taxes
  - ☐ 8260 · Real estate taxes
  - ☐ 8270 · Deprec, land improvements
  - ☐ 8275 · Deprec, buildings
  - ☐ 8280 · Amort, leasehold improvements
- ☐ S.829 · Utilities
  - ☐ 8291 · Gas
  - ☐ 8292 · Electric
  - ☐ 8293 · Water
  - ☐ 8294 · Trash removal
  - ☐ 8295 · Security system
  - ☐ S.829 · Utilities - Other
- ☐ S.830 · Office expenses
  - ☐ 8310 · Office supplies
  - ☐ 8320 · Printing & copying
  - ☐ 8330 · Postage & delivery
  - ☐ 8340 · Bank service charges
  - ☐ 8370 · Misc. office expense
- ☐ S.838 · Communications
  - ☐ 8380 · Telephone
  - ☐ 8382 · Cellular/digital phone services
  - ☐ 8384 · Pagers
  - ☐ 8386 · Answering services/voice mail
  - ☐ 8388 · On-line services
- ☐ S.839 · Office equipment, general
  - ☐ 8390 · Equipment rental/lease
  - ☐ 8392 · Repairs & maint, office equipmt
  - ☐ 8398 · Deprec, furn/fixt/equip't
- ☐ S.840 · Computer expense
  - ☐ 8410 · Computer supplies
  - ☐ 8420 · Computer rental/lease
  - ☐ 8430 · Computer repairs/maintenance
  - ☐ 8440 · Computer software support
  - ☐ 8490 · Software amortization
  - ☐ 8495 · Computer depreciation
- ☐ S.850 · General & admin expense
  - ☐ 8510 · Dues & memberships
  - ☐ 8530 · Books & subscriptions
  - ☐ 8540 · Training/education fees/mat'ls
  - ☐ 8550 · Amort of organization cost
  - ☐ 8555 · Amort of goodwill & NC covenant
  - ☐ 8560 · Bad debts & collection expense
- ☐ S.858 · Professional and service fees
  - ☐ 8580 · Accounting
  - ☐ 8582 · Payroll services
  - ☐ 8584 · Legal services
  - ☐ 8586 · Consulting services
  - ☐ 8590 · Other professional fees
- ☐ S.860 · Vehicle expenses
  - ☐ 8600 · Mileage reimbursement
  - ☐ 8610 · Vehicle lease/rental
  - ☐ 8620 · Vehicle gas & oil
  - ☐ 8630 · Vehicle licenses & insurance
  - ☐ 8640 · Vehicle repairs & maintenance
  - ☐ 8690 · Depreciation, vehicles
- ☐ S.870 · Travel & meals/entertainment
  - ☐ 8700 · Travel expense
  - ☐ 8710 · Lodging expense
  - ☐ 8720 · Meals/entertainment-50% ded'ble
- ☐ S.890 · Other taxes & licenses
  - ☐ 8920 · License fees
  - ☐ 8930 · State taxes (misc non-income)
  - ☐ 8990 · Other taxes
- ☐ S.89P · (Less) G&A allocation to COGS

Other Income

- ☐ S.910 · Interest and other income
  - ☐ 9100 · Interest income, banks/invstmts
  - ☐ 9110 · Finance charge income, clients
  - ☐ 9140 · Net inc-prtnrships/joint vent
  - ☐ 9150 · Rental property income
  - ☐ 9160 · Gain on sale of assets
  - ☐ 9190 · Income from other sources

Other Expense

- ☐ S.920 · Interest & other financing cost
  - ☐ 9210 · Interest, credit cards
  - ☐ 9212 · Annual credit fees
  - ☐ 9214 · Interest, loans (non-direct)
  - ☐ 9216 · Interest, stockholder loans
  - ☐ 9280 · Interest, late tax payments
  - ☐ 9290 · Discounts lost & late fees
- ☐ S.930 · Non-deductible expense
  - ☐ 9300 · N/D dues/penalties/gifts/contrb
- ☐ S.93P · Loss on sale of assets
- ☐ S.950 · Federal & State taxes
  - ☐ 9500 · Federal income taxes
  - ☐ 9550 · State taxes on earnings

APPENDIX B

Items to code to accounts

INCOMEXXXXXXXXXXXXXXX
Design and consulting services
CUSTOM HOME BUILDING SERVICESXXX
Custom Home-Draw A
Custom Home-Draw B
Custom Home-Draw C
Custom Home-Draw D
Custom Home-Draw E
Custom Home-Final Draw
Additional work authorizations
REMODELING-RESIDENTIALXXXXXX
Progress invoice A
Progress invoice B
Progress invoice C
Progress invoice D
Progress invoice E
Final invoice
Additional work authorizations
Commercial remodeling services
Handyman and repair services
Insurance restoration services
Services tradecontracted to other builders
Land or site development
Project management services
Custom home building services
Services for prebuilt or modular homes
Other services
Rental income
LOT/HOME SALEXXXXXXXXXX
Lot-Gross sales price
Home-Gross sales price
Lot & Home-Gross sales price
Income from extras
LOT/HOME SALEXXXXXXXXXXXXXXXX
Customer deposits
Commissions paid on sale of unit
Recording fees, title policy, county and state transfer fees, etc.
Pro-rated tax credit
Association dues proration
Payment on land contract
Payment on development loan
Payment on construction note
Interest charges/(credits)
Miscellaneous services
SIMPLE COSTXXXXXXXXXX
Direct labor-regular pay
Direct labor-overtime
Materials and supplies
Trade contractor costs
OTHER JOB-SPECIFIC EXPENSESXXXXXX
Permits
Blueprints & copies
Site utilities
Rentals for job
Travel and/or auto expense
Trash removal
Job promotion/client support (flowers, etc.)
Association dues
Builder's risk insurance
Other job specific misc expenses
NON JOB-SPECIFIC COSTSXXXXXX
General liability insurance
Small tools & supplies
Gas/oil on construc equipment/vehicles
Licenses/insurance on construction equipmt/vehicles
Repair/maint on constrn equipt/vehicles
Uniforms & laundry
Cell phones-field
Deprec-constrn equipt/tools/vehicles
EMPLOYEE BURDENXXXXXXXXXXXX
Workers' comp insurance-D/L portion
Health insurance-D/L portion
Private disability insurance-D/L portion
Life insurance-D/L portion
Retirement benefits-D/L portion
Employee assistance program-D/L portion
Gifts/promotion-D/L portion
100% deductible meals-D/L portion
Recruiting and hiring costs-D/L portion
BUILDING PROCESSXXXXXXXXXX
PROJECT MANAGEMENTXXXXXXXX
Project management-D/L-pay
Project management-trades
Project management-materials
Project management-other costs
Project management-variance from P.O.
Prepare job estimates-D/L-pay
General job supervison-D/L-pay
Gen'l admin (change orders, ordering materials, etc.)-D/L-pay
Filing & misc paperwork for job-D/L-pay
Review actual vs estimates/followup-D/L-pay
CLIENT MANAGEMENTXXXXXXXXXX
Client management-D/L-pay
Client management-trades
Client management-materials
Client management-other costs
Client management-variance from P.O.
Time spent with client-updates, changes, etc.-D/L time
Progress reports/progress billings for client/bank, etc.-D/L-pay
Misc client P/R time-D/L-pay
Misc client P/R costs (flowers, meals, etc.)-other
PERMITS, FEES, INSPECTIONSXXXXX
Permits, fees, inspections-D/L-pay
Permits, fees, inspections-trades
Permits, fees, inspections-materials
Permits, fees, inspections-other costs
Permits, fees, inspections-variance from P.O.
Building permits-costs
Third-party warranty fees
Impact fees
Misc inspection fees
Other permits & fees
DESIGN & ENGINEERINGXXXXXXXXX
Design & engineering-D/L-pay
Design & engineering-trades
Design & engineering-materials
Design & engineering-other costs
Design & engineering-variance from P.O.
Professional design fees-D/L-pay
Professional design fees-trades
Blueprints-trades
Blueprints-costs/copies
Engineering services-trades
Surveying-D/L-pay
Surveying-trades
Plan copies-D/L-pay
Plan copies-costs
EXTERIOR SITE PREPARATIONXXXXX
Exterior site preparation-D/L-pay
Exterior site preparation-trades
Exterior site preparation-materials
Exterior site preparation-other costs
Exterior site preparation-variance from P.O.
Site clearing/excavation-D/L-pay
Site clearing/excavation-trades
Site clearing/excavation-materials
Site clearing/excavation-other costs
Site clearing/excavation-variance from P.O.
Fill dirt/gravel/other fill materials-D/L-pay
Fill dirt/gravel/other fill materials-trades
Fill dirt/gravel/other fill materials-materials
Fill dirt/gravel/other fill materials-other costs
Fill dirt/gravel/other fill-variance from P.O.
Shoring/pilings/retaining walls-D/L
Shoring/pilings/retaining walls-trades
Shoring/pilings/retaining walls-materials
Shoring/pilings/retaining walls-other costs

Shoring/pilings/retaining walls-variance from P.O.
Rough grading-D/L-pay
Rough grading-trades
Rough grading-materials
Rough grading-other costs
Rough grading-variance from P.O.
Soil testing-D/L-pay
Soil testing-trades
Soil testing-materials
Soil testing-other costs
Soil testing-variance from P.O.
Equipt rental-exterior site prep-D/L-pay
Equipt rental-exterior site prep-trades
Equipt rental-exterior site prep-materials
Equipt rental-exterior site prep-rental cost
Equipt rental-exterior site prep-variance from P.O.
DEMO & INTERIOR PREPXXXXXXXXXX
Demo & interior prep-D/L-pay
Demo & interior prep-trades
Demo & interior prep-materials
Demo & interior prep-other costs
Demo & interior prep-variance from P.O.
Tear out/off/down-D/L-pay
Tear out/off/down-trades
Tear out/off/down-materials
Tear out/off/down-other costs
Tear out/off/down-variance from P.O.
Hazardous materials-tests, inspection, abatement (asbestos/lead/radon, etc.)-D/L-
Hazardous materials-tests, inspection, abatement (asbestos/lead/radon, etc.)-trade
Hazardous materials-tests, inspection, abatement (asbestos/lead/radon, etc.)-mate
Hazardous materials-tests, inspection, abatement (asbestos/lead/radon, etc.)-othe
Hazardous materials-tests, inspection, abatement (asbestos/lead/radon, etc.)-varia
Dust protection-D/L-pay
Dust protection-trades
Dust protection-materials
Dust protection-other costs
Dust protection-variance from P.O.
Equipt rental-demo & interior prep-D/L-prep
Equipt rental-demo & interior prep-trades
Equipt rental-demo & interior prep-materials
Equipt rental-demo & interior prep-other costs
Equipt rental-demo & interior prep-other costs
TEMPORARY FACILITIESXXXXXXXXXX
Temporary facilities-D/L-pay
Temporary facilities-trades
Temporary facilities-materials
Temporary facilities-other costs
Temporary facilities-variance from P.O.
Temporary electric-D/L-pay
Temporary electric-trades
Temporary electric-materials
Temporary electric-other costs
Temporary electric-variance from P.O.
Temporary toilets-D/L-pay
Temporary toilets-trades
Temporary toilets-materials
Temporary toilets-other costs
Temporary toilets-variance from P.O.
Storage trailer-D/L-pay
Storage trailer-trades
Storage trailer-materials
Storage trailer-rental
Storage trailer-variance from P.O.
Temporary security-D/L-pay
Temporary security-trades
Temporary security-materials
Temporary security-other costs
Temporary security-variance from P.O.
TRASH COLLECTION & REMOVALXXX
Trash collection & removal-D/L-pay
Trash collection & removal-trades
Trash collection & removal-materials
Trash collection & removal-other costs
Trash collection & removal-variance from P.O.
Dumpster rental-D/L-pay
Dumpster rental-trades
Dumpster rental-materials
Dumpster rental-cost
Dumpster-variance from P.O.
Hauling and dumping-D/L-pay
Hauling and dumping-trades
Hauling and dumping-materials
Hauling and dumping-other costs
Hauling and dumping-variance from P.O.
Dump fees
UTILITY SYSTEMSXXXXXXXXXXXXXX
Utility systems-D/L-pay
Utility systems-trades
Utility systems-materials
Utility systems-other costs
Utility systems-variance from P.O.
Individual well-D/L-pay
Individual well-trades
Individual well-materials
Individual well-other costs
Individual well-variance from P.O.
Water/sewer hookups-D/L-pay
Water/sewer hookups-trades
Water/sewer hookups-materials
Water/sewer hookups-other costs
Water/sewer hookups-variance from P.O.
Septic system-D/L-pay
Septic system-trades
Septic system-materials
Septic system-other costs
Septic system-variance from P.O.
Utility connections-D/L-pay
Utility connections-trades
Utility connections-materials
Utility connections-other costs
Utility connections-variance from P.O.
FOOTINGS & FOUNDATIONSXXXXXXXX
Footings & foundations-D/L-pay
Footings & foundations-trades
Footings & foundations-materials
Footings & foundations-other costs
Footings & foundations-variance from P.O.
Footings-D/L-pay
Footings-trades
Footings-materials
Footings-other costs
Footings-variance from P.O.
Foundation-D/L-pay
Foundation-trades
Foundation-materials
Foundation-other costs
Foundation-variance from P.O.
Waterproofing-D/L-pay
Waterproofing-trades
Waterproofing-materials
Waterproofing-other costs
Waterproofing-variance from P.O.
Dewatering-D/L-pay
Dewatering-trades
Dewatering-materials
Dewatering-other costs
Dewatering-variance from P.O.
Drainage-D/L-pay
Drainage-trades
Drainage-materials
Drainage-other costs
Drainage-variance from P.O.
Termite protection-D/L-pay
Termite protection-trades
Termite protection-materials
Termite protection-other costs
Termite protection-variance from P.O.

Continued

Winter protection-D/L-pay
Winter protection-trades
Winter protection-materials
Winter protection-rental/other costs
Winter protection-variance from P.O.
CONCRETE & FORMS PREPXXXXXXXX
Concrete & forms prep-D/L-pay
Concrete & forms prep-trades
Concrete & forms prep-materials
Concrete & forms prep-other costs
Concrete & forms prep-variance from P.O.
Slab-D/L-pay
Slab-trades
Slab-materials
Slab-other costs
Slab-variance from P.O.
Garage/carport-D/L-pay
Garage/carport-trades
Garage/carport-materials
Garage/carport-other costs
Garage/carport-variance from P.O.
Stairs & structures-D/L-pay
Stairs & structures-trades
Stairs & structures-materials
Stairs & structures-other costs
Stairs & structures-variance from P.O.
Concrete special finish-D/L-pay
Concrete special finish-trades
Concrete special finish-materials
Concrete special finish-other costs
Concrete special finish-variance from P.O.
Concrete tests/inspections-other costs
STRUCTURAL STEELXXXXXXXXXXXX
Structural steel-D/L-pay
Structural steel-trades
Structural steel-materials
Structural steel-other costs
Structural steel-variance from P.O.
Piers/posts/columns-D/L-pay
Piers/posts/columns-trades
Piers/posts/columns-materials
Piers/posts/columns-other costs
Piers/posts/columns-variance from P.O.
Beams-D/L-pay
Beams-trades
Beams-materials
Beams-other costs
Beams-variance from P.O.
Other structural steel-D/L-pay
Other structural steel-trades
Other structural steel-materials
Other structural steel-other costs
Other structural steel-variance from P.O.
FRAMING (ROUGH CARPENTRY)XXXXX
Framing (rough carpentry)-D/L-pay
Framing (rough carpentry)-trades
Framing (rough carpentry)-materials
Framing (rough carpentry)-other costs
Framing (rough carpentry)-variance from P.O.
Floor/sheathing framing-D/L-pay
Floor/sheathing framing-trades
Floor/sheathing framing-materials
Floor/sheathing framing-other costs
Floor/sheathing framing-variance from P.O.
Walls, partitions & stairs framing-D/L-pays
Walls, partitions & stairs framing-trades
Walls, partitions & stairs framing-materials
Walls, partitions & stairs framing-other costs
Walls, partitions & stairs framing-variance from P.O.
Roof & sheathing framing-D/L-pay
Roof & sheathing framing-trades
Roof & sheathing framing-materials
Roof & sheathing framing-other costs
Roof & sheathing framing-variance from P.O.
Basement framing-D/L-pay
Basement framing-trades
Basement framing-materials
Basement framing-other costs
Basement framing-variance from P.O.
Structures framing-D/L-pay
Structures framing-trades
Structures framing-materials
Structures framing-other costs
Structures framing-variance from P.O.
Special beams/girders/trusses-D/L-pay
Special beams/girders/trusses-trades
Special beams/girders/trusses-materials
Special beams/girders/trusses-other costs
Special beams/girders/trusses-variance from P.O.
Equipment rental-framing-D/L-pay
Equipment rental-framing-trades
Equipment rental-framing-materials
Equipment rental-framing-other costs
Equipment rental-framing-variance from P.O.
OTHER ROUGH WORKXXXXXXXXXXXX
Other rough work-D/L-pay
Other rough work-trades
Other rough work-materials
Other rough work-other costs
Other rough work-variance from P.O.
Rough sheet metal-D/L-pay
Rough sheet metal-trades
Rough sheet metal-materials
Rough sheet metal-other costs
Rough sheet metal-variance from P.O.
Rough plumbing-D/L-pay
Rough plumbing-trades
Rough plumbing-materials
Rough plumbing-other costs
Rough plumbing-variance from P.O.
Electrical wiring-D/L-pay
Electrical wiring-trades
Electrical wiring-materials
Electrical wiring-other costs
Electrical wiring-variance from P.O.
Duct work-D/L-pay
Duct work-trades
Duct work-materials
Duct work-other costs
Duct work-variance from P.O.
Venting-D/L-pay
Venting-trades
Venting-materials
Venting-other costs
Venting-variance from P.O.
ROOFINGXXXXXXXXXXXXXXXX
Roofing-D/L-pay
Roofing-trades
Roofing-materials
Roofing-other costs
Roofing-variance from P.O.
Shingles-D/L-pay
Shingles-trades
Shingles-materials
Shingles-other costs
Shingles-variance from P.O.
Tiles-D/L-pay
Tiles-trades
Tiles-materials
Tiles-other costs
Tiles-variance from P.O.
Other roof coverings-D/L-pay
Other roof coverings-trades
Other roof coverings-materials
Other roof coverings-other costs
Other roof coverings-variance from P.O.
Ventilation-D/L-pay

Ventilation-trades
Ventilation-materials
Ventilation-other costs
Ventilation-variance from P.O.
Skylights-D/L-pay
Skylights-trades
Skylights-materials
Skylights-other costs
Skylights-variance from P.O.
MASONRY & PREPXXXXXXXXXXXXXXX
Masonry & prep-D/L-pay
Masonry & prep-trades
Masonry & prep-materials
Masonry & prep-other costs
Masonry & prep-variance from P.O.
Masonry walls-D/L-pay
Masonry walls-trades
Masonry walls-materials
Masonry walls-other costs
Masonry walls-variance from P.O.
Chimney & cap-D/L-pay
Chimney & cap-trades
Chimney & cap-materials
Chimney & cap-other costs
Chimney & cap-variance from P.O.
Fireplace-D/L-pay
Fireplace-trades
Fireplace-materials
Fireplace-other costs
Fireplace-variance from P.O.
Veneer-D/L-pay
Veneer-trades
Veneer-materials
Veneer-other costs
Veneer-variance from P.O.
Hearth-D/L-pay
Hearth-trades
Hearth-materials
Hearth-other costs
Hearth-variance from P.O.
Enclosure-D/L-pay
Enclosure-trades
Enclosure-materials
Enclosure-other costs
Enclosure-variance from P.O.
Piers/curtain wall-D/L-pay
Piers/curtain wall-trades
Piers/curtain wall-materials
Piers/curtain wall-other costs
Piers/curtain wall-variance from P.O.
Other masonry work-D/L-pay
Other masonry work-trades
Other masonry work-materials
Other masonry work-other costs
Other masonry work-variance from P.O.
WINDOWSXXXXXXXXXXXXXXXXXXX
Windows-D/L-pay
Windows-trades
Windows-materials
Windows-other costs
Windows-variance from P.O.
Windows (incl storm & screen)-D/L-pay
Windows (incl storm & screen)-trades
Windows (incl storm & screen)-materials
Windows (incl storm & screen)-other costs
Windows (incl storm & screen)-variance from P.O.
Special windows & glass-D/L-pay
Special windows & glass-trades
Special windows & glass-materials
Special windows & glass-other costs
Special windows & glass-variance from P.O.
Awnings-D/L-pay
Awnings-trades
Awnings-materials
Awnings-other costs
Awnings-variance from P.O.
DOORSXXXXXXXXXXXXXXXXXX
Doors-D/L-pay
Doors-trades
Doors-materials
Doors-other costs
Doors-variance from P.O.
Exterior doors (incl storms & screens)-D/L-pay
Exterior doors (incl storms & screens)-trades
Exterior doors (incl storms & screens)-materials
Exterior doors (incl storm & screens)-other costs
Exterior doors (incl storms & screens)-variance from P.O.
Sliding glass/french doors-D/L-pay
Sliding glass/french doors-trades
Sliding glass/french doors-materials
Sliding glass/french doors-other costs
Sliding glass/french doors-variance from P.O.
Garage doors & openers-D/L-pay
Garage doors & openers-trades
Garage doors & openers-materials
Garage doors & openers-other costs
Garage doors & openers-variance from P.O.
Special doors & entrances-D/L-pay
Special doors & entrances-trades
Special doors & entrances-materials
Special doors & entrances-other costs
Special doors & entrances-variance from P.O.
Interior/closet doors/pocket doors-D/L-pay
Interior/closet doors/pocket doors-trades
Interior/closet doors/pocket doors-materials
Interior/closet doors/pocket doors-other costs
Interior/closet doors/pocket doors-variance from P.O.
Window and door hardware-D/L-pay
Window and door hardware-trades
Window and door hardware-materials
Window and door hardware-other costs
Window and door hardware-variance from P.O.
INSULATIONXXXXXXXXXXXXXXXX
Insulation-D/L-pay
Insulation-trades
Insulation-materials
Insulation-other costs
Insulation-variance from P.O.
Foundation/basement insulation-D/L-pay
Foundation/basement insulation-trades
Foundation/basement insulation-materials
Foundation/basement insulation-other costs
Foundation/basement insulation-variance from P.O.
Roof and ceiling insulation-D/L-pay
Roof and ceiling insulation-trades
Roof and ceiling insulation-materials
Roof and ceiling insulation-other costs
Roof and ceiling insulation-variance from P.O.
Wall insulation-D/L-pay
Wall insulation-trades
Wall insulation-materials
Wall insulation-other costs
Wall insulation-variance from P.O.
Floor insulation-D/L-pay
Floor insulation-trades
Floor insulation-materials
Floor insulation-other costs
Floor insulation-variance from P.O.
Vapor barrier & weatherstripping-D/L-pay
Vapor barrier & weatherstripping-trades
Vapor barrier & weatherstripping-materials
Vapor barrier & weatherstripping-other costs
Vapor barrier & weatherstripping-variance from P.O.
Soundproofing-D/L-pay
Soundproofing-trades
Soundproofing-materials
Soundproofing-other costs

Continued

Soundproofing-variance from P.O.
Fireproofing-D/L-pay
Fireproofing-trades
Fireproofing-materials
Fireproofing-other costs
Fireproofing-variance from P.O.
EXTERIOR TRIMXXXXXXXXXXXXXX
Exterior trim-D/L-pay
Exterior trim-trades
Exterior trim-materials
Exterior trim-other costs
Exterior trim-variance from P.O.
Flashings-D/L-pay
Flashings-trades
Flashings-materials
Flashings-other costs
Flashings-variance from P.O.
Deck trim-D/L-pay
Deck trim-trades
Deck trim-materials
Deck trim-other costs
Deck trim-variance from P.O.
Exterior wall finishes-D/L-pay
Exterior wall finishes-trades
Exterior wall finishes-materials
Exterior wall finishes-other costs
Exterior wall finishes-variance from P.C.
Siding-D/L-pay
Siding-trades
Siding-materials
Siding-other costs
Siding-variance from P.O.
Stucco-D/L-pay
Stucco-trades
Stucco-materials
Stucco-other costs
Stucco-variance from P.O.
Brick veneer-D/L-pay
Brick veneer-trades
Brick veneer-materials
Brick veneer-other costs
Brick veneer-variance from P.O.
Stone work-D/L-pay
Stone work-trades
Stone work-materials
Stone work-other costs
Stone work-variance from P.O.
Cornices/rake trim/facia/soffit-D/L-pay
Cornices/rake trim/facia/soffit-trades
Cornices/rake trim/facia/soffit-materials
Cornices/rake trim/facia/soffit-other costs
Cornices/rake trim/facia/soffit-variance from P.O.
Exterior stairs-D/L-pay
Exterior stairs-trades
Exterior stairs-materials
Exterior stairs-other costs
Exterior stairs-variance from P.O.
Columns/railings/posts-D/L-pay
Columns/railings/posts-trades
Columns/railings/posts-materials
Columns/railings/posts-other costs
Columns/railings/posts-variance from P.O.
Shutters/window box-D/L-pay
Shutters/window box-trades
Shutters/window box-materials
Shutters/window box-other costs
Shutters/window box-variance from P.O.
Gutters & downspouts-D/L-pay
Gutters & downspouts-trades
Gutters & downspouts-materials
Gutters & downspouts-other costs
Gutters & downspouts-variance from P.O.
Mailbox, house number, etc.-D/L-pay
Mailbox, house number, etc.-trades
Mailbox, house number, etc.-materials
Mailbox, house number, etc.-other costs
Mailbox, house number, etc.-variance from P.O.
Other exterior trim-D/L-pay
Other exterior trim-trades
Other exterior trim-materials
Other exterior trim-other costs
Other exterior trim-variance from P.O.
INTERIOR WALLSXXXXXXXXXXXXXX
Interior walls-D/L-pay
Interior walls-trades
Interior walls-materials
Interior walls-other costs
Interior walls-variance from P.O.
Drywall hanging-D/L-pay
Drywall hanging-trades
Drywall hanging-materials
Drywall hanging-other costs
Drywall hanging-variance from P.O.
Specialty drywall-D/L-pay
Specialty drywall-trades
Specialty drywall-materials
Specialty drywall-other costs
Specialty drywall-variance from P.O.
Plaster-D/L-pay
Plaster-trades
Plaster-materials
Plaster-other costs
Plaster-variance from P.O.
Textures-D/L-pay
Textures-trades
Textures-materials
Textures-other costs
Textures-variance from P.O.
Veneer-D/L-pay
Veneer-trades
Veneer-materials
Veneer-other costs
Veneer-variance from P.O.
Panels (fiberglass, cultured marble, etc.)-D/L-pay
Panels (fiberglass, cultured marble, etc.)-trades
Panels (fiberglass, cultured marble, etc.)-materials
Panels (fiberglass, cultured marble, etc.)-other costs
Panels (fiberglass, cultured marble, etc.)-variance from P.O.
CEILINGSXXXXXXXXXXXXXXX
Ceilings-D/L-pay
Ceilings-trades
Ceilings-materials
Ceilings-other costs
Ceilings-variance from P.O.
Accoustical/drop ceiling-D/L-pay
Accoustical/drop ceiling-trades
Accoustical/drop ceiling-materials
Accoustical/drop ceiling-other costs
Accoustical/drop ceiling-variance from P.O.
Other ceiling finishes-D/L-pay
Other ceiling finishes-trades
Other ceiling finishes-materials
Other ceiling finishes-other costs
Other ceiling finishes-variance from P.O.
FLOORINGXXXXXXXXXXXXXXXXX
Allowance for flooring
Flooring-D/L-pay
Flooring-trades
Flooring-materials
Flooring-other costs
Flooring-variance from P.O.
Resilient flooring-D/L-pay
Resilient flooring-trades
Resilient flooring-materials
Resilient flooring-other costs
Resilient flooring-variance from P.O.
Carpeting-D/L-pay

Carpeting-trades
Carpeting-materials
Carpeting-other costs
Carpeting-variance from P.O.
Hardwood floor and/or re-finish-D/L-pay
Hardwood floor and/or re-finish-trades
Hardwood floor and/or re-finish-materials
Hardwood floor and/or re-finish-other costs
Hardwood floor and/or re-finish-variance from P.O.
Ceramic tile-D/L-pay
Ceramic tile-trades
Ceramic tile-materials
Ceramic tile-other costs
Ceramic tile-variance from P.O.
Natural (marble/slate)-D/L-pay
Natural (marble/slate)-trades
Natural (marble/slate)-materials
Natural (marble/slate)-other costs
Natural (marble/slate)-variance from P.O.
Other flooring/finishes-D/L-pay
Other flooring/finishes-trade
Other flooring/finishes-materials
Other flooring/finishes-other costs
Other flooring/finishes-variance from P.O.
TRIM CARPENTRYXXXXXXXXXXXX
Trim carpentry-D/L-pay
Trim carpentry-trades
Trim carpentry-materials
Trim carpentry-other costs
Trim carpentry-variance from P.O.
Moldings-D/L-pay
Moldings-trades
Moldings-materials
Moldings-other costs
Moldings-variance from P.O.
Baseboards-D/L-pay
Baseboards-trades
Baseboards-materials
Baseboards-other costs
Baseboards-variance from P.O.
Chair-rail-D/L-pay
Chair-rail-trades
Chair-rail-materials
Chair-rail-other costs
Chair-rail-variance from P.O.
Windows and doors-D/L-pay
Windows and doors-trades
Windows and doors-materials
Windows and doors-other costs
Windows and doors-variance from P.O.
Built-ins-D/L-pay
Built-ins-trades
Built-ins-materials
Built-ins-other costs
Built-ins-variance from P.O.
Fireplace mantle/trim-D/L-pay
Fireplace mantle/trim-trades
Fireplace mantle/trim-materials
Fireplace mantle/trim-other costs
Fireplace mantle/trim-variance from P.O.
Paneling-D/L-pay
Paneling-trades
Paneling-materials
Paneling-other costs
Paneling-variance from P.O.
Shelving-D/L-pay
Shelving-trades
Shelving-materials
Shelving-other costs
Shelving-variance from P.O.
Stairs/handrails-D/L-pay
Stairs/handrails-trades
Stairs/handrails-materials
Stairs/handrails-other costs
Stairs/handrails-variance from P.O.
Trim hardware-D/L-pay
Trim hardware-trades
Trim hardware-materials
Trim hardware-other costs
Trim hardware-variance from P.O.
Custom millwork-D/L-pay
Custom millwork-trades
Custom millwork-materials
Custom millwork-other costs
Custom millwork-variance from P.O.
Other trim carpentry-D/L-pay
Other trim carpentry-trades
Other trim carpentry-materials
Other trim carpentry-other costs
Other trim carpentry-variance from P.O.
OTHER TRIM/SPECIALTYXXXXXXXXXX
Other trim speciality-D/L-pay
Other trim speciality-trades
Other trim speciality-materials
Other trim speciality-other costs
Other trim specialty-variance from P.O.
Trim tilework-D/L-pay
Trim tilework-trades
Trim tilework-materials
Trim tilework-other costs
Trim tilework-variance from P.O.
Mirrors-D/L-pay
Mirrors-trades
Mirrors-materials
Mirrors-other costs
Mirrors-variance from P.O.
Glass block specialty work-D/L-pay
Glass block specialty work-trades
Glass block specialty work-materials
Glass block specialty work-other costs
Glass block specialty work-variance from P.O.
Steel stairs-D/L-pay
Steel stairs-trades
Steel stairs-materials
Steel stairs-other costs
Steel stairs-variance from P.O.
CABINETS/VANITIES/STORAGE SYSTEMSXXXX
Allowance for cabinetry
Cabinets-D/L-pay
Cabinets-trades
Cabinets-materials
Cabinets-other costs
Cabinets-variance from P.O.
Kitchen cabinets-D/L-pay
Kitchen cabinets-trades
Kitchen cabinets-materials
Kitchen cabinets-other costs
Kitchen cabinets-variance from P.O.
Countertops-D/L-pay
Countertops-trades
Countertops-materials
Countertops-other costs
Countertops-variance from P.O.
Bathroom cabinets/vanities-D/L-pay
Bathroom cabinets/vanities-trades
Bathroom cabinets/vanities-materials
Bathroom cabinets/vanities-other costs
Bathroom cabinets/vanities-variance from P.O.
Cabinet hardware-D/L-pay
Cabinet hardware-trades
Cabinet hardware-materials
Cabinet hardware-other costs
Cabinet hardware-variance from P.O.
Cabinet storage systems-D/L-pay
Cabinet storage systems-trades
Cabinet storage systems-materials
Cabinet storage systems-other costs

Continued

Cabinet storage systems-variance from P.O.
Closet storage systems-D/L-pay
Closet storage systems-trades
Closet storage systems-materials
Closet storage systems-other costs
Closet storage systems-variance from P.O.
Bath accessories/finishes-D/L-pay
Bath accessories/finishes-trades
Bath accessories/finishes-materials
Bath accessories/finishes-other costs
Bath accessories/finishes-variance from P.O.
Bulkheads-D/L-pay
Bulkheads-trades
Bulkheads-materials
Bulkheads-other costs
Bulkheads-variance from P.O.
APPLIANCESXXXXXXXXXXXXXXXXXX
Allowance for appliances
Appliances-D/L-pay
Appliances-trades
Appliances-materials
Appliances-other costs
Appliances-variance from P.O.
Ranges/grills/hoods/microwaves-D/L-pay
Ranges/grills/hoods/microwaves-trades
Ranges/grills/hoods/microwaves-materials
Ranges/grills/hoods/microwaves-other costs
Ranges/grills/hoods/microwaves-variance from P.O.
Refrigerators & freezers-D/L-pay
Refrigerators & freezers-trades
Refrigerators & freezers-materials
Refrigerators & freezers-other costs
Refrigerators & freezers-variance from P.O.
Dishwasher-D/L-pay
Dishwasher-trades
Dishwasher-materials
Dishwasher-other costs
Dishwasher-variance from P.O.
Garbage disposal-D/L-pay
Garbage disposal-trades
Garbage disposal-materials
Garbage disposal-other costs
Garbage disposal-variance from P.O.
Trash compactor-D/L-pay
Trash compactor-trades
Trash compactor-materials
Trash compactor-other costs
Trash compactor-variance from P.O.
Hot water dispenser-D/L-pay
Hot water dispenser-trades
Hot water dispenser-materials
Hot water dispenser-other costs
Hot water dispenser-variance from P.O.
Washer and/or dryer-D/L-pay
Washer and/or dryer-trades
Washer and/or dryer-materials
Washer and/or dryer-other costs
Washer and/or dryer-variance from P O.
Other appliances-D/L-pay
Other appliances-trades
Other appliances-materials
Other appliances-other costs
Other appliances-variance from P.O.
FINISH PLUMBINGXXXXXXXXXXXXX
Finish plumbing-D/L-pay
Finish plumbing-trades
Finish plumbing-materials
Finish plumbing-other costs
Finish plumbing-variance from P.O.
Tub and shower-D/L-pay
Tub and shower-trades
Tub and shower-materials
Tub and shower-other costs
Tub and shower-variance from P.O.
Toilets/bidet-D/L-pay
Toilets/bidet-trades
Toilets/bidet-materials
Toilets/bidet-other costs
Toilets/bidet-variance from P.O.
Sinks-D/L-pay
Sinks-trades
Sinks-materials
Sinks-other costs
Sinks-variance from P.O.
Whirlpool-D/L-pay
Whirlpool-trades
Whirlpool-materials
Whirlpool-other costs
Whirlpool-variance from P.O.
Hot water heater-D/L-pay
Hot water heater-trades
Hot water heater-materials
Hot water heater-other costs
Hot water heater-variance from P.O.
Plumbing hardware-D/L-pay
Plumbing hardware-trades
Plumbing hardware-materials
Plumbing hardware-other costs
Plumbing hardware-variance from P.O.
Other plumbing fixtures-D/L-pay
Other plumbing fixtures-trades
Other plumbing fixtures-materials
Other plumbing fixtures-other costs
Other plumbing fixtures-variance from P.O.
FINISH ELECTRICALXXXXXXXXXXXXX
Allowance for lighting
Finish electrical-D/L-pay
Finish electrical-trades
Finish electrical-materials
Finish electrical-other costs
Finish electrical-variance from P.O.
Electrical fixtures (lights/fans/chimes, etc.)-D/L-pay
Electrical fixtures (lights/fans/chimes, etc.)-trades..
Electrical fixtures (lights/fans/chimes, etc.)-materials
Electrical fixtures (lights/fans/chimes, etc.)-other costs
Electrical fixtures (lights/fans/chimes, etc.)-variance from P.O.
Security system-D/L-pay
Security system-trades
Security system-materials
Security system-other costs
Security system-variance from P.O.
Audio system-D/L-pay
Audio system-trades
Audio system-materials
Audio system-other costs
Audio system-variance from P.O.
Protection systems (fire/lightning, etc.)-D/L-pay
Protection systems (fire/lightning, etc.)-trades
Protection systems (fire/lightning, etc.)-materials
Protection systems (fire/lightning, etc.)-other costs
Protection systems (fire/lightning, etc.)-variance from P.O.
Other electrical systems (antenna, phone, computers, low voltage lighting, etc.)-D/
Other electrical systems (antenna, phone, computers, low voltage lighting, etc.)-tra
Other electrical systems (antenna, phone, computers, low voltage lighting, etc.)-ma
Other electrical systems (antenna, phone, computers, low voltage lighting, etc.)-oth
Other electrical systems (antenna, phone, computers, low voltage lighting, etc.)-va
FINISH HVACXXXXXXXXXXXXXX
Finish HVAC-D/L-pay
Finish HVAC-trades
Finish HVAC-materials
Finish HVAC-other costs
Finish HVAC-variance from P.O.
Furnace-D/L-pay
Furnace-trades
Furnace-materials
Furnace-other costs
Furnace-variance from P.O.

Air conditioning-D/L-pay
Air conditioning-trades
Air conditioning-materials
Air conditioning-other costs
Air conditioning-variance from P.O.
Filtering systems-D/L-pay
Filtering systems-trades
Filtering systems-materials
Filtering systems-other costs
Filtering systems-variance from P.O.
Exhaust fans-D/L-pay
Exhaust fans-trades
Exhaust fans-materials
Exhaust fans-other costs
Exhaust fans-variance from P.O.
Attic fan-D/L-pay
Attic fan-trades
Attic fan-materials
Attic fan-other costs
Attic fan-variance from P.O.
Prefab fireplace-D/L-pay
Prefab fireplace-trades
Prefab fireplace-materials
Prefab fireplace-other costs
Prefab fireplace-variance from P.O.
PAINTINGXXXXXXXXXXXXXXXXX
Painting-D/L-pay
Painting-trades
Painting-materials
Painting-other costs
Painting-variance from P.O.
Exterior painting-D/L-pay
Exterior painting-trades
Exterior painting-materials
Exterior painting-other costs
Exterior painting-variance from P.O.
Interior painting-D/L-pay
Interior painting-trades
Interior painting-materials
Interior painting-other costs
Interior painting-variance from P.O.
Wall painting-D/L-pay
Wall painting-trades
Wall painting-materials
Wall painting-other costs
Wall painting-variance from P.O.
Ceiling painting-D/L-pay
Ceiling painting-trades
Ceiling painting-materials
Ceiling painting-other costs
Ceiling painting-variance from P.O.
Trim painting-D/L-pay
Trim painting-trades
Trim painting-materials
Trim painting-other costs
Trim painting-variance from P.O.
Decorative/faux painting-D/L-pay
Decorative/faux painting-trades
Decorative/faux painting-materials
Decorative/faux painting-other costs
Decorative/faux painting-variance from P.O.
Wall coverings-D/L-pay
Wall coverings-trades
Wall coverings-materials
Wall coverings-other costs
Wall coverings-variance from P.O.
Other painting-D/L-pay
Other painting-trades
Other painting-materials
Other painting-other costs
Other painting-variance from P.O.
INTERIOR DECORATINGXXXXXXXXX
Interior decorating-D/L-pay
Interior decorating-trades
Interior decorating-materials
Interior decorating-other costs
Interior decorating-variance from P.O.
Window treatments-D/L-pay
Window treatments-trades
Window treatments-materials
Window treatments-other costs
Window treatments-variance from P.O.
Interior design-D/L-pay
Interior design-trades
Interior design-materials
Interior design-other costs
Interior design-variance from P.O.
DRIVEWAYXXXXXXXXXXXXXXXXXX
Driveway-D/L-pay
Driveway-trades
Driveway-materials
Driveway-other costs
Driveway-variance from P.O.
LANDSCAPINGXXXXXXXXXXXXXX
Landscaping-D/L-pay
Landscaping-trades
Landscaping-materials
Landscaping-other costs
Landscaping-variance from P.O.
Final fill dirt/top soil/grading-D/L-pay
Final fill dirt/top soil/grading-trades
Final fill dirt/top soil/grading-materials
Final fill dirt/top soil/grading-other costs
Final fill dirt/top soil/grading-variance from P.O.
Plants, trees, shrubs, etc.-D/L-pay
Plants, trees, shrubs, etc.-trades
Plants, trees, shrubs, etc.-materials
Plants, trees, shrubs, etc.-other costs
Plants, trees, shrubs, etc.-variance from P.O.
Sod/seed-D/L-pay
Sod/seed-trades
Sod/seed-materials
Sod/seed-other costs
Sod/seed-variance from P.O.
Ground covering-D/L-pay
Ground covering-trades
Ground covering-materials
Ground covering-other costs
Ground covering-variance from P.O.
Irrigation system-D/L-pay
Irrigation system-trades
Irrigation system-materials
Irrigation system-other costs
Irrigation system-variance from P.O.
Landscape lighting-D/L-pay
Landscape lighting-trades
Landscape lighting-materials
Landscape lighting-other costs
Landscape lighting-variance from P.O.
EXTERIOR STRUCTURESXXXXXXXX
Exterior structures-D/L-pay
Exterior structures-trades
Exterior structures-materials
Exterior structures-other costs
Exterior structures-variance from P.O.
Patio-D/L-pay
Patio-trades
Patio-materials
Patio-other costs
Patio-variance from P.O.
Sidewalks & walkways-D/L-pay
Sidewalks & walkways-trades
Sidewalks & walkways-materials
Sidewalks & walkways-other costs
Sidewalks & walkways-variance from P.O.
Fences-D/L-pay
Fences-trades

Continued

Fences-materials
Fences-other costs
Fences-variance from P.O.
Splashblocks & A/C pads-D/L-pay
Splashblocks & A/C pads-trades
Splashblocks & A/C pads-materials
Splashblocks & A/C pads-other costs
Splashblocks & A/C pads-variance from P.O.
Ornamental iron/concrete-D/L-pay
Ornamental iron/concrete-trades
Ornamental iron/concrete-materials
Ornamental iron/concrete-other costs
Ornamental iron/concrete-variance from P.O.
Deck-D/L-pay
Deck-trades
Deck-materials
Deck-other costs
Deck-variance from P.O.
ACCESSIBILITY ITEMSXXXXXXXXXXXX
Accessibility items-D/L-pay
Accessibility items-trades
Accessibility items-materials
Accessibility items-other costs
Accessibility items-variance from P.O.
Ramps-D/L-pay
Ramps-trades
Ramps-materials
Ramps-other costs
Ramps-variance from P.O.
Grab bars-D/L-pay
Grab bars-trades
Grab bars-materials
Grab bars-other costs
Grab bars-variance from P.O.
Other accessibility items-D/L-pay
Other accessibility items-trades
Other accessibility items-materials
Other accessibility items-other costs
Other accessibility items-variance from P.O.
WATER PURIFICATION SYSTEMSXXXX
Water purification systems-D/L-pay
Water purification systems-trades
Water purification systems-materials
Water purification systems-other costs
Water purification systems-variance from P.O.
LEISURE SPACESXXXXXXXXXXXXXXX
Leisure-D/L-pay
Leisure-trades
Leisure-materials
Leisure-other costs
Leisure-variance from P.O.
Sunrooms/screened rooms-D/L-pay
Sunrooms/screened rooms-trades
Sunrooms/screened rooms-materials
Sunrooms/screened rooms-other costs
Sunrooms/screened rooms-variance from P.O.
Pool/spa-D/L-pay
Pool/spa-trades
Pool/spa-materials
Pool/spa-other costs
Pool/spa-variance from P.O.
ADD-ONSXXXXXXXXXXXXXXXXXXX
ACCESSIBILITY ITEMSXXXXXXXXXX
Accessibility items-D/L pay
Accessibility items-trades
Accessibility items-materials
Accessibility items-other costs
Accessibility items-variance from P.O.
Ramps-D/L-pay
Ramps-trades
Ramps-materials
Ramps-other costs
Ramps-variance from P.O.
Grab bars-D/L-pay
Grab bars-trades
Grab bars-materials
Grab bars-other costs
Grab bars-variance from P.O.
Other accessibility items-D/L-pay
Other accessibility items-trades
Other accessibility items-materials
Other accessibility items-other costs
Other accessibility items-variance from P.O.
GARAGE DOOR OPENERXXXXXXXX
Garage door opener and installation-D/L-pay
Garage door opener and installation-trades
Garage door opener and installation-materials
Garage door opener and installation-other costs
Garage door opener and installation-variance from P.O.
GARAGE-FULL DRYWALL & TAPEXXX
Full drywall & tape garage-D/L-pay
Full drywall & tape garage-trades
Full drywall & tape garage-materials
Full drywall & tape garage-other costs
Full drywall & tape garage-variance from P.O.
AIR HUMIDIFIERXXXXXXXXXX
Air humidifier-D/L-pay
Air Humidifier-trades
Air Humidifier-materials
Air Humidifier-other costs
Air Humidifier-variance from P.O.
AIR CLEAN AIR FILTERXXXXX
Air clean air filter-D/L-pay
Air clean air filter-trades
Air clean air filter-materials
Air clean air filter-other costs
Air clean air filter-variance from P.O.
CLEAN AIR SYSTEMXX
Clean air system-D/L-pay
Clean air system-trades
Clean air system-materials
Clean air system-other costs
Clean air system-variance from P.O.
OAK TRIM AROUND KITCHEN COUNTERSXXXX
Oak trim around kitchen counters-D/L-pay
Oak trim around kitchen counters-trades
Oak trim around kitchen counters-materials
Oak trim around kitchen counters-other costs
Oak trim around kitchen counters-variance from P.O.
FIREPLACE REMOTE CONTROL & INSTALLATIONXXXX
Fireplace remote control & installation-D/L-pay
Fireplace remote control & installation-trades
Fireplace remote control & installation-materials
Fireplace remote control & installation-other costs
Fireplace remote control & installation-variance from P.O.
DOG DISPOSAL UNIT & INSTALLATIONXXXXXX
Dog disposal unit & installation-D/L-pay
Dog disposal unit & installation-trades
Dog disposal unit & installation-materials
Dog disposal unit & installation-other costs
Dog disposal unit & installation-variance from P.O.
CLEAN-UP & FINALIZE JOBXXXXXXXX
Clean-up & finalize job-D/L-pay
Clean-up & finalize job-trades
Clean-up & finalize job-materials
Clean-up & finalize job-other costs
Clean-up & finalize job-variance from P.O.
Daily cleanup-D/L-pay
Final cleanup-D/L-pay
Final cleanup-trades
Final cleanup-materials
Final cleanup-other costs
Final cleanup-variance from P.O.
Punchlist & resolution-D/L-pay
MISC JOB COSTSXXXXXXXXXXXXX
Misc job costs-D/L pay
Misc job costs-trades

Misc job costs-materials
Misc job costs-other costs
Misc job costs-variance from P.O.
Job timekeeping-D/L-pay
Staff meetings/updates re: job-D/L-pay
Order/locate/distribute materials, supplies, tools, equipment for job-D/L-pay
Trips for materials/supplies-D/L-pay
Trips to/from job site-D/L-pay
Trips to/from job site (mileage)-other costs
Photography-D/L-pay
Photography-trades
Photography-materials
Photography-other costs
Photography-variance from P.O.
Rework (our error)-D/L-pay
Rework (our error)-trades
Rework (our error)-materials
Rework (our error)-other costs
Rework (our error)-variance from P.O.
OTHER DIRECT CONSTRUCTION COSTSXXXX
Other direct construction costs-D/L-pay
Other direct construction costs-trades
Other direct construction costs-materials
Other direct construction costs-other costs
Other direct construction costs-variance from P.O.
Consumable small tools-time spent to purchase-D/L-pay
Consumable small tools-materials
Consumable small tools-other costs
Consumable small tools-variance from P.O.
Vehicle expense-truck/auto mileage, etc.-reimbursement to employees and/or trad
Direct interest & loan fees-construction loan
Bond expenses
Builder's risk insurance
Sales commissions
Bonuses-D/L
Bonuses-trade contractors
WARRANTYXXXXXXXXXXXXXXXXXX
Warranty work-D/L-pay
Warranty work-trades
Warranty work-materials
Warranty work-other costs
Warranty work-variance from P.O.
ALLOCATONS BETWEEN JOBS
Allocated land development costs
Allocate estimated development costs between jobs (from "development" job to oth
Allocated interest-from 'interest' job to all other jobs
G&A costs allocated to units
Allocated bonuses
Allocated warranty

TO BE DETERMINEDXXXXXXXXXX
MODEL HOME OPERATING EXPENSEXXXXXXXXXX
Property taxes
Utilities
Cleaning model home
Maintenance of premises-model home
Interior decoration-model home
Furnishings/deprec on furnishings for model
Association dues
Interest on model home
Other expenses-model home
INTERIOR DECORATINGXXXXXXXXXX
Interior decorating-D/L-pay
Interior decorating-trades
Interior decorating-materials
Interior decorating-other costs
Interior decorating-variance from P.O.
Window treatments-D/L-pay
Window treatments-trades
Window treatments-materials
Window treatments-other costs
Window treatments-variance from P.O.
Interior design-D/L-pay
Interior design-trades
Interior design-materials
Interior design-other costs
Interior design-variance from P.O.
LAND & DEVELOPMENTXXXXXXXXXX
HOLDING COSTS-DEVELOPMENTXXXX
Real estate taxes on homes in process or ready for sale
Direct interest-land/dev loans
Legal fees-trades
Legal fees-variance from P.O.
Maintenance/repair for development
Inflation contingency
DESIGN & ENGINEERING-DEVELOPMENTXXXXX
Design & engineering-D/L-pay
Design & engineering-trades
Design & engineering-materials
Design & engineering-other costs
Design & engineering-variance from P.O.
Professional design fees-D/L-pay
Professional design fees-trades
Professional design fees-variance from P.O.
Architectural fees-trades
Architectural fees-variance from P.O.
Surveying-D/L-pay
Surveying-trades
Surveying-variance from P.O.

Engineering services-trades
Engineering services-variance from P.O.
Blueprints-trades
Blueprints-costs/copies
Blueprints-variance from P.O.
Plan copies-D/L-pay
Plan copies-costs
EXCAVATING-DEVELOPMENTXXXXXX
Excavating-D/L-pay
Excavating-trades
Excavating-materials
Excavating-other costs
Excavating-variance from P.O.
UTILITIES-DEVELOPMENTXXXXXXXX
Sewer-D/L-pay
Sewer-trades
Sewer-materials
Sewer-other costs
Sewer-variance from P.O.
Water-D/L-pay
Water-trades
Water-materials
Water-other costs
Water-variance from P.O.
Natural gas-D/L-pay
Natural gas-trades
Natural gas-materials
Natural gas-other costs
Natural gas-variance from P.O.
Electric utilities installation-D/L-pay
Electric utilities installation-trades
Electric utilities installation-materials
Electric utilities installation-other costs
Electric utilities installation-variance from P.O.
PAVING/SIDEWALKS-DEVELOPMENTXXX
Paving-D/L-pay
Paving-trades
Paving-materials
Paving-other costs
Paving-variance from P.O.
Curb & gutter-D/L-pay
Curb & gutter-trades
Curb & gutter-materials
Curb & gutter-other costs
Curb & gutter-variance from P.O.
Sidewalks-D/L-pay
Sidewalks-trades
Sidewalks-materials
Sidewalks-other costs

Continued

Sidewalks-variance from P.O.
LIGHTS/SIGNS/BOXES-DEVELOPMENTXXXX
Entrance signs-D/L-pay
Entrance signs-trades
Entrance signs-materials
Entrance signs-other costs
Entrance signs-variance from P.O.
Street lights-D/L-pay
Street lights-trades
Street lights-materials
Street lights-other costs
Street lights-variance from P.O.
Street signs-D/L-pay
Street signs-trades
Street signs-materials
Street signs-other costs
Street signs-variance from P.O.
Postal boxes-D/L-pay
Postal boxes-trades
Postal boxes-materials
Postal boxes-other costs
Postal boxes-variance from P.O.
LEISURE PROVISIONS-DEVELOPMENTXXXX
Nature plantings-D/L-pay
Nature plantings-trades
Nature plantings-materials
Nature plantings-other costs
Nature plantings-variance from P.O.
Nature paths-D/L-pay
Nature paths-trades
Nature paths-materials
Nature paths-other costs
Nature paths-variance from P.O.
Nature structures (e.g. gazebos, benches, pavilions, etc.)-D/L-pay
Nature structures (e.g. gazebos, benches, pavilions, etc.)-trades
Nature structures (e.g. gazebos, benches, pavilions, etc.)-materials
Nature structures (e.g. gazebos, benches, pavilions, etc.)-other costs
Nature structures (e.g. gazebos, benches, pavilions, etc.)-variance from P.O.
Nature bridges, steps, etc.-D/L-pay
Nature bridges, steps, etc.-trades
Nature bridges, steps, etc.-materials
Nature bridges, steps, etc.-other costs
Nature bridges, steps, etc.-variance from P.O.
LANDSCAPING-DEVELOPMENTXXXXX
Final fill dirt/top soil/grading-D/L-pay
Final fill dirt/top soil/grading-trades
Final fill dirt/top soil/grading-materials
Final fill dirt/top soil/grading-other costs
Final fill dirt/top soil/grading-variance from P.O.
Plants, trees, shrubs, etc.-D/L-pay
Plants, trees, shrubs, etc.-trades
Plants, trees, shrubs, etc.-materials
Plants, trees, shrubs, etc.-other costs
Plants, trees, shrubs, etc.-variance from P.O.
Sod/seed-D/L-pay
Sod/seed-trades
Sod/seed-materials
Sod/seed-other costs
Sod/seed-variance from P.O.
Ground covering-D/L-pay
Ground covering-trades
Ground covering-materials
Ground covering-other costs
Ground covering-variance from P.O.
Irrigation system-D/L-pay
Irrigation system-trades
Irrigation system-materials
Irrigation system-other costs
Irrigation system-variance from P.O.
Landscape lighting-D/L-pay
Landscape lighting-trades
Landscape lighting-materials
Landscape lighting-other costs
Landscape lighting-variance from P.O.
Other development costs
TEMP ITEM USED TO CREATE OPENING INVOICES
Finance Charges on Overdue Balance
tradetotal
Less discount of xx%
Refunds to clients
Institutional (church, school, etc.)-no sales tax
Michigan sales tax
No sales tax
Reseller-no sales tax

APPENDIX C

Month-end procedures

Month-End Procedures

Month Ending ____________________

Date to be Completed	Date Complete

Income

Finalize invoicing for the month:

- Determine that all sales/closings have been entered (all invoices related to a closed unit should be dated the same day as the closing date - - use the memo portion to indicate order date.)
- Check to see that special orders for units with future closing dates are post-dated (use anticipated closing date). Indicate order date in memo field.
- Review unbilled costs by job if working on a time and materials (or "cost plus") basis.
- Review jobs in process for contracted payment agreements.
- Create remaining invoices as appropriate.
- Prepare client statements for outstanding balances. (Assign finance charges to clients if applicable)

Finalize deposits for the month:

- Receive payments
- Make (enter) deposits from payments received

Post earnings on change in officer's life insurance cash value

Post earnings on short and/or long-term investments

Review the "Unclassified Income" account. Determine appropriate classifications to bring balance to $0.

Expenses and liability payments

Owner receives sealed bank statement(s) at home and reviews cancelled checks for proper vendors.

- Owner reviews deposits and other bank account reductions for reasonableness.
- Owner gives bank statement(s) to accounting assistant for further reconciliation.

Owner receives sealed credit card statements at home and reviews charges for proper vendors.

- Owner gives credit card statements to accounting assistant for further reconciliation.

Finalize payroll for the month:

- Enter payroll timesheets.
- Create all payroll checks and balance against payroll service reports or review payroll report for current period and then print checks.
- Balance payroll liabilities against payroll service reports (e.g. Paychex). Correct liabilities as necessary.
- Create other special payroll-related reports for outside agencies (e.g., unemployment, insurance, unions, etc.)
- Balance payroll liabilities against calculated reports. Correct liabilities as necessary.
- Enter payroll liability payments. Make payment if necessary.

Determine if use tax expense/payable applies to current period. If so, recognize as a payable in accounts payable.

Prepare sales and/or use tax return if applicable.

Balance and proof sales tax payable account against the sales tax report.

Pay sales tax liability for prior period.

Determine that petty cash and cash box balances are correct. If petty cash is low on funds, write a check or enter a payable to replenish funds to working level. Split costs to correct categories based on current receipts. (You may wish to enter expenses as charge card slips, in order to record vendor, expense or item and job as accurately as possible. Then enter a bill to pay the "Reimbursements due petty cash" credit card.)

Enter credit card charges shown on the most recent statement (use transaction date shown on statement).

- Record vendor and choose "items" or "expenses" (the same as for bills) on charge card slips.
- Enter late fees as charge slips-post to "Discounts lost & late fees" account.
- Enter annual fees as charge slips-post to "Annual credit fees" account.

Reconcile charge account statements:

- Enter interest/finance charges as part of reconciliation process.
- Investigate any old outstanding items. Prepare AJE's to reverse in current period if appropriate.

Continued

Month-End Procedures

Month Ending ______________________

Date to be Completed	Date Complete

Expenses and liability payments (continued)

Enter remaining bills payable (use invoice dates, or day expense incurred). Be sure to include payments due for:
- Credit cards.
- Loan payments due and/or interest due.
- Employee expense reports.
  - (You may wish to enter expenses as charge card slips, in order to record vendor, expense or item and job as accurately as possible. Then enter a bill to pay the "Business exp reimbursements due" credit card.)

Review payables and jobs in process to be sure all applicable expenses have been included in payables.
Prepare an "Unpaid Bills Report." Reconcile outstanding payables to vendor statements. Resolve differences.
Change customer:job names to indicate which jobs have closed ($) and which show final net income (f)
(I.e., no more bills anticipated).
Finalize bills for current month:
- Prepare outstanding payables list for owner(s) to review.
- Pay bills as indicated by owner.
- Print checks, attach support.
- Hand-write, in pen, first and last number in the check series to a "Checks log".
- Owner reviews & initials support, signs checks, and approves "checks issued log".
- Owner, or third staff person, mails checks.

"Write checks" to record any checks that were written by hand, or not entered through the payables system.
Record any electronic or automatic payments that post to your checking account.
NSF checks:
- Record any NSF returns that have not been re-deposited within the same period (see memorized transactions)
- Record any re-deposits of NSF's not initially returned by bank during this period (see memorized transactions).
- Review items still in the "NSF's in transit" account. Write off uncollectibles (see memorized transactions).

Check the balance in the "Undeposited funds" account. It should only contain amounts waiting to be deposited.
Reconcile all checking and savings accounts (note outstanding checks and deposits for immediate follow-up).

Check to see that personal expenditures to, or for, the owner (or the owner's family) have been classified as "draws", "distributions", "dividends", or loan adjustments.
Review the "Unclassified COGS" account. Determine appropriate classifications to bring balance to $0.
Review the "Unclassified Expense" account. Determine appropriate classifications to bring balance to $0.

Accounts Receivable

Prepare Accounts Receivable Aging Summary and review.
Clear small outstanding debit and credit balances if no longer applicable.
Note credit balances that represent customer deposits--make period-end adjustment between receivables and customer deposits. Prepare reversing entry.
Prepare and review Percentage of Completion schedule. Post AJE by job to recognize any income earned, but not yet invoiced. Prepare and post reversing entry.
Note any old receivables for collection action.
Follow up on "due from employees" and "due from related parties" balances.

Notes & Loans Receivable

Determine that payments received have been properly split between principal and interest.
Post any interest receivable through current period.

Fixed Assets and Depreciation

Determine that fixed asset acquisitions have been so recorded.
Post entries for estimated depreciation and amortization (see memorized transactions-could be automatic).

Other Assets

Review balances for reasonableness.
Adjust inventory and Cost of Good Sold accounts if applicable (see memorized transactions).

Continued

Month-End Procedures

Month Ending ______________________

Date to be Completed	Date Complete

Accounts Payable

Prepare aging summary. Review for reasonableness. Remove any payables no longer applicable.

Other Current Liabilities

Determine that outstanding payroll tax or other payroll-related liabilities are accurate. (Note: If you have a payroll service that pays your payroll taxes, those liabilities should be $0.)

Prepare a payroll liabilities report as of the last day of the month. Prepare a standard balance sheet report for the same date. Verify that the payroll liabilities on the balance sheet match the Liabilities report. If not, research and correct.

Determine if Federal, State, real estate, or personal property taxes are due (may need to check with accountant.)

Follow up on amounts owed to related parties. Determine that balances are accurate and whether to make a payment.

Notes and Loans Payable

Determine that payments made have been properly split between principal and interest.

If new loans have been created, determine that they are recorded. Obtain an amortization schedule.

Accrue any interest payable if not paid for current period. (Reverse entry as of first day of following period.)

Equity

Investigate and error-correct any balances shown in the account "Opening balance equity".

Budget

Review the annual budget to determine that total budget has not changed.

Income and Expense Accounts-Review for Reasonableness

As you perform the following process, keep a notepad handy to jot down your questions and observations.

(Note: this perusal should be the most important part of your job! Keeping the owner informed is critical!)

a. Run a year-to-date Profit and Loss Report (print and also leave on screen):

Review all "miscellaneous" or "unclassified" accounts to determine proper classification.

Look for any accounts that end in "--Other" and show a balance. The "--other" means that a transaction posted to the "summary" account and needs to be corrected to post to a more specific account.

Customize on screen to a "by month" view. Look to see that YTD totals look correct, and review for missing entries (e.g. rent and/or utilities, and/or interest should appear every month. "Holes" indicate missing expenses.

b. Double click on the "total COGS" line to create a transaction report.

Customize the report to include the "source name" column, and the "item" column.

Correct each transaction that has not been recorded with an item.

Compare the "name" and "source name" columns. If they are the same for any transaction, correct the transaction to include the job name.

c. Run a year-to-date Profit and Loss Report by Job (income and COGS accounts only):

Compare the totals in the Income and COGS sections of this report against the previous report. The total for each account should be the same. If not, some transaction in the standard (total) P&L does not have a job assigned to it. Review transactions in the standard P&L--locate and correct missing job assignments.

Customize the report to include "% of income" columns. Change the date to the current period(s). Do you see a reasonable relationship between income and expense for each job? (If not, either income or expense is probably not recorded correctly.)

Continued

Month-End Procedures

Month Ending ________________

Date to be Completed	Date Complete

*d. **Run a year-to-date Profit and Loss Report by Class (if you use classes):***

Drill down into any amounts shown in the "unclassified" column and assign the appropriate class (journal entries may be necessary.

Then compare the totals by account against the first report. The total for each account should be the same. If not, some transaction in the standard (total) P&L does not have a class assignment. Review transactions in the standard P&L--locate and correct missing class assignments.

*e. **Run project detail reports as requested by owner (against estimate if applicable):***

Run a Profit and Loss Report by Job (income and COGS accounts only)-date should be "ALL". Filter this report to include only the jobs requested by the owner for reporting above.

Before these reports are shared with the owner and/or crew, each job cost report should be compared to the job total from the report above. If totals do not agree, some expense was entered as "expense" rather than as "item", and must be corrected. "No item" transactions should be corrected unless they net out to $0.

Allocations

Compute and post any inter-account, inter-job, or inter-class allocations or adjustments.

Work in Process adjustments

Run a P&L by Job Report. Review income and costs in process for unfinished units. Determine which should be moved to Work in Process, or to Lots/home ready for sale balance sheet accounts.

Prepare journal entry, by job, to move costs to balance sheet as of end of accounting period.

Prepare reversing journal entry, by job, to move costs back to income statement as of following day.

Prepare standard monthly reports (expanded and collapsed versions available)

Balance Sheet

Income Statement (Profit & Loss) with percentages.

Closed jobs/units only: Job cost report in Profit and Loss format, including percentages (determine time frame)

Interest- - Budget vs. Actual (use life-to-date report dates)

Individual project reports (actual vs. estimate as appropriate)--usually life to date, but could be current period.

Trial Balance (for historical reference)

Owner's review

Owner reviews bank reconciliation and compares to bank statement. Owner initials approval.

Owner reviews reports and notes questions for bookkeeper and for accountant.

Owner follows up with production employees regarding production issues.

Owner returns questions/corrections to bookkeeper.

Owner signs on to QBP, presses CTRL/1 to confirm that audit trail is enabled

Owner contacts accountant regarding remaining questions or requests system changes.

Respond to owner's questions - - make correcting entries and re-run reports if necessary.

Owner's final review of reports

Close your file for the month

Close all on-screen reports. Go to FILE/UTILITIES/VERIFY DATA. Run a file integrity check.

If your data has lost integrity, re-build your file. After re-build, re-run your file integrity check. If you get the same message, contact Intuit or a Certified QBP ProAdvisor.

Set your date password for the month ended mo/day/yr.

Prepare 2 QBP file backups. Title disks "for the month ended yr/mo/day" and store off site.

Create a third backup to your hard drive and name as "z_yr/mo/day.qbb".

Note: This is now your month-end back up file. The next time you back up on your hard drive, DO NOT use this same name! (The "z" will place it at the end of the backup list on your hard drive).

Go back to your normal backup name next time!

Owner signoff for the month

Give this checklist to owner for approval

Owner signs off and returns to bookkeeper.

File approved reports and these procedures in a permanent company notebook.